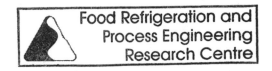
Food Refrigeration and
Process Engineering
Research Centre

KT-557-553

INTRODUCTION TO FOOD PROCESS ENGINEERING

INTRODUCTION TO FOOD PROCESS ENGINEERING

P. G. Smith

University of Lincoln
Lincoln, United Kingdom

Kluwer Academic / Plenum Publishers
New York, Boston, Dordrecht, London, Moscow

Library of Congress Cataloging-in-Publication Data

Smith, P. G.
    Introduction to food process engineering/P.G. Smith.
        p.   cm.
    Includes bibliographical references and index.
    ISBN 0-306-47397-6
        1. Food industry and trade.   I. Title.

TP370 .S585 2002
664—dc21

2002028686

ISBN: 0-306-47397-6

©2003 Kluwer Academic / Plenum Publishers, New York
233 Spring Street, New York, New York 10013

http://www.wkap.nl/

10 9 8 7 6 5 4 3 2

A C.I.P. record for this book is available from the Library of Congress

Printed in the United States of America

Contents

Contents

Preface

There are now a large number of food related first-degree courses offered at universities in Britain and elsewhere in the world which either specialise in, or contain a significant proportion of, food technology. This is a new book on food process engineering which treats the principles of processing in a scientifically rigorous yet concise manner, which can be used as a lead in to more specialised texts for higher study and which is accessible to students who do not necessarily possess a traditional science A-level background. It is equally relevant to those in the food industry who desire a greater understanding of the principles of the processes with which they work. Food process engineering is a quantitative science and this text is written from a quantitative and mathematical perspective and is not simply a descriptive treatment of food processing. The aim is to give readers the confidence to use mathematical and quantitative analyses of food processes and most importantly there are a large number of worked examples and problems with solutions. The mathematics necessary to read this book is limited to elementary differential and integral calculus and the simplest kind of differential equation.

This book is the result of fifteen years experience of teaching food processing technology and food engineering to students on a variety of diploma, first degree and postgraduate courses. It is designed, *inter alia*, to:

- emphasise the importance of thermodynamics and heat transfer as key elements in food processing
- stress the similarity of heat, mass and momentum transfer and make the fundamentals of these essential concepts readily accessible
- develop the theory of mass transfer, which is under used in studies of food processing and little understood, in a useful and readily applicable way
- widen the usual list of unit operations treated in text books for undergraduates to include the use of membranes
- introduce a proper treatment of the characterisation of food solids and of solids processing and handling

Chapters one and two set out the background for a quantitative study of food processing by defining the objectives of process engineering; by describing the mathematical and analytical approach to the design and operation of processes; and by establishing the use of SI units. Much of what follows in the book is made easier by a thorough understanding of the SI system. Important thermodynamics concepts are introduced in chapter three which underpin the sections on energy balances and heat transfer, itself so central to food processing. Chapter four is concerned with material and energy balances and concentrates upon the techniques required to solve problems. Most of the chapter is devoted to numerical examples drawn from a wide range of operations.

In chapter five the concepts of heat, mass and momentum transfer are introduced; the similarity between heat, mass and momentum transfer is stressed. This acts as an introduction to the material of the following three chapters. These cover: firstly, the flow of food fluids, in which the importance of laminar flow in food processing is emphasised, and food rheology where the objective is to enable the reader to apply rheological models to experimental data and to understand their significance in mechanistic and structural terms. Secondly, heat transfer, which is at the heart of many food processing operations. The basic principles are covered in detail and illustrated with numerous worked examples. Thirdly, mass transfer, which is often perceived as a difficult topic and indeed is poorly treated in many food texts. As a consequence mass transfer theory is under used in the analysis of food processes. Chapter eight is intended to redress this imbalance and the treatment of mass transfer is extended in chapter nine where the principles of psychrometry are explained.

The principal preservation operations are covered in chapters ten, eleven and twelve. These include the commercial sterilisation of foods, where the bases of the general and mathematical models are outlined and emphasis is given to a clear explanation of calculation procedures; low temperature preservation, including coverage of the principles of the refrigeration cycle; evaporation and drying. The processing of food particulates is often overlooked and chapter thirteen is an attempt to address this oversight. It considers the characterisation of individual particles and the development of relationships for particle-fluid interaction; fluidisation is included at this point because it is a fundamental processing technique with wide application to many unit operations. Finally chapter fourteen covers mixing and physical separation processes including the increasingly important area of separation using ultrafiltration and reverse osmosis.

INTRODUCTION TO FOOD
PROCESS ENGINEERING

1

An Introduction to Food Process Engineering

A process may be thought of as a sequence of operations which take place in one or more pieces of equipment, giving rise to a series of physical, chemical, or biological changes in the feed material and which results in a useful or desirable product. More traditional definitions of the concept of *process* would not include the term *biological* but, because of the increasing sophistication, technological advance and economic importance of, the food industry, and the rise of the biotechnology industries, it is ever more relevant to do so.

Process engineering is concerned with developing an understanding of these operations and with the prediction and quantifying of the resultant changes to feed materials (such as composition and physical behaviour). This understanding leads in turn to the specification of the dimensions of process equipment and the temperatures, pressures and other conditions required to achieve the necessary output of product. It is a quantitative science in which accuracy and precision, measurement, mathematical reasoning, modelling and prediction are all important. Food process engineering is about the operation of processes in which food is manufactured, modified, and packaged. Two major categories of process might be considered; those which ensure food safety, that is the preservation techniques such as freezing or sterilisation, which usually involve the transfer of heat and induce changes to microbiological populations, and those which may be classified as food manufacturing steps. Examples of the latter include the addition of components in mixing, the separation of components in filtration or centrifugation, or the formation of particles in spray drying. Classification in this way is rather artificial and by no means conclusive but serves to illustrate the variety of reasons for processing food materials.

Although foods are always liquid or solid in form, many foods are aerated (e.g., ice cream), many processes utilise gases or vapours (e.g., steam as a heat source) and many storage procedures require gases of a particular composition. Thus it is important for the food technologist or the food engineer to understand in detail the properties and behaviour of gases, liquids, and solids. In other words the transfer of heat, mass, and momentum in fluids and an understanding of the behaviour of solids, especially particulate solids, form the basis of food processing technology. At the heart of process engineering is the concept of the unit operation. Thus the principles which underlie drying, extraction, evaporation, mixing and sterilisation are independent of the material which is being processed. Once understood these principles can be applied to a wide range of products.

The overall purpose of food process engineering then is to design processes which result in safe food products with specific properties and structure. Foods, of course, have their own particular and peculiar properties: most food liquids are non-Newtonian; structures are often complex and multi-phase; non-isotropic properties are common. In addition to this, hygiene is of paramount importance in all manufacturing steps. The correct design of such processes is possible only as a result of the development of mathematical models which incorporate the relevant mechanisms. Thus it is important to understand the chemical, structural, and microbiological aspects of food in so far as they contribute to an understanding of the process; that is, how to develop, design, operate and improve the process to give better performance at reduced cost and, above all, improved safety and quality.

The first step in the design of a process is the conception stage. What is the product to be manufactured? What steps will be needed in order to manufacture it? In some cases the necessary steps may be very well known and there is no particular innovation required. As an example take the manufacture of ice cream. Whilst individual products may be innovative to a degree, the essential production steps are well known. There will be a mixing step in which the solid and liquid ingredients are added to the batch, followed by pasteurisation, storage or ageing, freezing and finally filling and packaging. For many food products there is an established way of doing things and there may be no realistic alternative. In other cases it may be far less obvious what the final process design will look like. In each case a simple flow sheet of the process should be prepared.

At this point it is likely to become apparent whether the process is to be batch or continuous. A batch process is one in which a given mass of material is subject to a series of operations in a particular sequence. For example, a batch of liquid may be heated, a second component added, the mixture agitated and then the resultant liquid cooled all within a single vessel. Alternatively the sequence of operations may involve a number of pieces of equipment. In a simple mixing operation, or where a chemical reaction occurs, the composition of the batch changes with time. If a liquid is heated in a stirred vessel the temperature of the liquid will be uniform throughout the vessel, provided the agitation is adequate, but will change with time. Batch processes generally have two disadvantages. Firstly they are labour intensive because of the bulk handling of material involved and the large number of individual operations which are likely to be used. Secondly the quality of the product may well vary from batch to batch. These problems are largely overcome if the process becomes continuous. Here, material flows through a series of operations and individual items of equipment undergoing a continuous change without manual handling. Once running, a continuous process should run for a long period under steady-state conditions; that is the composition, flow rate, temperature or any other measurable quantity should remain constant at any given point in the process. In this way a continuous process gives a more consistent product.

The mathematical analysis of a process also highlights an important difference between batch and continuous operation. Continuous, steady-state processes are usually considerably simpler to analyse than are unsteady-state batch processes because the latter involve changes in composition or temperature with time. However the difference between batch and continuous may not always be clear cut; many individual operations in the food industry are batch (often because of the scale of operation required) but are placed between other continuous operations. Thus the entire process, or a major section of it, is then best described as either semi-batch or semi-continuous.

The second stage of the design process may be called process analysis and this entails establishing both a material balance and an energy balance. The material balance aims to answer the question: what quantities of material are involved? What flow rates of ingredients are needed? In many cases this will be simply a case of establishing the masses of components to be added to a batch mixer. In others it will require the determination of flow rates of multi-component streams at several points in a complex process covering a large factory unit. In food processing the energy or enthalpy balance assumes enormous significance; sterilisation, pasteurisation, cooking, freezing, drying, and evaporation all involve the addition of heat to or removal of heat from the product. Establishing the necessary heat flows with accuracy is therefore of crucial importance both for reasons of food safety and of process efficiency.

A third stage comprises the specification of each operation and the design of individual pieces of equipment. In order to do this the prevailing physical mechanisms must be understood as well as the nature and extent of any chemical and biochemical reaction and the kinetics of microbiological growth and death. Specification of the size of heat transfer process equipment depends upon being able to predict the rate at which heat is transferred to a food stream being sterilised. In turn this requires a knowledge of the physical behaviour of the fluid, in short an understanding of fluid flow and rheology. This allows judgements to be made about how best to exploit the flow of material; for example, whether the flow should be co-current or counter-current.

Crucial to any process design is a knowledge of equilibrium and kinetics. Equilibrium sets the boundaries of what is possible. For example, in operations involving heat transfer a knowledge of thermal equilibrium (the heat capacity and the final temperatures required) allows the quantity of heat to be removed or added to be calculated. Equipment and processes can be sized only if the rate at which heat is transferred is known. Each rate process encountered in food engineering follows the same kind of law: where molecular diffusion is responsible for transfer, the rate of transfer of heat, mass or of momentum is dependant upon the product of a gradient in temperature, concentration or velocity respectively, and a diffusivity — a physical property which characterises the particular system under investigation. Where artificial convection currents are introduced, by the use of deliberate agitation, then an empirical coefficient must be used in conjunction with gradient term; little progress can be made in the application of heat transfer in food processing without a knowledge of the relevant heat transfer coefficient.

The overall design of the food process now moves onto the specification of instrumentation and process control procedures, to detailed costing and economic calculations, to detailed mechanical design and to plant layout. However all of these latter stages are beyond the scope of this book.

Dimensions, Quantities, and Units

2.1. DIMENSIONS AND UNITS

The dimensions of all physical quantities can be expressed in terms of the four basic dimensions: mass, length, time, and temperature. Thus velocity has the dimensions of length per unit time and density has the dimensions of mass per unit length cubed. A system of units is required so that the magnitudes of physical quantities may be determined and compared one with another. The internationally agreed system which is used for science and engineering is the Systeme International d'Unites, usually abbreviated to SI. Table 2.1 lists the SI units for the four basic dimensions together with those for electrical current and plane angle which, although strictly are derived quantities, are usually treated as basic quantities. Also included is the unit of molar mass.

The SI system is based upon the general metric system of units which itself arose from the attempts during the French Revolution to impose a more rational order upon human affairs. Thus the metre was originally defined as one ten-millionth part of the distance from the North Pole to the equator along the meridian which passes through Paris. It was subsequently defined as the length of a bar of platinum–iridium maintained at a given temperature and pressure at the Bureau International des Poids et Measures (BIPM) in Paris, but is defined now by the wavelength of a particular spectral line emitted by a krypton 86 atom.

TABLE 2.1

Dimensions and SI Units of the Four Basic Quantities
and Some Derived Quantities

Dimension	Symbol	SI unit	Symbol
Mass	M	Kilogram	kg
Length	L	Metre	m
Time	T	Second	s
Temperature	Θ	Degree Kelvin	K
Plane angle	—	Radian	rad
Electrical current	—	Ampere	A
Molar mass	—	Kilogram-molecular weight	kmol

The remaining units in Table 2.1 are defined as follows:

kilogram	The mass of a cylinder of platinum–iridium kept under given conditions at BIPM, Paris.
second	A particular fraction of a certain oscillation within a caesium 133 atom.
degree Kelvin	The temperature of the triple point of water, on an absolute scale, divided by 273.16. The degree Kelvin is the unit of *temperature difference* as well as the unit of thermodynamic temperature.
radian	The angle subtended at the centre of a circle by an arc equal in length to the radius.
Ampere	The electrical current which if maintained in two straight parallel conductors of infinite length and negligible cross-section, placed one metre apart in a vacuum, produces a force between them of 2×10^{-7} Newtons per metre length.
kmol	The amount of substance containing as many elementary units (atoms or molecules) as there are in 12 kg of carbon 12.

The SI system is very logical and, in a scientific and industrial context, has a great many advantages over previous systems of units. However it is usually criticised on two counts. First, that the names given to certain derived units, such as the Pascal for the unit of pressure, of themselves mean nothing and that it would be better to remain with, for example, the kilogram per square metre. This is erroneous; the definitions of the Newton, Joule, Watt, and Pascal are simple and straightforward if the underlying principles are understood. Derived units which have their own symbols, and which are encountered in this book, are listed in Table 2.2.

The second criticism concerns the magnitude of many units and the resulting numbers which are often inconveniently large or small. This problem would occur with any system of units and is not peculiar to SI. However there are instances when strictly non-SI units may be preferred. For example, the wavelengths of certain kinds of electromagnetic radiation may be more conveniently written in terms of the angstrom, one angstrom being equal to 10^{-10} m. Flow rates and production figures, when expressed in $\mathrm{kg\,h}^{-1}$ or even in tonnes day^{-1}, may be more convenient than in $\mathrm{kg\,s}^{-1}$. Pressures are still often quoted in bars or standard atmospheres rather than Pascals simply because the Pascal is a very small unit. Many of these latter disadvantages

TABLE 2.2
Some Derived SI Units

Name	Symbol	Quantity represented	Basic units
Newton	N	Force	$\mathrm{kg\,m\,s}^{-2}$
Joule	J	Energy or work	$\mathrm{N\,m}$
Watt	W	Power	$\mathrm{J\,s}^{-1}$
Pascal	Pa	Pressure	$\mathrm{N\,m}^{-2}$
Hertz	Hz	Frequency	s^{-1}
Volt	V	Electrical potential	$\mathrm{W\,A}^{-1}$

can be overcome by using prefixes. Thus a pressure of 10^5 Pa might better be expressed as 100 kPa. A list of prefixes is given in Appendix A.

It must be stressed that whatever shorthand methods are used to present data, the strict SI unit must be used in calculations. Mistakes are made frequently by using, for example, $kW\,m^{-2}\,K^{-1}$ for the units of heat transfer coefficients in place of $W\,m^{-2}\,K^{-1}$. Although such errors ought to be obvious it is often the case that compound errors of this kind result in plausible values based upon erroneous calculations.

A further note on presentation is appropriate at this point. I believe firmly that the use of negative indices, as in $W\,m^{-2}\,K^{-1}$, avoids confusion and is to be preferred to the solidus as in $W/m^2\,K$. The former method is used throughout this book. Appendix B gives a list of conversion factors between different units.

2.2. DEFINITIONS OF SOME BASIC PHYSICAL QUANTITIES

2.2.1. Velocity and Speed

Velocity and speed are both defined as the rate of change of distance with time. Thus, speed u is given by:

$$u = \frac{dx}{dt} \tag{2.1}$$

where x is the distance and t is the time. Average speed is then the distance covered in unit elapsed time. Velocity and speed differ in that speed is a scalar quantity (its definition requires only a magnitude together with the relevant units) but velocity is a vector quantity and requires a direction to be specified. A process engineering example would be the velocity of a fluid flowing in a pipeline; the velocity must be specified as the velocity in the direction of flow as opposed to, for example, the velocity perpendicular to the direction of flow. Thus velocity in the x direction might be designated u_x where

$$u_x = \frac{dx}{dt} \tag{2.2}$$

In practice the term velocity is used widely without specifying direction explicitly because the direction is obvious from the context. The SI unit of velocity is $m\,s^{-1}$.

2.2.2. Acceleration

Acceleration is the rate of change of velocity with time. Thus the acceleration at any instant, a, is given by

$$a = \frac{du}{dt} \tag{2.3}$$

It should be noted that the term acceleration does not necessarily indicate an increase in velocity with time. Acceleration may be positive or negative and this should be indicated along with the magnitude. The SI unit of acceleration is $m\,s^{-2}$.

2.2.3. Force and Momentum

The concept of force can only be understood by reference to Newton's laws of motion.

First law   A body will continue in its state of rest or uniform motion in a straight line unless acted upon by an impressed force.

Second law   The rate of change of momentum of the body with time is proportional to the impressed force and takes place in the direction of the force.

Third law   To each force there is an equal and opposite reaction.

These laws cannot be proved but they have never been disproved by any experimental observation. The momentum of an object is the product of its mass m and velocity u:

$$\text{momentum} = mu \tag{2.4}$$

From Newton's second law, the magnitude of a force F acting on a body may be expressed as

$$F \propto \frac{d(mu)}{dt} \tag{2.5}$$

If the mass is constant, then

$$F \propto m\frac{du}{dt} \tag{2.6}$$

and

$$F \propto ma \tag{2.7}$$

In other words, force is proportional to the product of mass and acceleration. A suitable definition of the unit of force will result in a constant of proportionality in Equation (2.7) of unity. In the SI system the unit of force is the Newton (N). One Newton is that force, acting upon a body with a mass of 1 kg, which produces an acceleration of $1\,\text{m}\,\text{s}^{-2}$. Hence

$$F = ma \tag{2.8}$$

2.2.4. Weight

Weight is a term for the localised gravitational force acting upon a body. The unit of weight is therefore the Newton and not the kilogram. The acceleration produced by gravitational force varies with the distance from the centre of the earth and at sea level the standard value is $9.80665\,\text{m}\,\text{s}^{-2}$ which is usually approximated to $9.81\,\text{m}\,\text{s}^{-2}$. The acceleration due to gravity is normally accorded the symbol g.

The magnitude of a Newton can be gauged by considering the apple falling from a tree which was observed supposedly by Isaac Newton before he formulated the theory of universal gravitation. The force acting upon an average-sized apple with a mass of, say, 0.10 kg, falling under gravity, would be, using Equation (2.8):

$$F = 0.10 \times 9.81\,\text{N}, \quad F = 0.981\,\text{N}, \quad F \approx 1.0\,\text{N}$$

The fact that an average-sized apple falls with a force of about 1.0 N is nothing more than an interesting coincidence. However this simple illustration serves to show that the Newton is a very small unit.

2.2.5. Pressure

A force F acting over a specified surface area A gives rise to a pressure P. Thus

$$P = \frac{F}{A} \tag{2.9}$$

In the SI system the unit of pressure is the Pascal (Pa). One Pascal is that pressure generated by a force of $1\,N$ acting over an area of $1\,m^2$. Note that standard atmospheric pressure is 1.01325×10^5 Pa. There are many non-SI units with which it is necessary to become familiar; the bar is equal to 10^5 Pa and finds use particularly for pressures exceeding atmospheric. A pressure can be expressed in terms of the height of a column of liquid which it would support which leads to many common pressure units, derived from the use of the simple barometer for pressure measurement. Thus atmospheric pressure is approximately $0.76\,m$ of mercury or $10.34\,m$ of water. Very small pressures are sometimes expressed as mm of water or 'mm water gauge'. Unfortunately it is still common to find imperial units of pressure, particularly the pound per square inch or psi.

Consider a narrow tube held vertically so that the lower open end is below the surface of a liquid (Figure 2.1). The pressure of the surrounding atmosphere P forces a column of liquid up the tube to a height h. The pressure at the base of the column (point B) is given by the weight of liquid in the column acting over the cross-sectional area of the tube.

$$\text{mass of liquid in tube} = \frac{\pi d^2}{4} h \rho \tag{2.10}$$

where ρ is the density of the liquid. The weight of liquid (force acting on the cross-section) F is

$$F = \frac{\pi d^2}{4} h \rho g \tag{2.11}$$

The pressure at B must equal the pressure at A, therefore:

$$P = \frac{\pi d^2}{4} h \rho g \frac{4}{\pi d^2} \tag{2.12}$$

and

$$P = \rho g h \tag{2.13}$$

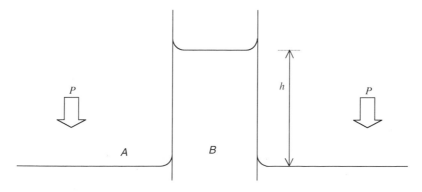

Figure 2.1. Pressure at the base of a column of liquid.

Example 2.1

What pressure will support a column of mercury (density $= 13,600\,\mathrm{kg\,m^{-3}}$) 80 cm high?

From Equation (2.13) the pressure at the base of the column is

$$P = 13,600 \times 9.81 \times 0.80\,\mathrm{Pa}$$

and therefore

$$P = 1.0673 \times 10^5\,\mathrm{Pa}$$

2.2.6. Work and Energy

Work and energy are interchangeable quantities; work may be thought of as energy in transition. Thus in an internal combustion engine, for example, chemical energy in the fuel is changed into thermal energy which in turn produces expansion in a gas and then motion, first of a piston within a cylinder and then of a crankshaft and of a vehicle. The SI unit of all forms of energy and of mechanical work is the Joule (J). A Joule is defined as the work done when a force of 1 N moves through a distance of 1 m. For example, if an apple of weight 1 N (see section 2.2.4) is lifted through a vertical distance of 1 m then the net work done on the apple is 1 J. This is the energy required simply to lift the apple against gravity and does not take into account any inefficiencies in the device — be it a mechanical device or the human arm. Clearly the Joule is a small quantity. A further illustration of the magnitude of the Joule is that approximately 4,180 J of thermal energy are required to increase the temperature of 1 kg of water by 1 K (see section 3.1).

Example 2.2

A mass of 250 kg is to be raised by 5 m against gravity. What energy input is required to achieve this?

From Equation (2.8) the gravitational force acting on the mass is

$$F = 250 \times 9.81\,\mathrm{N} \quad\text{or}\quad F = 2452.5\,\mathrm{N}$$

The work done (or energy required) W in raising the mass against this force is

$$W = F \times \text{distance}$$

and therefore

$$W = 2452.5 \times 5\,\mathrm{J} \quad\text{or}\quad W = 1.23 \times 10^4\,\mathrm{J}$$

2.2.7. Power

Power is defined as the rate of working or the rate of usage or transfer of energy and thus it involves time. The apple may be lifted slowly or quickly. The faster it is lifted through a given distance the greater is the power of the mechanical device. Alternatively the greater the mass lifted (hence the greater the force overcome) within the same period, the greater will be the power. A rate of energy usage of 1 Joule per second ($1\,\mathrm{J\,s^{-1}}$) is defined as 1 Watt (W).

Example 2.3

The mass in Example 2.2 is lifted in 5 s. What power is required to do this? What power is required to lift it in 1 min?

The power required is equal to the energy used (or work done) divided by the time over which the energy is expended. Therefore,

$$\text{power} = \frac{1.23 \times 10^4}{5} \text{W}$$

$$= 2452.5 \text{ W or } 2.45 \text{ kW}.$$

If now the mass is lifted over a period of 1 min,

$$\text{power} = \frac{1.23 \times 10^4}{60} \text{W}$$

$$= 204.4 \text{ W}$$

Whilst most examples in this book are concerned with rates of thermal energy transfer, it is important to understand (and may be easier to visualise) the definitions of work and power in mechanical terms. In addition, students of food technology and food engineering do require some basic knowledge of electrical power supply and usage, although that is outside the scope of this book. Suffice it to say that the electrical power consumed (in watts) when a current flows in a wire is given by the product of the current (in amperes) and the electrical potential (in volts).

2.3. DIMENSIONAL ANALYSIS

2.3.1. Dimensional Consistency

All mathematical relationships which are used to describe physical phenomena should be dimensionally consistent. That is, the dimensions (and hence the units) should be the same on each side of the equality. Take Equation (2.13) as an example.

$$P = \rho g h \tag{2.13}$$

Using square brackets to denote dimensions or units, the dimensions of the terms on the *right*-hand side are as follows:

$$[\rho] = ML^{-3}, \quad [g] = LT^{-2}, \quad [h] = L$$

Thus the dimensions on the *right*-hand side of Equation (2.13) are:

$$[\rho g h] = (ML^{-3})(LT^{-2})(L)$$

and

$$[\rho g h] = ML^{-1}T^{-2}$$

Pressure is a force per unit area, force is given by the product of mass and acceleration and therefore the dimensions on the *left*-hand side of Equation (2.13) are:

$$[P] = \frac{[\text{mass}][\text{acceleration}]}{[\text{area}]}$$

or

$$[P] = \frac{(\text{M})(\text{LT}^{-2})}{\text{L}^2}, \qquad [P] = \text{ML}^{-1}\text{T}^{-2}$$

Similarly the units must be the same on each side of the equation. This is simply a warning not to mix SI and non-SI units and to be aware of prefixes. The units of the various quantities in Equation (2.13) are

$$[P] = \text{Pa}, \qquad [\rho] = \text{kg}\,\text{m}^{-3} \qquad [g] = \text{m}\,\text{s}^{-2}, \qquad [h] = \text{m}$$

and therefore the *right*-hand side of Equation (2.13) becomes

$$[\rho g h] = (\text{kg}\,\text{m}^{-3}) \times (\text{m}\,\text{s}^{-2}) \times (\text{m})$$

or

$$[\rho g h] = \text{kg} \times \text{m}\,\text{s}^{-2} \times \text{m}^{-2}$$

As a force of one Newton, acting upon a body of mass 1 kg, produces an acceleration of $1\,\text{m}\,\text{s}^{-2}$, the units on the *right*-hand side of Equation (2.13) are now

$$[\rho g h] = \text{N}\,\text{m}^{-2}$$

and thus

$$[\rho g h] = \text{Pa}$$

2.3.2. Dimensional Analysis

The technique of dimensional analysis is used to rearrange the variables which represent a physical relation to give a relationship that can more easily be determined by experimentation. For example, in the case of the transfer of heat to a food fluid in a pipeline, the rate of heat transfer might depend upon the density and viscosity of the food, the velocity in the pipeline, the pipe diameter and other variables. Dimensional analysis, by using the principle of dimensional consistency, suggests a more detailed relationship which is then tested experimentally to give a working equation for predictive or design purposes. It is important to stress that this procedure must always consist of dimensional analysis followed by detailed experimentation. The theoretical basis of dimensional analysis is too involved for this text but an example, related to heat transfer (chapter seven), is set out in Appendix C.

NOMENCLATURE

a Acceleration
A Area
d Diameter
F Force
g Acceleration due to gravity
h Height
m Mass
P Pressure
t Time
u Speed or velocity
u_x Velocity in the x direction
W Work
x Distance

GREEK SYMBOLS

ρ Density

PROBLEMS

2.1. What is the pressure at the base of a column of water 20 m high? The density of water is $1,000 \, \text{kg m}^{-3}$.

2.2. What density of liquid is required to give standard atmospheric pressure (101.325 k Pa) at the base of a 12 m high column of that liquid?

2.3. A person of mass 75 kg climbs a vertical distance of 25 m in 30 s. Ignoring inefficiencies and friction, (a) how much energy does the person expend, and (b) how much power must the person develop?

2.4. Determine the dimensions of the following:

(a) linear momentum (mass × velocity)
(b) work
(c) power
(d) pressure
(e) weight
(f) moment of a force (force × distance)
(g) angular momentum (linear momentum × distance)
(h) pressure gradient
(i) stress
(j) velocity gradient

3

Thermodynamics and Equilibrium

3.1. INTRODUCTION

Thermodynamics takes its name from the Greek for 'movement of heat' and is the science concerned with the interchange of energy, particularly that between thermal energy and mechanical work. Thermodynamics is concerned with systems which have come to equilibrium and not with the rate at which equilibrium is achieved. Two examples can be used to illustrate this point. Chemical thermodynamics describes the extent to which a chemical reaction proceeds based upon a knowledge of the total quantity of energy involved, in contrast to chemical kinetics which attempts to describe and predict the rate at which a chemical reaction takes place. Perhaps more pertinent to the aim of this book, thermodynamics is able to specify the thermal energy changes required to bring about certain physical changes in a system: an increase or decrease in temperature; a change of state from liquid to vapour or from solid to liquid. Thus thermodynamics predicts the heat input required to raise the temperature of a food to evaporate water from an aqueous food solution or to thaw a block of frozen food. It does not, however, have anything to say about the rate at which the transfer of thermal energy should or can take place; that is the province of heat transfer which is covered in chapters five and seven. Thus thermodynamics is concerned with the initial and final states of a process and not with how the movement between those states is achieved. It is perhaps, cosmology and quantum mechanics apart, the most philosophical of all the sciences and indeed it underpins even those subjects.

This chapter is concerned with defining and explaining the major thermodynamic quantities and making practical use of the laws of thermodynamics and the gas laws. This will give the reader an understanding of the behaviour of vapours and gases, a basis from which to carry out energy balances across processes and sufficient theory to move on to the study of heat transfer. It is not concerned with applied thermodynamics in the mechanical engineering sense, nor with chemical thermodynamics as such, but with the theory necessary for understanding a range of food processing operations.

3.1.1. Temperature and the Zeroeth Law of Thermodynamics

We know from experience that it is possible to detect, by a sense of touch, that bodies can be either 'hotter' or 'colder' than others. It is also a matter of experience that if a 'hotter' and a 'colder' body are in close contact, and preferably isolated from their surroundings as much as is possible, then the hotter body will become cooler and the colder body will become warmer. In the same way, if the state of a hot body is maintained constant by the continuous supply of energy, then a colder body in contact with it will, over a period, approach the degree of hotness of the first body. Of course the preceding description could have employed the word 'temperature' but there is no *a priori* reason why that should be the case. It is part of the function of the subject of thermodynamics to define what is meant by temperature.

It is also part of everyday experience that the transfer of heat between two bodies is linked to the perceived difference in their 'hotness'. However there is a fundamental question to be resolved here. Is the difference in 'hotness' independent from the transfer of energy (heat) and which is observed to occur with a difference in 'hotness'? In other words, is the temperature difference something distinct and significant?

If two bodies are each in thermal equilibrium (i.e., they are in contact with each other such that heat can be transferred and all observable change has come to an end) with a third body then it follows that the first two bodies are in thermal equilibrium with each other. This result is known as the Zeroeth law of thermodynamics and forms the basis for the concept of temperature. The Zeroeth law (so-called because, although it is fundamental, it was formulated only after the first and second laws) establishes that temperature itself is a meaningful concept and thus it can be measured and assigned values. Temperature can be measured because there is an observed correlation between the 'sense of hotness' and physical phenomena such as the expansion of alcohol or liquid mercury in a glass tube which forms the basis of the device we call a thermometer.

3.1.2. Temperature Scale

Having established the concept of temperature and some physical means of registering a quantitative measure of temperature, a temperature scale is required which is inalterable and via which comparisons can be made. Fahrenheit made a worthy attempt to obtain the lowest temperature possible (with a mixture of water, ice and various salts) which he labelled $0°F$. For unknown reasons he set the temperature of the human body at $96°F$ and, having established this scale, the freezing and boiling points of water became $32°F$ and $212°F$, respectively. (Subsequently, more accurate measurements gave the human body temperature to be $98.4°F$.) Arguably, the Swedish physicist Elvins was more rational in setting the freezing point of water to $0°C$ and the boiling point to $100°C$. His centigrade scale has become known as the Celsius scale (after a later Swedish physicist) in which the freezing and boiling points of pure water are $0°C$ and $100°C$, respectively.

William Thomson (Lord Kelvin) showed that the lowest possible temperature, absolute zero, is $-276.15°C$. An absolute temperature scale has considerable theoretical significance and is used extensively in the physical sciences. Thus, absolute zero on the Kelvin scale is $0\,K$ and the freezing point of water is $273.15\,K$, which is usually approximated to $273\,K$. Note that the symbol for degree Kelvin is K and not $°K$. Heat transfer depends upon temperature difference which should properly be quoted in degree Kelvin and not degree Celsius, but of course they have the same magnitude. There is an absolute temperature scale based upon the Fahrenheit scale on which absolute zero is zero degree Rankine ($0°R$), $0°F$ is equal to $460°R$ and water freezes at $492°R$. The relationship between these scales is shown diagrammatically in Figure 3.1.

3.1.3. Heat, Work, and Enthalpy

A number of terms must now be defined before applying thermodynamic reasoning to the transfer of work and heat and arriving at a series of working relationships.

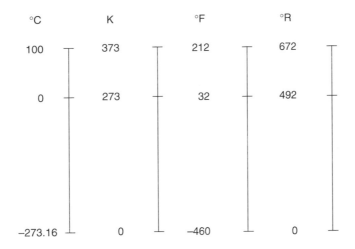

Figure 3.1. Relationship between temperature scales.

Heat is a form of energy, which is transferred from one body to another body due to a temperature difference between them. Heat cannot be contained in a body or possessed by a body. The term *sensible heat* is used to describe heat which results in a temperature change within a substance, that is the addition or removal of sensible heat can be sensed because of the temperature change. *Latent heat* refers to the heat which accompanies a change of phase, for example, from a liquid to a vapour (latent heat of vaporisation) or from a solid to a liquid (latent heat of fusion). The addition or removal of latent heat does not involve a temperature change, so that the exact quantity of heat equal to the latent heat of vaporisation, when added to pure water at $100°C$ and atmospheric pressure, will produce vapour at $100°C$.

Work, like heat, is energy in transition and similarly it is never contained in a body or possessed by a body. However work can be done *on* a thermodynamic fluid or *by* a thermodynamic fluid on the surroundings.

A *thermodynamic fluid* is fluid which is subjected to thermodynamic processes (e.g., the addition and removal of heat or work) for some useful purpose. An example of a thermodynamic fluid would be the refrigerant (perhaps ammonia) which is circulated around a refrigeration system to transfer heat from a cold body to a hot body. The state of a fluid can be defined by its properties (e.g., temperature and pressure). Two other properties must now be defined: Any body or substance at a temperature above absolute zero has a positive energy content. The energy of a fluid may be increased by adding heat or by doing work on it. Such work might result in a fluid gaining kinetic energy. However, *internal energy* is the intrinsic energy content of a fluid which is not in motion. It is a function of temperature and pressure.

Enthalpy is defined as the sum of internal energy and the work done on the fluid. It may be thought of as 'energy content', and is explained more fully in section 3.4 in the context of the first law of thermodynamics. Thus, while bodies do not contain heat or work they do possess internal energy and/or enthalpy.

3.1.4. Other Definitions

The term *isothermal* refers to a line of constant temperature or to a process in which temperature is constant. An *adiabatic* process is one in which no heat is transferred to or from the thermodynamic fluid. *Reversibility* is defined thus: 'when a fluid undergoes a *reversible* process, both the fluid and the surroundings can always be restored to their original state'. In practice no process is truly reversible, but a close approach to *internal* reversibility is possible when the fluid returns to its original state, although the surroundings do not.

3.2. THE GASEOUS PHASE

Studying the nature of gases is important for two reasons. First, students of food engineering need to understand the behaviour of gases and vapours in order to understand many food processing operations. Second, a gas is a more convenient thermodynamic fluid than is a liquid with which to understand thermodynamic principles.

3.2.1. Kinetic Theory of Gases

The kinetic theory of gases assumes that a gas is composed of molecules with each molecule behaving as if it were a separate particle (rather like a billiard ball) and free to move in space according to Newton's laws. Each gas 'particle' is constantly in motion and has an inherent kinetic energy. When heat is added to a gas the result is that the kinetic energy of the gas increases and it is this average kinetic energy which defines the temperature of the gas. Thus an increase in gas temperature signifies an increase in average kinetic energy and an increase in the average particle velocity.

In this simple treatment individual gas molecules impinge upon the wall of the container within which the gas is held (rather as billiard balls hit the cushions on the side of a billiard table) and thus exert a force on the wall. This force averaged over the surface area of the wall is the pressure of the gas within the container. Now a rise in temperature produces an increase in average particle velocity and a corresponding increase in gas pressure as molecules hit the walls of the container more frequently and with higher velocities.

3.2.2. Perfect Gases

(*a*) *The Gas Laws*   Robert Boyle was the first to discover that, at a constant temperature, the volume of a gas V varies inversely with its pressure P. Thus,

$$V \propto \frac{1}{P} \tag{3.1}$$

or

$$PV = \text{constant} \tag{3.2}$$

Equation (3.2) is the equation of a hyperbola which is represented in Figure 3.2. This curve therefore represents a line of constant temperature and is known as an isotherm. This simple relationship holds for many common gases at moderate pressures; those gases which follow the relationship exactly are known as perfect or ideal gases. Working much later than Boyle, Gay-Lussac investigated the variation of the volume of a gas with temperature and showed that

$$V = V_0(1 + \alpha T) \tag{3.3}$$

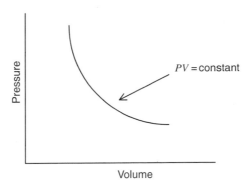

Figure 3.2. Boyle's law: variation in pressure and volume of a gas.

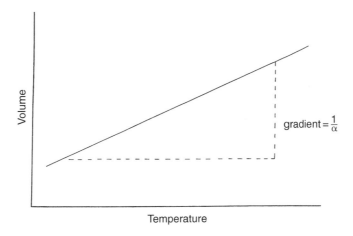

Figure 3.3. Determination of absolute zero: volume–temperature relationship for a perfect gas.

where V_0 is the volume at $0°C$ and α is a coefficient of expansion. His measurements gave a value for α of $1/267$. More accurate work with nitrogen, hydrogen and helium, amongst other gases, gave α equal to $1/273.15$. Therefore, by extrapolation, the value of absolute zero becomes $-273.15°C$. This is shown graphically in Figure 3.3. It should be pointed out that the inability to agree precisely the value for absolute zero led to its definition by convention in terms of the triple point of water ($0.01°C$), hence the definition of the degree Kelvin in section 2.1.

(b) *The Ideal Gas Law* A perfect gas, or ideal gas, is by definition one which obeys the ideal gas law

$$PV = nRT \qquad (3.4)$$

where T is absolute temperature, n is the molar mass of the gas in kmol and R is the universal gas constant. Equation (3.4) contains within it the laws and observations of Boyle, Gay-Lussac, and Avogadro, the third of which may be summarised as 'one mole of any gas contains the same number of atoms or molecules and occupies the same volume at a given temperature'.

If the gas is held at standard temperature and pressure (STP), 101.325 kPa and 273.15 K, then 1 kmol will occupy 22.4 m^3. This corresponds to a value for R of 8,314 J kmol^{-1} K^{-1}.

Example 3.1

A sample of 1 kmol of an ideal gas is held in a container with a volume of 22.4 m^3 and at a temperature of 0°C. What is the pressure in the container?

Taking care to use absolute temperature, that is $T = 273.15$ K, the ideal gas law, Equation (3.4), gives

$$P = \frac{1.0 \times 8314 \times 273.15}{22.4} \text{ Pa}, \qquad P = 1.014 \times 10^5 \text{ Pa}$$

Example 3.2

What volume will be occupied by 56 kg of nitrogen at a pressure of 1.3 bar and a temperature of 25°C? Assume that nitrogen behaves as an ideal gas.

The absolute temperature of the nitrogen is 298.15 K and its pressure is 1.3×10^5 Pa. The molecular weight of nitrogen is 28 and therefore, 56 kg represents $\frac{56}{28}$ or 2 kmol. Equation (3.4) can be rearranged to give

$$V = \frac{nRT}{P}$$

and therefore,

$$V = \frac{2.0 \times 298.15 \times 8314}{1.3 \times 10^5} \text{ m}^3, \qquad V = 38.14 \text{ m}^3$$

Example 3.3

A fermenter requires an oxygen flow of 1.7×10^{-5} kmol s^{-1} at 37°C. The oxygen is supplied as air (which may be assumed to be a perfect gas) at 200 kPa. Determine the necessary volumetric flow rate of air.

Assuming that the volumetric (and therefore the molar) composition of air is 21% oxygen, the required molar flow rate of air is $(1.7 \times 10^{-5}/0.21)$ or 8.10×10^{-5} kmol s^{-1}. The ideal gas law can now be used to find the volumetric flow rate of air from the molar flow rate. Thus,

$$V = \frac{nRT}{P}$$

and therefore,

$$V = \frac{8.1 \times 10^{-5} \times 8314 \times 310}{200 \times 10^3} \text{ m}^3, \qquad V = 1.04 \times 10^{-3} \text{ m}^3\text{s}^{-1}$$

If a perfect gas is originally at a pressure P_1, temperature T_1 and has a volume V_1, and the conditions are changed such that the pressure, temperature, and volume become P_2, T_2, and V_2 respectively, then

$$P_1 V_1 = nRT_1 \qquad\qquad (3.5)$$

and

$$P_2 V_2 = n R T_2 \tag{3.6}$$

where n is of course constant. Rearranging Equations (3.5) and (3.6) gives

$$n R = \frac{P_1 V_1}{P_2 V_2} = \frac{T_1}{T_2} \tag{3.7}$$

Therefore if the original conditions are known, together with two of the temperature, pressure and volume at the new condition, then the third can be found.

Example 3.4

Nitrogen is stored in a cylinder which has a working volume of 50 l under a pressure of 230 bar. What will be the volume of nitrogen released at standard atmospheric pressure, if the temperature remains constant on expansion?

Nitrogen may be assumed to behave as an ideal gas and therefore the relationship between pressure, volume, and temperature is given by Equation (3.7). However the temperature remains constant on expansion, that is $T_1 = T_2$, and thus

$$V_2 = \frac{P_1 V_1}{P_2}$$

where state 1 refers to conditions in the cylinder and state 2 to the expanded gas at atmospheric pressure. Hence

$$V_2 = \frac{230 \times 10^5 \times 50 \times 10^{-3}}{1.013 \times 10^5} \, \mathrm{m}^3, \qquad V_2 = 11.35 \, \mathrm{m}^3$$

Example 3.5

A can containing air, which may be assumed to be ideal, is heated to $80°\mathrm{C}$ and then sealed. The can is then left to cool in water at $15°\mathrm{C}$. Calculate the new pressure within the can.

The volume of the can remains constant ($V_1 = V_2$) and thus the pressure–volume–temperature relationship for a perfect gas [Equation (3.7)] reduces to

$$P_2 = \frac{T_2}{T_1} P_1$$

If the original pressure P_1 is atmospheric (101.3 kPa) then the new pressure in the can is

$$P_2 = \frac{288}{353} \times 1.013 \times 10^5 \, \mathrm{Pa}$$

and

$$P_2 = 8.26 \times 10^4 \, \mathrm{Pa} \quad \text{or} \quad 0.816 \text{ of an atmosphere.}$$

(c) *Density and Molar Concentration*   The ideal gas law can be used to determine the concentration of a gas. A further rearrangement of Equation (3.4) yields

$$\frac{n}{V} = \frac{P}{RT} \tag{3.8}$$

where n/V is the molar concentration (C) of the gas, that is the total number of moles per unit volume of gas. Thus

$$C = \frac{P}{RT} \tag{3.9}$$

Density, that is mass per unit volume, is simply a measure of concentration. For an ideal gas of molecular weight M, the density ρ is given by

$$\rho = \frac{nM}{V} \tag{3.10}$$

and therefore

$$\rho = \frac{PM}{RT} \tag{3.11}$$

Example 3.6

Determine the density of air at atmospheric pressure and 20°C. Take the mean molecular weight of air to be 28.96.

Using Equation (3.11) and an absolute temperature of 293.15 K,

$$\rho = \frac{1.013 \times 10^5 \times 28.96}{8314 \times 293.15} \text{ kg m}^{-3}, \qquad \rho = 1.204 \text{ kg m}^{-3}$$

3.2.3. Pure Component Vapour Pressure

In theory it is possible to change the state of all elements and compounds. So, for example, water can exist in the form of liquid water, as ice or as a vapour. Consider a beaker of water at atmospheric pressure and room temperature. Because there is a distribution of energy within the molecules of water in the beaker, some molecules of water will be able to change phase from liquid to vapour. Equally some vapour molecules will condense back from the vapour phase into the liquid phase. At a given temperature, the rates of vaporisation and condensation will be in equilibrium. The pressure exerted by a vapour (which has been generated from a pure substance) at equilibrium is called the pure component vapour pressure. This is usually abbreviated to vapour pressure or saturated vapour pressure (SVP).

All solids and liquids exert a vapour pressure however small; for solids the vapour pressure will usually be negligible. At low temperatures the number of molecules with sufficient energy to escape from the liquid phase to the vapour phase will be relatively small; for example, the pure component vapour pressure of water is only 2.34 kPa at 20°C. However vapour pressure is usually a strong function of temperature (see Figure 3.4) and at 100°C the rates of vaporisation and condensation are significantly greater such that, at equilibrium, the vapour pressure above a sample of pure water exerts a pressure of 101.325 kPa. This of course is atmospheric pressure and the example serves to demonstrate the significance of boiling point. The boiling point of

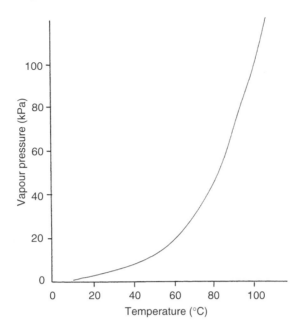

Figure 3.4. Vapour pressure of water as a function of temperature.

a substance (sometimes called bubble point) is that temperature at which the pure component vapour pressure is equal to atmospheric pressure. Thus at standard atmospheric pressure water boils at 100°C. However if ambient pressure is reduced to 80 kPa the boiling point of water falls to 93.5°C. Similarly an increase in pressure produces an increase in boiling point. This phenomenon has a particular practical consequence; namely that the temperature of steam (water vapour) used for process heating is a function of pressure. For example, steam at 300 kPa has a temperature of 133.5°C whilst at 500 kPa the temperature rises to 151.8°C. The availability and use of this kind of data is covered under the heading of energy balances in chapter four.

3.2.4. Partial Pressure and Pure Component Volume

In a mixture of gases, for example, air, each component is distributed uniformly throughout the volume of the mixture and the total pressure of the mixture is given by the sum of the pressures exerted by the molecules of each component. Thus, if for the purposes of this example air consists of 21% by volume oxygen, 78% nitrogen and 1% argon, the oxygen will be present at a uniform concentration (21%) throughout the volume of the container. Partial pressure is then the pressure which would be exerted by a single component occupying the same volume, and held at the same temperature, as the mixture. This is summarised by Dalton's law which states that the sum of the partial pressures in a mixture of ideal gases equals the total pressure of the mixture. This may be expressed as

$$P = p_A + p_B + p_C + \cdots \tag{3.12}$$

where P is the total pressure of the mixture and p_A, p_B and p_C are the partial pressures of components A, B, C, and so on.

It should be noted that Equation (3.9) can be applied to concentrations of components in a mixture of gases. For example, in a binary gaseous mixture of A and B, the molar concentrations of A and B are given by Equations (3.13) and (3.14), respectively.

$$C_A = \frac{p_A}{RT} \tag{3.13}$$

$$C_B = \frac{p_B}{RT} \tag{3.14}$$

Now from Dalton's law the total pressure is the sum of the partial pressures, thus it follows that

$$C = C_A + C_B \tag{3.15}$$

In other words the total molar concentration is equal to the sum of the concentrations of the individual components. Pure component volume is the volume occupied by a single component of a mixture of ideal gases if it is held at the same temperature and pressure as the mixture. The total volume of the mixture is then the sum of the individual pure component volumes. This is a statement of Amagat's law. Thus

$$V = v_A + v_B + v_C + \cdots \tag{3.16}$$

where V is the total volume of the mixture and v_A, v_B, and v_C are the pure component volumes of components A, B, C, and so on. Now, if ideality is obeyed, the ideal gas law can be applied to each component of the mixture in turn, and

$$p_A V = n_A RT \tag{3.17}$$

where n_A is the number of moles of A. For the mixture

$$PV = (n_A + n_B + n_C + \cdots)RT \tag{3.18}$$

Eliminating R, T, and V from Equations (3.17) and (3.18) gives a further statement of Dalton's law

$$p_A = \frac{n_A}{(n_A + n_B + n_C + \cdots)} P \tag{3.19}$$

or

$$p_A = y_A P \tag{3.20}$$

where y_A is the mole fraction of component A. Thus the partial pressure exerted in a mixture of ideal gases is given by the product of the mole fraction of that component and the total pressure.

A similar treatment yields a further statement of Amagat's law

$$v_A = y_A V \tag{3.21}$$

These relationships for ideal gases can now be summarised as:

$$\text{partial pressure}\% = \text{mol}\% = \text{volume }\% \tag{3.22}$$

noting that % can always be replaced by 'fraction'.

Example 3.7

A modified atmosphere for the packaging of chicken contains 30% carbon dioxide, 40% nitrogen, and 30% oxygen by volume. For a total pressure of 102 kPa and a temperature of 15°C, tabulate the mole fraction, partial pressure and mass of each component per m^3 of gas mixture.

The molar mass of $1\,m^3$ of the gas mixture can be obtained from the ideal gas law [Equation (3.4)], thus

$$n = \frac{1.02 \times 10^5 \times 1}{8314 \times 288}\,\text{kmol}, \qquad n = 0.0426\,\text{kmol}$$

For an ideal gas the mole fraction is equal to the volume fraction [Equation (3.22)] and hence for carbon dioxide the molar mass is 0.3×0.0426 or $0.0128\,\text{kmol}$. Taking the molecular weight of carbon dioxide as 44, the mass of carbon dioxide is then 0.0128×44 or $0.562\,\text{kg}$. The masses and molar masses of nitrogen and oxygen can be calculated in the same way. The partial pressure of carbon dioxide is found from Dalton's law [Equation (3.20)]:

$$p_{CO_2} = y_{CO_2} P$$

Hence

$$p_{CO_2} = 0.3 \times 1.02 \times 10^5\,\text{Pa} \qquad p_{CO_2} = 3.06 \times 10^4\,\text{Pa}$$

Consequently, the following table can be generated:

Component	Molecular weight	Mole fraction y	Partial pressure p_A (kPa)	Molar mass (kmol)	Mass (kg)
Carbon dioxide	44	0.30	30.6	0.0128	0.562
Nitrogen	28	0.40	40.8	0.0170	0.477
Oxygen	32	0.30	30.6	0.0128	0.409
Total		1.00	102.0	0.426	1.448

Example 3.8

Tomatoes are stored at 5°C in an atmosphere containing oxygen, nitrogen, and carbon dioxide. The partial pressure of oxygen must not exceed 2.6 kPa when the total pressure is 105 kPa. If 0.10 kg of carbon dioxide is charged to the gas mixture, then determine the volumetric composition.

Assuming ideality and using Equation (3.4), the total molar mass of the gas mixture, for a volume of $1\,m^3$, is

$$n = \frac{105 \times 10^3 \times 1}{8314 \times 278}\,\text{kmol}, \qquad n = 0.0454\,\text{kmol}$$

The molar mass of carbon dioxide added to the mixture is $0.10/44 = 2.273 \times 10^{-3}\,\text{kmol}$ and thus the mole fraction (equal to the volume fraction) of carbon dioxide is $(2.273 \times 10^{-3}/0.0454) = 0.05$. The mole fraction of oxygen can be deduced from the partial pressure fraction and thus,

$$y_{O_2} = \frac{2.6}{105} \quad \text{or} \quad y_{O_2} = 0.0248$$

By difference the mole fraction of nitrogen is 0.9252.

3.3. THE LIQUID–VAPOUR TRANSITION

3.3.1. Vaporisation and Condensation

Most substances can exist in each phase, however there is a tendency to think of materials in terms of the equilibrium phase which exists at ambient pressure and temperature. For example, nitrogen, oxygen, and carbon dioxide are thought of as gases but it is possible to liquefy each of them and liquid nitrogen is used in certain food-freezing applications. Mercury is thought of as a liquid but it can be vaporised. It is more common to consider water in each of its different phases (ice, liquid water, and water vapour) but water does also become a gas at very high temperatures. The phase change from a liquid to a vapour can be represented on a $P–V$ (pressure–volume) diagram an example of which is shown in Figure 3.5.

If heat is added to a liquid, which is maintained at a constant pressure P_1, then there is a fixed temperature at which vapour bubbles appear. This is known as the boiling point or bubble point (point Q in Figure 3.5). Further heating of the liquid at constant pressure leads to a phase change from liquid to vapour but both the temperature and pressure will remain constant. The resulting vapour occupies a fixed volume V_R at the point where vaporisation is just complete (point R) which is considerably greater than the volume occupied by the boiling liquid V_Q. At a higher pressure P_2 the boiling point of the liquid (represented by S) will be

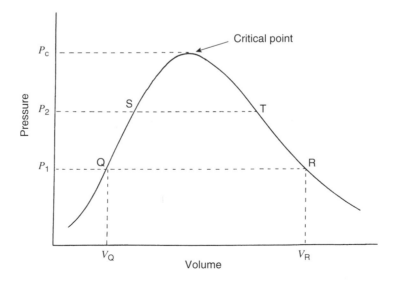

Figure 3.5. Pressure–volume diagram for a fluid.

higher and the volume occupied by the boiling liquid will have increased slightly (in other words its density will have decreased slightly) because of its expansion with temperature. If the pressure is increased sufficiently, then a continuous curve is formed; the turning point of the curve is called the critical point and the pressure at which this occurs is the critical pressure P_c.

The almost linear left-hand part of the curve in Figure 3.5 is called the saturated liquid line and represents all the possible combinations of pressure and volume for a (given) boiling liquid. The right-hand part of the curve is called the saturated vapour line and represents all possible combinations of pressure and volume for a saturated vapour. As we have seen in earlier sections of this chapter, the variation in the volume of a vapour with pressure is considerable. Note that the term 'saturation' here refers to energy saturation. The saturation state is therefore the state at which a phase change will occur without a change in temperature or pressure. In other words, a very slight addition of heat to a boiling liquid (i.e. a liquid on the saturated liquid line) gives rise to some vapour. Similarly the saturated vapour line represents the point where a liquid has been completely changed into vapour and if a substance on this line is very slightly cooled then droplets of liquid will begin to form. The $P-V$ diagram can now be used as a kind of phase diagram (Figure 3.6); to the left of the saturated liquid line the substance exists only in the liquid phase whilst to the right of the saturated vapour line only vapour exists. Inside the loop the fluid is a mixture of liquid and vapour (or 'wet vapour'). The heat which must be supplied to bring about the change of phase from saturated liquid to saturated vapour is the latent heat of vaporisation which is represented by the magnitude of the gap between the two saturation lines. Note that at the critical point the latent heat is zero. Sensible heat which is added to a saturated vapour is known as superheat. Condensation is simply the reverse of vaporisation. The latent heat of condensation is exactly equal in magnitude to the latent heat of vaporisation, but it is more usual to refer to the latter term. Removal of the latent heat from a saturated vapour will produce a saturated liquid, at the same temperature and pressure.

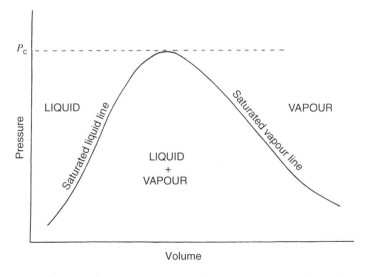

Figure 3.6. Phase diagram for a fluid at the vapour–liquid transition.

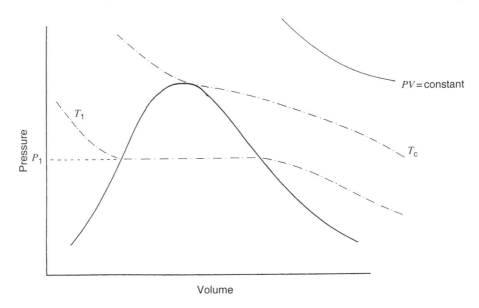

Figure 3.7. Isotherms and the critical temperature.

3.3.2. Isotherms and Critical Temperature

Isotherms or lines of constant temperature can be plotted on the $P-V$ diagram and there is a saturation temperature for each saturation pressure (Figure 3.7). Thus at the pressure P_1 the temperature is T_1 and this remains constant as the latent of vaporisation is added or removed from a saturated fluid. Because both the temperature and pressure are constant between the two curves, some other property (such as the specific volume) must be specified in order to define the state of the fluid, which could be a vapour, a liquid or a mixture of the two. The critical temperature T_c is the isotherm which is tangential to the turning point.

3.3.3. Definition of Gas and Vapour

The distinction between a gas and a vapour is not a rigid one. However at high degrees of superheat, well beyond the saturated vapour line, the isotherms become hyperbolic, that is, they are described by the equation

$$PV = \text{constant} \tag{3.23}$$

Now a perfect gas obeys Equation (3.4) which may be rearranged to give

$$\frac{PV}{T} = \frac{n}{R} = \text{constant} \tag{3.24}$$

and therefore when an isotherm (constant T) obeys Equation (3.23) Equation (3.24) is satisfied. Thus the shape of the isotherm plotted on a $P-V$ diagram will indicate whether the fluid behaves as a vapour or as a perfect gas.

All substances tend towards the perfect gas state at very high degrees of superheat but the common gases such as nitrogen and oxygen are already superheated at ambient conditions.

TABLE 3.1
Critical Pressure and Temperature of Some
Common Substances

	T_c (K)	T_c (°C)	P_c (kPa)
Water	647	374	2.15
Oxygen	154	−119	0.49
Hydrogen	33	−240	0.12
Carbon dioxide	304	31	0.72
Mercury	1,765	1,492	11.2

Substances which are normally vapours at ambient must be raised to much higher temperatures before they behave as perfect gases. Referring to Figures 3.6 and 3.7, it should be noted that there is no discontinuity of phase above the critical temperature and that it is impossible to liquefy a gas, no matter how high the pressure, if the temperature is above T_c. Table 3.1 gives the critical pressures and temperature of some common substances.

From this data it can be seen that water (at temperatures well above its freezing point) is always experienced as a vapour–liquid mixture at atmospheric pressure because $T_c = 374°C$ and ambient temperature is well below this.

3.3.4. Vapour–Liquid Equilibrium

In examining the transition between the vapour and liquid phases we have so far considered only a pure component with the obvious consequence that the composition of each phase is 100% of that component in all cases. However, in a binary system (e.g., ethanol and water) different concentrations of each component can be expected in the vapour and liquid phases because of the difference in volatility (or 'ease of vaporisation') of the two components. These concentrations can be predicted using the laws of Raoult, Henry, and Dalton. In the treatment presented here only binary systems are considered.

(a) *Raoult's Law* Raoult's law applies to ideal solutions and may be stated as follows:

The equilibrium partial pressure exerted by a component in solution is proportional to the mole fraction of that component in the liquid phase.

Now, the surface of an homogenous solution contains molecules of each component and the number of molecules of a given component (say of the more volatile component (MVC) of the two) per unit surface inevitably will be smaller than in the case of a pure liquid. Therefore the rate of vaporisation of the MVC will be correspondingly smaller from a solution than from a pure liquid. However, the rate of condensation of molecules of MVC is unaffected by the presence of the other component at the solution surface. In other words, it is more difficult for a molecule of the MVC in solution to move into the vapour phase and remain there than is the case with a pure liquid. Consequently the equilibrium partial pressure exerted by a component in solution is less than the pure component vapour pressure. Raoult's law may be written

$$p_A = x_A p_A'$$ (3.25)

where p_A is the partial pressure of component A, x_A is the mole fraction of A in the liquid phase and p_A' is the pure component vapour pressure of A at the relevant temperature. Unfortunately

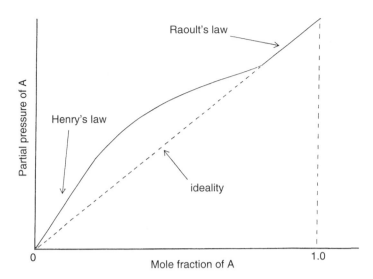

Figure 3.8. Vapour–liquid equilibrium for a binary system.

Raoult's law is obeyed only at high liquid phase concentrations, that is at high values of x_A. This is shown diagrammatically in Figure 3.8. Below mole fractions of approximately 0.80 most solutions do not demonstrate ideality and the actual partial pressure is greater than that predicted by Raoult. However at low values of x_A the relationship between partial pressure and mole fraction once again becomes linear. This region of the curve can be described by Henry's law.

(*b*) *Henry's Law*   This may be stated as follows:

> The equilibrium concentration of a gas solute dissolved in a liquid is directly proportional to the concentration of that component in the vapour space above the liquid surface.

An alternative wording would be:

> The equilibrium partial pressure in the vapour phase exerted by a component in solution is directly proportional to the concentration of that component in solution.

Mathematically this becomes

$$p_A = x_A H \tag{3.26}$$

where H is Henry's constant. This is not a true constant but is material specific and is a function of temperature and pressure. Further, because the gas (vapour) and liquid concentrations can be expressed in a variety of ways, Henry's law may also be written as:

$$p_A = C_A H' \tag{3.27}$$

or as

$$y_A = x_A H'' \tag{3.28}$$

where C_A is the molar concentration of A in the liquid phase (kmol m^{-3}) and y_A is the mole fraction in the vapour phase. Henry's constant (H' and H'') must therefore take on both different units and different numerical values depending upon the choice of concentration terms. The use of Henry's law and the values of Henry's constant are considered in greater detail in the treatment of mass transfer in chapter eight.

(c) *Prediction of Vapour Liquid Equilibrium*   If ideality is assumed then the laws of Dalton and Raoult can be combined to predict the equilibrium vapour and liquid concentrations for a binary system. Equating the expressions for the partial pressure of component A from Equations (3.20) and (3.25) (the laws of Dalton and Raoult) gives

$$y_A P = x_A p'_A \tag{3.29}$$

and thus

$$y_A = \frac{x_A p'_A}{P} \tag{3.30}$$

This applies equally to any other component in the system and thus for B

$$y_B = \frac{x_B p'_B}{P} \tag{3.31}$$

However, in a binary system, the sum of the mole fractions of A and B must equal unity and therefore

$$y_A + y_B = 1 \tag{3.32}$$

and

$$\frac{x_A p'_A}{P} + \frac{x_B p'_B}{P} = 1 \tag{3.33}$$

Similarly the sum of the liquid phase mole fractions of A and B must equal unity and

$$x_A + x_B = 1 \tag{3.34}$$

Substituting from Equation (3.34) into Equation (3.33) gives

$$\frac{x_A p'_A}{P} + \frac{(1 - x_A) p'_B}{P} = 1 \tag{3.35}$$

which can be rearranged to give an expression explicit in x_A

$$x_A = \frac{P - p'_B}{p'_A - p'_B} \tag{3.36}$$

Thus Equations (3.30) and (3.36) can be used to predict the liquid concentration x_A which is in equilibrium with the vapour concentration y_A at any temperature and pressure, for an ideal binary system. This requires only a knowledge of the pure component vapour pressures (p'_A and p'_B), which are of course a function of temperature, together with the total system pressure P.

Example 3.9

Find the equilibrium concentrations in both liquid and vapour phases for a binary mixture of ethanol and water held at 363 K and standard atmospheric pressure. The pure component vapour pressures of ethanol and water at 363 K are 157.34 and 70.11 kPa, respectively.

Assuming ideality and that the laws of Dalton and Raoult apply, Equation (3.36) gives the mole fraction of the more volatile component (ethanol) in the liquid phase as

$$x_{\text{ethanol}} = \frac{101.3 - 70.11}{157.34 - 70.11}$$

hence

$$x_{\text{ethanol}} = 0.358$$

Consequently the mole fraction of ethanol in the vapour phase [Equation (3.30)] is

$$y_{\text{ethanol}} = \frac{0.358 \times 157.34}{101.3}, \qquad y_{\text{ethanol}} = 0.556$$

Example 3.9 shows that there is a difference between the concentration of an ideal solution and the concentration of the vapour above that solution at equilibrium. It is this difference which is the basis for distillation as a separation process for mixtures of volatile liquids. Successive stages of partial vaporisation followed by condensation allow a distillate to be produced which is progressively richer in the more volatile component. Plotting the equilibrium values of the molecule fraction in vapour and liquid phases respectively, for an ideal solution, against temperature gives rise to a temperature-composition or phase diagram such as the example in Figure 3.9. The lower curve (values of x_A), known as the bubble point or boiling point curve, and equivalent to the saturated liquid line on a $P-V$ diagram, gives the boiling points of mixtures of any composition. Below this curve the mixture is always liquid. The upper curve (values of y_A) is known as the dew point curve and is equivalent to the saturated vapour line. Above the dew point curve the mixture is always a vapour and the space between the two lines represents a two-phase region of liquid plus vapour. The boiling points of the more volatile and less volatile components are represented by the intersections of the two curves at $x_A = 0$ and $x_A = 1$, respectively.

In fact the ethanol/water system of Example 3.9 is not ideal but shows what is known as a positive deviation from Raoult's law and gives rise to a temperature-composition diagram

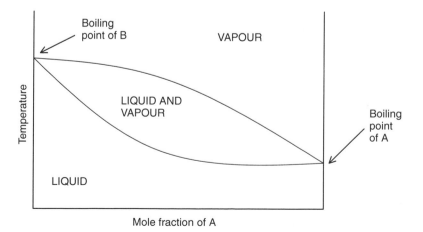

Figure 3.9. Vapour–liquid equilibrium: temperature-composition diagram for an ideal solution.

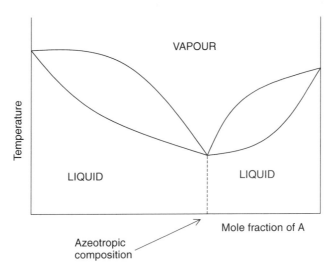

Figure 3.10. Temperature-composition diagram: positive deviation from ideality.

like the one in Figure 3.10. Ethanol and water form an azeotrope, or constant boiling mixture, where the vapour and liquid concentrations are equal. At atmospheric pressure the concentration of the azeotropic mixture is 89.4 mol% ethanol and 10.6 mol% water and its boiling point is 78.2 °C. This means that it is impossible to separate ethanol and water by simple distillation alone to give ethanol concentrations higher than 89.4 mol%. However the composition of azeotropic mixtures is usually a strong function of pressure and in the case of ethanol and water no azeotrope forms below a pressure of 9.3 kPa.

3.4. FIRST LAW OF THERMODYNAMICS

The simplest statement of the first law of thermodynamics is:

In all processes the total energy is conserved.

In other words this is the principle of the conservation of energy; energy cannot be created nor can it be destroyed, it is merely converted from one form to another. For example, in a large canning factory, fuel will be burnt in the boiler house to generate process steam which in turn will be used in a retort to transfer heat to individual cans in order that the can contents may be sterilised. The energy transferred to the cans is only a fraction of the energy content of the fuel which is burnt; waste gases from the boiler and leakage from steam pipes all represent losses of energy which cannot be used in sterilisation. Although this energy is lost to the canning process, it is not destroyed and may be accounted for.

A more useful statement of the first law is the 'non-flow energy equation' which refers to a body of fluid which is not in motion:

gain in internal energy = heat supplied − work output

This is expressed mathematically as

$$du = dQ' - dW' \qquad (3.37)$$

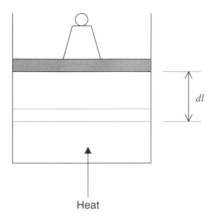

Figure 3.11. Addition of heat to a frictionless gas in a cylinder of cross section A.

or, in terms of finite changes,

$$\Delta u = \Delta Q' - \Delta W' \tag{3.38}$$

where Q' is the heat added to unit mass of the thermodynamic fluid, u is the internal energy per unit mass of the fluid and W' is the work done, per unit mass, by the fluid on the surroundings. The quantities dW' and $\Delta W'$ are *positive* because of the convention of defining work done *by* a fluid on the external environment as positive and work done *on* the fluid as negative.

We must now use the first law to develop the concept of enthalpy in detail. Imagine an ideal frictionless gas contained in a cylinder by a piston (Figure 3.11) with no friction between the piston and the cylinder and a source of energy able to heat the gas. The pressure of the gas will be kept constant by means of the weight of a block placed on the piston. The cylinder is enclosed in a vacuum so that the only force exerted upon the cylinder contents is the weight of the block. Now, a quantity of heat Q' is added to the gas so that its temperature rises and it expands, keeping the pressure constant. The energy which is supplied can be accounted for in two ways. First, by the increase in internal energy of the gas, which is a function of the gas temperature, and second the energy supplied must provide the work to lift the piston against the weight. This can also be thought of as the work of expansion of the gas. If, instead of a weight placed on the piston, the cylinder is placed in air then the work of expansion will done against the pressure of the surrounding atmosphere.

Now, let A be the cross-sectional area of the cylinder and P be the pressure of the gas. The force exerted by the gas on the piston is given by pressure $\times$ area. If, on expansion of the gas, the piston moves a distance dl, then the total work done by the gas (force $\times$ distance) is given by

$$W = (PA)\,dl \tag{3.39}$$

and, because area $\times$ distance represents the increase in volume of the gas,

$$W = P\,dV \tag{3.40}$$

where V is the volume of the fluid. Equation (3.40) can be written on a mass specific basis as:

$$W = P\,dv \tag{3.41}$$

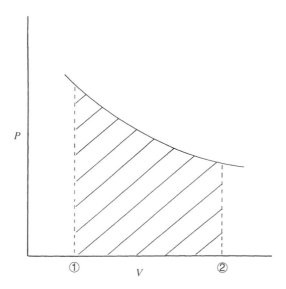

Figure 3.12. Pressure–volume diagram for a fluid: representation of reversible work.

where υ is the specific volume of the fluid (volume per unit mass). Thus for reversible work between the two states 1 and 2, the work done is given by the integral

$$W = \int_1^2 P \, d\upsilon \tag{3.42}$$

which is equal to the shaded area under the curve in Figure 3.12. The quantity which is represented by internal energy plus the work of expansion is called enthalpy, h. It is defined by Equation (3.43)

$$h = u + P\upsilon \tag{3.43}$$

which takes the general form of the First law as expressed in Equation (3.37). Enthalpy may be thought of as 'energy content' and it is most important in food processes to be able to follow the changes in enthalpy of materials as heat is added or removed. In practice enthalpy changes usually manifest themselves as changes in sensible or latent heat. Only rarely are we concerned with mechanical work.

3.5. HEAT CAPACITY

In sensible heat changes, that is those where a change of temperature is registered and where there is no phase change, the determination of the quantities of heat involved requires a value of heat capacity. Heat capacity may be defined as the amount of heat required to raise the temperature of a body by one unit under specified conditions. For example, the heat capacity of a fruit juice may be quoted as $3800 \, J \, kg^{-1} \, K^{-1}$. This means that $3800 \, J$ of thermal energy must be supplied to 1 kg of the fruit juice to raise its temperature by 1 K. If the temperature is to be raised by 10 K, then $3800 \times 10 = 38,000 \, J$ of energy must be supplied. If the mass of the juice sample is increased to 5 kg, then a total of $3800 \times 10 \times 5 = 190,000 \, J$ will be needed for a 10 K change in temperature. However, a problem exists if the value of heat capacity itself varies with temperature. In practice this is the case for very many substances. Heat capacity must now be re-defined as the limiting value of the quantity (dQ'/dT), where dQ' is an infinitesimal change in heat and dT is the resultant temperature change, as dT approaches zero. In other words this is a way of examining the relationship between the value of the heat capacity and the quantity of heat added over the smallest possible temperature increase. In considering heat added to a fluid, there are two possible ways of measuring heat capacity: at constant volume and at constant pressure.

3.5.1. Heat Capacity at Constant Volume

The definition of heat capacity, when the volume of the fluid is kept constant, now becomes

$$dQ' = c_v \, dT \tag{3.44}$$

where c_v is the heat capacity at constant volume, per unit mass. The work done *on* the fluid is given by

$$dW' = -P_{ext} \, dv \tag{3.45}$$

where P_{ext} is the pressure external to the system and the negative sign arises from the sign convention referred to earlier. If the fluid is held and heated in a rigid container, such that its volume cannot change, then, by definition,

$$dv = 0 \tag{3.46}$$

and consequently, from Equation (3.45),

$$dW' = 0 \tag{3.47}$$

In other words no work is done against the external surroundings. Substituting from Equation (3.46) into the first law gives

$$du = dQ' \tag{3.48}$$

and substituting back into the definition of heat capacity in Equation (3.44) yields the result

$$du = c_v \, dT \tag{3.49}$$

or

$$c_v = \left(\frac{du}{dT} \right)_v \tag{3.50}$$

Therefore the heat capacity at constant volume is given by the rate of change of internal energy with temperature when the volume of the fluid remains constant and thus changes in internal energy Δu can be calculated by evaluating

$$\Delta u = \int c_v \, dT \tag{3.51}$$

3.5.2. Heat Capacity at Constant Pressure

For the case where heat is added to a fluid which is not rigidly contained, the heat capacity at constant pressure c_p is now defined by

$$dQ' = c_p \, dT \tag{3.52}$$

However, the pressure is now kept constant and therefore the volume of the fluid must change as it expands and

$$dv \neq 0 \tag{3.53}$$

Further, the external pressure and the pressure of the fluid must be equal if the fluid is not restrained and hence

$$dW' = P \, dv \tag{3.54}$$

Substituting Equations (3.52) and (3.54) into the first law [Equation (3.37)] results in

$$du = c_p dT - P \, dv \tag{3.55}$$

Now enthalpy is defined by Equation (3.43)

$$h = u + Pv \tag{3.43}$$

and on differentiation this becomes

$$dh = du + P \, dv + v \, dP \tag{3.56}$$

Of course the pressure is constant (and changes in pressure are equal to zero) thus

$$dP = 0 \tag{3.57}$$

and Equation (3.56) becomes

$$dh = du + P \, dv \tag{3.58}$$

or

$$du = dh - P \, dv \tag{3.59}$$

Substituting this last result into Equation (3.55) gives

$$dh - P \, dv = c_p \, dT - P \, dv \tag{3.60}$$

which simplifies to

$$dh = c_p dT \tag{3.61}$$

Heat capacity at constant pressure can now be defined as the rate of change of enthalpy with temperature when the pressure of the fluid remains constant. The mathematical expression of this is

$$c_p = \left(\frac{dh}{dT} \right)_p \tag{3.62}$$

Thus heat capacity can be measured either at constant volume or at constant pressure. The latter is more generally of interest in food technology, there being few examples of constant volume processes. Integrating Equation (3.61) now gives changes in enthalpy between two states (1 and 2) of temperature T_1 and T_2, respectively:

$$h_2 - h_1 = \int_{T_1}^{T_2} c_p \, dT \tag{3.63}$$

which for a constant heat capacity would become

$$h_2 - h_1 = c_p(T_2 - T_1) \tag{3.64}$$

Thus changes in enthalpy are given by the product of heat capacity at constant pressure and change in temperature. Enthalpy is genuinely zero only at a temperature of absolute zero and has a positive value at any temperature above this. However, it is more convenient to define enthalpy as zero at a temperature which is within everyday experience. Accordingly, by international convention, enthalpy is taken to be zero at the triple point of water ($0.01°C$ and $611.2\,Pa$); the unique conditions at which water exists as liquid, solid and vapour. Thus, substances above $0.01°C$ have positive enthalpies but this does not mean that below $0.01°C$ enthalpy is 'negative'. Rather it emphasises that [as is implicit in Equation (3.64)] it is changes in enthalpy (due either to changes in temperature or to changes of phase) that are important, especially when considering the operation of food processes.

The definition of heat capacity now becomes:

> the amount of heat required to raise the temperature of unit mass of material by one unit under the specified conditions.

An important note on terminology at this point: the quantities c_v and c_p (which of course are physical properties) are properly described as *specific heat capacity*, but are usually described simply as *heat capacity* without the prefix 'specific,' although it is assumed that they are mass specific, and such terminology is adopted throughout this book. The term 'specific heat' has no meaning and should always be avoided.

The SI units of heat capacity are now $J\,kg^{-1}\,K^{-1}$ but for convenience are often quoted in $kJ\,kg^{-1}\,K^{-1}$. If molar quantities are being used the units become $J\,kmol^{-1}\,K^{-1}$ or $kJ\,kmol^{-1}\,K^{-1}$. It is important to realise that heat capacity almost always varies with temperature and that it is usually necessary to use mean values relevant to the temperature range over which the substance is being heated or cooled. This is dealt with in some detail in chapter four where examples of the use of heat capacity will be found.

3.5.3. The Relationship between Heat Capacities for a Perfect Gas

For a perfect gas it is possible to derive a simple relationship between the heat capacity at constant pressure and the heat capacity at constant volume. Rewriting the ideal gas law as

$$Pv = RT \tag{3.65}$$

where v is now the *molar* specific volume and writing the definition of enthalpy

$$h = u + Pv \tag{3.43}$$

in *molar* specific terms, that is u and h have units of $J\,kmol^{-1}$ rather than $J\,kg^{-1}$, allows them to be combined giving the relationship

$$h = u + RT \tag{3.66}$$

Since, for a perfect gas, internal energy is a function only of temperature and R is a constant (the universal gas constant) then it follows, from inspection of Equation (3.66), that enthalpy is also a function only of temperature. Now differentiating Equation (3.66) yields

$$dh = du + R\,dT \tag{3.67}$$

Changes in internal energy and enthalpy are given by Equations (3.49) and (3.61) respectively and substituting these into Equation (3.67) and equating the coefficients of the dT terms gives

$$c_p\,dT = c_v\,dT + R\,dT \tag{3.68}$$

which can be simplified to yield the result

$$c_p - c_v = R \tag{3.69}$$

3.5.4. The Pressure, Volume, Temperature Relationship for Gases

The ratio of heat capacities is defined by

$$\gamma = \frac{c_p}{c_v} \tag{3.70}$$

For most gases c_p is a strong function of temperature. Some exceptions are the noble gases and diatomic molecules for which the variation with temperature is small. For these gases

$$P\,V^\gamma = \text{constant} \tag{3.71}$$

and therefore,

$$P_1\,V_1^\gamma = P_2\,V_2^\gamma \tag{3.72}$$

where the subscripts 1 and 2 refer to two separate states of the gas. Now equating the expressions for nR in Equations (3.5) and (3.6) for a perfect gas yields

$$\frac{V_2}{V_1} = \frac{T_2}{T_1}\frac{P_1}{P_2} \tag{3.73}$$

but from Equation (3.71)

$$\left(\frac{V_2}{V_1}\right)^\gamma = \frac{P_1}{P_2} \tag{3.74}$$

and therefore

$$\frac{V_2}{V_1} = \left(\frac{P_1}{P_2}\right)^{1/\gamma} \tag{3.75}$$

Now substituting Equation (3.75) into Equation (3.73) gives

$$\left(\frac{P_1}{P_2}\right)^{1/\gamma} = \frac{T_2}{T_1}\frac{P_1}{P_2} \tag{3.76}$$

which may be rearranged to give the result

$$\frac{T_2}{T_1} = \left(\frac{P_2}{P_1}\right)^{(\gamma-1)/\gamma} \tag{3.77}$$

It should be noted that the heat capacity at constant pressure must always be greater than the heat capacity at constant volume because of Equation (3.69) and that the ratio of heat capacities γ is always greater than 1. In practice γ has the following values:

$$\gamma \approx 1.6 \quad \text{for monatomic gases (He, Ar)}$$
$$\gamma \approx 1.6 \quad \text{for diatomic gases (O}_2\text{, N}_2\text{, H}_2)$$
$$\gamma \approx 1.3 \quad \text{for triatomic gases (CO}_2\text{, SO}_2)$$

3.6. SECOND LAW OF THERMODYNAMICS

In essence the first law of thermodynamics states that the net quantity of heat transferred in a process is equal to the net quantity of work done, that is energy is conserved. However, the second law of thermodynamics states that the gross quantity of heat supplied to a process must be larger than the amount of work done. This is best understood by reference to a heat engine (see Figure 3.13). This is a thermodynamic system operating in a cycle which develops work from a supply of heat and is the basis for the steam engine, the internal combustion engine and industrial power plant. The heat engine consists of three components: a heat source supplying heat at a rate Q_1, a thermodynamic fluid undergoing a cycle which results in a rate of work output W and a heat sink to which heat is rejected at a rate Q_2. This now illustrates Planck's statement of the second law which might be summarised as:

It is impossible to construct a system which will operate in a cycle, extract heat from a reservoir and do an equivalent amount of work on the surroundings.

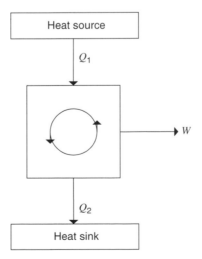

Figure 3.13. Schematic diagram of a heat engine.

The heat source, of course, must be at a higher temperature than the heat sink and thus an alternative statement of the second law is:

> It is impossible for a heat engine to produce net work in a complete cycle if it exchanges heat only at a single fixed temperature.

A simplified mathematical statement of the first law (the conservation of energy principle) for a thermodynamic cycle of the kind illustrated in Figure 3.13 is

$$Q_1 - Q_2 = W \tag{3.78}$$

and for the second law

$$Q_1 > W \tag{3.79}$$

Thus, the practical consequence of the second law for the heat engine is that inevitably a proportion of the thermal energy input Q_1 is wasted and is not used to provide useful work.

3.6.1. The Heat Pump and Refrigeration

The second law can also be applied to a cycle working in the reverse direction to a heat engine. This device is a heat pump which forms the basis of the refrigerator. Figure 3.14 is a schematic diagram of a heat pump in which net work (W) is done on the system and which according to the first law must be equal to the net heat rejected by the system. Thus heat is supplied at a rate Q_2 from a cold reservoir and is rejected to a heat sink at a rate Q_1.

From the first law:

$$Q_2 - Q_1 = W \tag{3.80}$$

and from the second law:

$$W > O \tag{3.81}$$

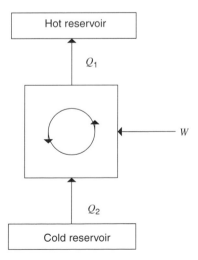

Figure 3.14. Schematic diagram of a heat pump.

In other words a work input is necessary in order to transfer heat from a low temperature to a higher temperature. Exactly how this is put into practice in a domestic refrigerator or industrial freezing plant is covered in chapter eleven. The illustration of the heat pump now leads to a further statement of the second law, due to Clausius:

> It is impossible to construct a device that, operating in a cycle, will produce no effect other than the transfer of heat from a cooler to a hotter body.

A heat pump does indeed transfer heat from cold to hot but there *is* another effect; the work input to the cycle. In other words heat is never observed to flow from a cold body to a hot body and therefore a refrigerator requires an energy input (work) to extract heat from a cold chamber and reject it at a higher temperature. In a heat pump the important quantity is the flow of heat which is rejected (Q_1) and used for heating. A refrigerator works on exactly the same principle: a thermodynamic fluid is circulated through a series of processes with a continuous input of mechanical work. However, in this case, it is the flow of heat (Q_2) removed from a cold chamber, and consequently the rate of heat removal from frozen or chilled food, which is the important quantity.

3.6.2. Consequences of the Second Law

The second law of thermodynamics suggests that work is a more valuable form of energy than heat. For a heat engine the value of Q_1, the heat rejected from the cycle, cannot be zero. Thus it is impossible to convert continuously a supply of heat Q_1 into mechanical work W. However, for a heat pump the value of Q_2 may well be zero; mechanical work can be converted completely (and continuously) into heat.

The thermal efficiency of a heat engine η may be defined as the net work done on the surroundings during the cycle (i.e. the net work produced compared to the gross heat supplied. Thus,

$$\eta = \frac{W}{Q_1} \tag{3.82}$$

Substituting for W from the first law [Equation (3.78)] gives

$$\eta = \frac{Q_1 - Q_2}{Q_1} \tag{3.83}$$

or

$$\eta = 1 - \frac{Q_2}{Q_1} \tag{3.84}$$

Now Q_2 will be smaller than Q_1 and η becomes a positive fractional number. In other words the thermal efficiency of a heat engine must always be less than 100%.

NOMENCLATURE

A	Area
C	Molar concentration of gas
C_A	Molar concentration of component A in the gas phase
c_p	Heat capacity at constant pressure
c_v	Heat capacity at constant volume
h	Enthalpy
H	Henry's constant
H'	Henry's constant
H''	Henry's constant
l	Length
M	Molecular weight
n	Molar mass of gas
n_A	Molar mass of A
P	Pressure
P_c	Critical pressure
P_{ext}	External pressure of surroundings
p	Partial pressure
p_A	Partial pressure of component A
p'	Pure component vapour pressure
p'_A	Pure component vapour pressure of A
Q	Heat; flow rate of heat
Q'	Heat per unit mass of fluid
R	Universal gas constant
T	Absolute temperature
T_c	Critical temperature
u	Internal energy
V	Volume
V_0	Volume at $0°C$
W	Work; rate of working
W'	Work per unit mass of fluid
x_A	Mole fraction of A in the liquid phase
y_A	Mole fraction of component A in vapour phase

GREEK SYMBOLS

α	Coefficient of expansion
γ	Ratio of heat capacities
η	Thermal efficiency
v	Pure component volume
v_A	Pure component volume of A
ρ	Density
υ	Specific volume

PROBLEMS

3.1. Assuming that air behaves as an ideal gas, calculate the temperature of 0.2 kg of air held in a 50 l container at a pressure of 300 kPa.

3.2. Determine the pressure in a $0.10\,m^3$ container which holds a gas mixture at $25°C$ consisting of 0.09 kg nitrogen, 0.01 kg ethylene and 0.15 kg carbon dioxide.

3.3. An ideal gas is expanded at constant pressure such that its volume is doubled. If the initial temperature is $25°C$, calculate the final temperature of the gas.

3.4. A gas with an initial volume of $0.5\,m^3$ is heated from $50°C$ to $200°C$. If the gas behaves ideally what is its final volume? The pressure remains constant on heating.

3.5. A gas, initially at $50°C$, is compressed to one-fifth of its original volume as the pressure is increased from 101.3 to 400 kPa. Determine the final temperature of the gas.

3.6. Calculate the density, at 1.5 bar and 298 K, of a gas mixture with a mass composition of 70% nitrogen and 30% oxygen.

3.7. Fruit is to be ripened at 1.04 bar in an atmosphere containing ethylene with a partial pressure of 3.5 kPa and carbon dioxide at a partial pressure of 2 kPa. The balance will be air. If 0.08 kg of ethylene is charged to the storage container, determine the total mass of gas used. If the temperature of the store is maintained at $15°C$ what is the density of the gas mixture?

3.8. A binary mixture of water and acetic acid forms an ideal solution at standard atmospheric pressure. Find the equilibrium mole fractions of water in the liquid and vapour phases at a temperature of $110°C$. The pure component vapour pressures of water and acetic acid at $110°C$ are 1.433×10^5 and 7.88×10^4 Pa, respectively.

3.9. Assuming that the laws of Raoult and Dalton apply, construct a vapour–liquid equilibrium phase diagram for ethanol and acetic acid at a pressure of 101.3 kPa. The pure component vapour pressures of ethanol and acetic acid are as follows:

	Vapour pressure (kPa)	
Temperature ($°C$)	Ethanol	Acetic acid
78.3	101.3	25.7
80.0	107.5	27.5
90.0	157.3	39.9
100.0	222.6	56.7
110.0	311.7	78.8
115.0	369.8	92.3
117.9	405.1	101.3

3.10. Using the phase diagram from Problem 3.10, determine the bubble point of a mixture containing 40 mol% ethanol and 60 mol% acetic acid.

FURTHER READING

T.D. Eastop and A. McConkey, *Applied Thermodynamics for Engineering Technologists*, Longman (1993).

G.F.C. Rogers and Y.R. Mayhew, *Engineering Thermodynamics, Work and Heat Transfer*, Longman (1992).

G.F.C. Rogers and Y.R. Mayhew, *Thermodynamic and Transport Properties of Fluids, SI units*, Blackwell (1995).

F. Tyler, *Heat and Thermodynamics*, Edward Arnold (1973).

J.R.W. Warn, *Concise Chemical Thermodynamics*, Chapman and Hall (1990).

4

Material and Energy Balances

4.1. PROCESS ANALYSIS

Before a food manufacturing process can be installed and commissioned into a factory, it must be designed and the size and nature of the equipment to be used must be specified. However before this stage an analysis of the process must be undertaken to determine, inter alia, the quantities of product and raw materials to be handled. This *process analysis* step consists largely of a material balance and an energy balance. Process analysis itself is preceded by a concept stage at which the broad outline of the food manufacturing process is decided upon. For example the concept may involve, for a given product and a given rate of production, a series of operations such as mixing, evaporation, pasteurisation, filtration, drying, freezing, and packaging with specified values of important quantities at each stage. In other words an answer must be found to the question — *how* is the product to be manufactured? When the concept has been established the very first step in the process analysis is a material balance. This allows the mass flow rate and composition of the various process streams to be determined. Thus the necessary inputs for a given production rate can be calculated or alternatively the output rate for given raw material feed rates can be determined. It is often the case in food processes that the material balance is relatively trivial, perhaps because relatively few components and few process steps are involved, although in such cases the procedure described in the next section is still extremely useful. Even where the material balance is trivial the energy balance (often reduced to a 'heat' balance) is usually very important; few food processes do not employ heat at some stage. In many instances (freezing, evaporation, pasteurisation) the addition or removal of heat is the substance of the manufacturing process and may involve very considerable quantities of energy. Only when the material and energy balances are complete can the detailed design or specification of process equipment begin.

4.2. MATERIAL BALANCES

Consider a process in which there are a number of input and output streams, as in Figure 4.1. Imagine an envelope (represented by the dotted line) around the process. A material balance may be written in the form:

$$\text{input} + \text{generation} - \text{consumption} = \text{output} + \text{accumulation} \qquad (4.1)$$

That is, the mass entering the envelope plus the mass generated within the envelope must equal the sum of the mass leaving, the mass consumed and the mass which accumulates

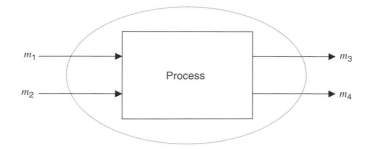

Figure 4.1. Material balance: a process with four streams.

within the envelope. In food manufacturing processes there is no generation or consumption of mass and these terms may be omitted. Indeed only in nuclear processes need these terms be considered. If we are dealing with a continuous process operated at steady state (i.e., all flow rates and compositions are constant with time) then the accumulation term may be set equal to zero. The material balance now reduces to:

$$input = output \tag{4.2}$$

In other words all mass entering the envelope must leave it. Note that in a batch process it is not important to distinguish between 'output' and 'accumulation'.

4.2.1. Overall Material Balances

If the streams in Figure 4.1 have mass flow rates m_1, m_2, m_3, and m_4, respectively, then because the sum of the mass flow rates of all streams entering the process must equal the sum of the flow rates of all streams leaving the process,

$$m_1 + m_2 = m_3 + m_4 \tag{4.3}$$

Equation (4.3) is an example of an overall material balance. The SI unit of mass flow rate is $kg\,s^{-1}$ but it may often be convenient to use different units depending on the magnitude of flow rates in a particular process.

4.2.2. Concentration and Composition

(*a*) *Fractions*   The concentration or composition of a stream can be expressed in a number of ways. A mass fraction is simply the mass of a given component expressed as a fraction of the total mass of the mixture containing that component. Thus if a mixture consists of masses m_A and m_B of components A and B respectively the mass fraction of A is given by

$$x_A = \frac{m_A}{m_A + m_B} \tag{4.4}$$

Similarly a mole fraction is defined as the number of moles of a specified component expressed as a fraction of the total number of moles in the mixture. Hence if the mixture contains n_A moles of A and n_B moles of B, the mole fraction is given by

$$x_A = \frac{n_A}{n_A + n_B} \tag{4.5}$$

Note that the symbol x is used to denote mass or mole fraction in the liquid phase and the symbol y is used for the vapour or gaseous phase. This is an almost universal convention and is adopted throughout this book.

Example 4.1

A solution contains 2.5 kg of ethanol in 3.75 kg of water. Express the concentration of ethanol as both a mass fraction and a mole fraction.

The total mass of solution $= 2.5 + 3.75$ kg, therefore the mass fraction of ethanol x is

$$x = \frac{2.5}{2.5 + 3.75}, \qquad x = 0.40$$

Taking the molecular weights of water and ethanol to be 18 and 46, respectively, the molar masses of water and ethanol are 2.5/18 kmol and 3.75/46 kmol, respectively. Thus the mole fraction becomes

$$x = \frac{(2.5/46)}{(2.5/46) + (3.75/18)}, \qquad x = 0.207$$

A mass percentage is simply the mass fraction multiplied by 100. Therefore, in the above example, the concentration of ethanol is 40% by mass and the molar concentration is 20.7%.

(b) *Ratios*   As an alternative to mass fraction (or mass percentage) the concentration of a given component may be expressed as the ratio of the mass of that component to the mass of a second component. This is particularly appropriate in a binary system, or in a system where a number of components can be treated together and which therefore approximates to a binary mixture, and where the mass of the second component remains constant. Thus in a binary mixture of A and B the mass ratio of A is defined by

$$X_A = \frac{m_A}{m_B} \tag{4.6}$$

Note that upper case symbols (X for liquid and Y for vapour or gas) are used for ratios. Again this is an almost universal convention. Mass ratios are used frequently to express the concentration of water in solid foods (i.e., moisture content) and of water vapour in air (humidity). However in the case of moisture content the use of both mass fractions and mass ratios can lead to considerable confusion. Consider the next example.

Example 4.2

Ten kilograms of water is added to 30 kg of dry solid food. What is the resulting moisture content?

Expressing the moisture content as a mass fraction gives

$$x = \frac{10}{10 + 30} = 0.25$$

Thus the mass fraction of water is 0.25 and the mass percentage of water is 25%. However using a mass ratio gives

$$X = \frac{10}{30} = 0.333$$

That is, the moisture content as a percentage is 33.3%.

Clearly mass fractions and mass ratios give very different values for the same concentration. In the particular case of water, the mass fraction is often referred to as moisture content on a *wet weight basis* and in the case of mass ratio as a *dry weight basis*. Obviously care should be taken not to confuse the two. It should be noted that ratios can have values greater than unity (whereas fractions have values only between zero and unity). For ratios greater than unity it is inappropriate, and misleading, to refer to moisture content in terms of a percentage.

The relationship between fraction and ratio may be formalised as follows

$$x = \frac{X}{1 + X} \tag{4.7}$$

and

$$X = \frac{x}{1 - x} \tag{4.8}$$

Using the data of Example 4.2, from Equation (4.7) the mass fraction is

$$x = \frac{0.333}{1 + 0.333} = 0.25$$

and from Equation (4.8) the mass ratio is

$$X = \frac{0.25}{1 - 0.25} = 0.333$$

4.2.3. Component Material Balances

Suppose that each of the four streams in Figure 4.1 contain water. Clearly, at steady state, the combined mass flow rate of water in all streams entering the process must equal the combined mass flow rates of water leaving the process. Now if x is the mass fraction of water in a stream of total mass flow rate m then the mass flow rate of water is xm. Hence,

$$x_1 m_1 + x_2 m_2 = x_3 m_3 + x_4 m_4 \tag{4.9}$$

Equation (4.9) is known as a component material balance. Any material balance problem, other than the most trivial, will require the simultaneous solution of an overall balance and at least one component balance equation.

Example 4.3

Determine the respective masses of 10% and 50% (by weight) aqueous sucrose solution which must be mixed to prepare 100 kg of a 22% sucrose solution.

Let the masses of 10% and 50% solutions be A and B, respectively. The overall material balance is then

$$A + B = 100$$

where each term has units of kg. The component material balance (for sucrose) is

$$0.10A + 0.50B = 0.22 \times 100$$

Note that mass fractions are used rather than the percentages given in the question. Substituting for B from the overall balance now gives

$$0.10A + 0.50(100 - A) = 22.0$$

which can be solved to give

$$A = 70 \, \text{kg}$$

and therefore

$$B = 30 \, \text{kg}$$

Example 4.4

Tomato juice containing 7% solids by mass is fed to an evaporator and water is removed at a rate of $500 \, \text{kg h}^{-1}$. If the concentrate is to contain 35% solids, then determine the necessary feed rate.

Let F and L be the feed rate and product rate (kg h^{-1}), respectively. The overall material balance can be written as

$$F = 500 + L$$

In other words, the total mass in the feed stream must appear either as concentrated product or as evaporated water. Now, because the water vapour removed from the evaporator cannot contain solids, the component balance for solids becomes

$$0.07F = 0 + 0.35L$$

On substitution from the overall balance this becomes

$$0.07F = 0 + 0.35(F - 500)$$

from which

$$F = 625 \, \text{kg h}^{-1}$$

Examples 4.3 and 4.4 made use of mass fractions simply because the concentration information was supplied in that form. In cases where it is natural to use mass ratios for concentrations it is often more convenient to work throughout with ratios as in Example 4.5.

Example 4.5

Diced potato is dehydrated in a drier from a moisture content of 85% (wet weight basis) to 20% using warm air with an absolute humidity of 0.010 kg water/kg dry air. Dry air enters the drier at a mass flow rate of 500 times that of the dry potato. Calculate the absolute humidity of the outlet air.

This problem is best solved by working in terms of mass ratios and the unknown flow rate of dry solid. Thus the moisture content of the inlet potato, on a dry weight basis, is 0.85/0.15 kg water per kg dry solid. The component balance for water is therefore

$$\frac{0.85}{0.15}P + (0.010 \times 500P) = \frac{0.20}{0.80}P + (H_2 \times 500P)$$

where H_2 is the outlet humidity and P is the flow rate of dry potato. The left hand side of this equation represents the combined mass flow rate of water into the drier and the right hand side the flow of water leaving the drier. P, of course, is constant, and can be cancelled from each term. Because there is only a single unknown it is not necessary to write an overall balance and therefore the component balance can be solved directly to give

$$H_2 = 0.0208 \text{ kg water/kg dry air}$$

Often it is necessary to decide upon a *basis* for a material balance calculation which does not represent an actual mass or flow rate but which allows compositions or proportions to be determined. For example, in the following problem the mass of product is not quoted but can be selected arbitrarily.

Example 4.6

Skim milk with a fat content of 0.4% by mass is produced by centrifuging whole milk (containing 3.5% fat). The cream layer, which separates from the skim milk in the centrifuge, contains 50% fat. What will be the proportion of skim milk to cream?

As a basis for the calculation, let the feed rate of whole milk be 1.0 kg s^{-1} and the mass flow rates of skim milk and cream S and $C \text{ kg s}^{-1}$, respectively. The overall material balance therefore becomes

$$1.0 = S + C$$

and the component material balance (for fat) is

$$0.035 \times 1.0 = 0.004S + 0.50C$$

Substitution now yields:

$$0.035 = 0.004S + 0.50(1.0 - S)$$

which is solved to give

$$S = 0.9375 \text{ kg s}^{-1}, \qquad C = 0.0625 \text{ kg s}^{-1}$$

Thus the *proportion* of skim milk is 93.75% of the feed and cream 6.25%; alternatively the ratio of skim milk to cream is $15 : 1$.

It is important to realise that the solution to this problem does not depend upon the magnitude of the basis for the calculation; the question asked about proportions not about actual flow rates. The proportions will remain the same whatever flow rate of whole milk is chosen.

A component balance can be, and may need to be, written for all components in the system. If n unknowns (flow rates or compositions) are to be determined then n independent equations must be solved simultaneously: an overall balance together with $(n - 1)$ component balances.

Example 4.7

Cheese whey containing 4% by mass lactose and 2% protein, with the balance being water, is fed to an experimental ultrafiltration unit at a rate of $100 \, \text{kg h}^{-1}$. The membrane is chosen so that the protein concentration in the permeate (effectively the waste stream passing through the membrane) falls to 0.02% whilst the lactose concentration rises slightly to 4.5%. If the feed is concentrated by a factor of 2, then determine the composition of the retentate stream.

Let P be the permeate mass flow rate (kg h^{-1}) and x_p, x_l, and x_w the mass fractions of protein, lactose, and water, respectively, in the retentate. The feed is concentrated by 50% and therefore the retentate flow rate is $50 \, \text{kg h}^{-1}$. The overall material balance is then

$$100 = 50 + P$$

and a component balance for protein gives

$$0.02 \times 100 = x_p 50 + 0.0002 P$$

Clearly the permeate rate is $50 \, \text{kg h}^{-1}$ which allows the component balance to be solved to yield

$$x_p = 0.0398$$

A second component balance is necessary to find the concentration of lactose

$$0.04 \times 100 = x_l 50 + 0.045 P$$

from which

$$x_l = 0.035$$

The mass fraction of water is then $1.0 - (0.0398 + 0.035) = 0.9252$ and the retentate composition is therefore: 92.52% water, 3.98% protein, and 3.5% lactose.

Mass balances can be written for a process, for part of a process, for an individual piece of equipment or even for an entire factory. In order to solve more complex problems it may be necessary to consider several mass balance envelopes together to generate sufficient equations to solve for the required number of unknowns.

4.2.4. Recycle and By-pass

Figure 4.2(a) shows a process in which part of the output stream is returned to join the feed stream. Recycle may be used for a number of reasons:

   (i)   to increase the yield from a chemical reaction by recycling unreacted species,
   (ii) to conserve energy by recycling high temperature streams which would otherwise be wasted (e.g., the recycle of air in a drier)
   (iii) to reduce the inlet concentration of a given component to a particular level by diluting that component (e.g., the moisture content of a feed to a drier may be reduced by recycling a proportion of the dried product).

In processes employing recycle [or by-pass streams as in Figure 4.2(b)] mass balances over more than one envelope are usually needed to solve for the required number of unknowns. The flow rate of the recycle stream is often expressed as a dimensionless recycle ratio. This is usually, but not always, defined as the mass ratio of recycle to fresh feed.

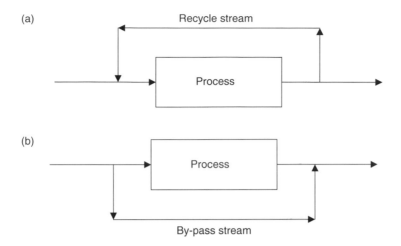

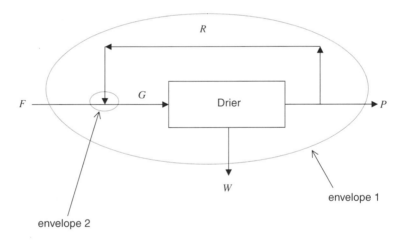

Figure 4.2. Material balance: (a) recycle, (b) by-pass.

Figure 4.3. Material balance for a drier.

Example 4.8

A continuously operated rotary drier is used to dry 12 kg min^{-1} of a starch-based food containing 25% moisture (wet weight basis) to give a product containing 10% moisture. However the drier cannot handle feed material with a moisture content greater than 15% and therefore a proportion of dry product must be recycled and mixed with the fresh feed. Calculate the evaporation rate and the recycle ratio.

Let W, R, and P represent the mass flow rates of evaporated water, recycle and product respectively. Referring to Figure 4.3, an overall balance over the entire process (envelope 1) gives

$$12 = W + P$$

The component balance for water is

$$0.25 \times 12 = W + 0.10P$$

Combining these gives

$$3.0 = W + 0.10\,(12 - W)$$

From which the evaporation rate is

$$W = 2\,\mathrm{kg\,s^{-1}}$$

The recycle stream combines with the fresh feed to give a combined feed rate to the drier G. However the overall balance over the entire process tells us nothing about the recycle within the envelope and a second overall material balance around envelope 2 is necessary, thus

$$12 + R = G, \qquad (0.25 \times 12) + 0.1R = 0.15G$$

or

$$3.0 + 0.10R = 0.15(12 + R), \qquad R = 24\,\mathrm{kg\,s^{-1}}$$

The recycle ratio is thus equal to $\frac{24}{12}$ or 2.

4.3. THE STEADY-FLOW ENERGY EQUATION

In the same way that a material balance accounts for all mass entering and leaving an envelope around a process, an energy balance accounts for energy added to or removed from a process in whatever form it may exist. Figure 4.4 represents the flow of a fluid in a pipeline, at a constant mass flow rate m, through a process in which heat is added to the fluid at a rate Q and work is done by the fluid on the surroundings at a rate W. The total energy of the fluid at any point consists of the following contributions:

(i) internal energy, u
(ii) potential energy due to the height Z of the pipeline above a datum,

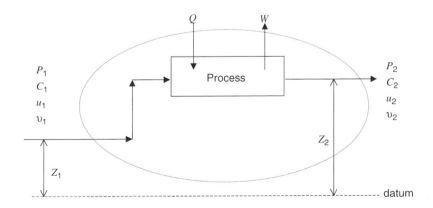

Figure 4.4. Steady flow of a fluid through a process.

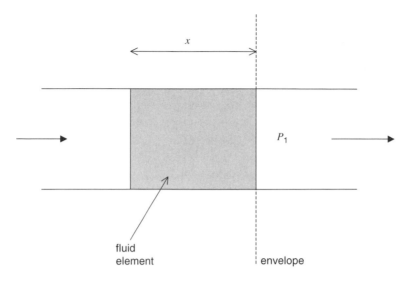

Figure 4.5. Energy requirement at process inlet.

(iii)   kinetic energy, $mC^2/2$ where C is the velocity of the fluid, and

(iv)   the energy required to push the fluid across the boundary against the pressure P.

This latter quantity can be understood by referring to Figure 4.5 which depicts an element of fluid passing along the pipeline at point 1. The force acting on the fluid element is equal to $P_1 A$ where A is the cross-sectional area of the pipeline. Now the work done in moving the element through a distance x is $P_1 A x$ which is equal to $P_1 \times$ volume of the element. Consequently, the work done per unit mass of fluid is $P_1 v_1$ where v_1 is the specific volume of the fluid ($\mathrm{m}^3\,\mathrm{kg}^{-1}$).

Ignoring changes in chemical or electrical energy, an energy balance may now be written in which terms on the left-hand side represents energy entering the envelope and terms on the right-hand side represent energy leaving

$$mu_1 + \frac{mC_1^2}{2} + mZ_1 g + mP_1 v_1 + Q = mu_2 + \frac{mC_2^2}{2} + mZ_2 g + mP_2 v_2 + W \qquad (4.10)$$

Each term in Equation (4.10) represents the *rate* of energy addition to (or removal from) the process and has units of W. Dividing Equation (4.10) by the mass flow rate m reduces each term to energy per unit mass of fluid ($\mathrm{J\,kg}^{-1}$).

Now in most food processes the change in potential energy is relatively trivial and $Z_1 = Z_2$. Similarly, the change in kinetic energy of the food fluid will be negligible compared to enthalpy changes and the inlet and outlet velocities can be equated, and therefore $C_1 = C_2$. Third, the work done by the fluid (or done on the fluid) will very likely be zero. In many 'processes' such as the generation of useful work in a turbine W will be significantly large, but here $W = 0$. The steady flow energy equation now reduces to:

$$mu_1 + mP_1 v_1 + Q = mu_2 + mP_2 v_2 \qquad (4.11)$$

or

$$m(u_1 + P_1 v_1) + Q = m(u_2 + P_2 v_2) \qquad (4.12)$$

Of course, the term $u + Pv$ is the enthalpy of the fluid h [Equation (3.42)] and therefore Equation (4.12) can be simplified further to give

$$mh_1 + Q = mh_2 \qquad (4.13)$$

and

$$Q = m(h_2 - h_1) \qquad (4.14)$$

The significance of Equation (4.14) is that an energy balance for a food process, in the vast majority of cases, reduces to an enthalpy balance, which is often referred to as a 'heat balance'. The latter term, however, is not rigorous and should be avoided. In practice, in a constant volume process, enthalpy is a function of the heat capacity at constant pressure and the temperature at states 1 and 2. Thus, from Equation (3.64), the rate of heat addition to a process where the temperature changes from T_1 to T_2 is given by

$$Q = mc_p(T_2 - T_1) \qquad (4.15)$$

In processes where there is a change of phase between saturated liquid and saturated vapour the rate at which heat is supplied or removed is given by

$$Q = mh_{fg} \qquad (4.16)$$

where h_{fg} is the latent heat of vaporisation. For a continuous process with a constant flow rate m (kg s^{-1}), Q in Equations (4.14) and (4.16) is the rate of heat addition (W). However, for a batch process, Q becomes the quantity of heat added (J) to a batch of mass m (kg).

4.4. THERMOCHEMICAL DATA

It is clear from Equations (4.14)–(4.16) that energy balances require enthalpy, heat capacity and latent heat data. Some attention must now be given to the availability of this kind of data for foods.

4.4.1. Heat Capacity

(a) *Gases*   Heat capacity at constant pressure is usually a strong function of temperature. For gases the variation with temperature can be described by a polynomial of the form

$$c_p = a + bT + cT^2 \qquad (4.17)$$

where a, b, and c are constants and T is the absolute temperature. It is possible to add further terms to Equation (4.17) but this is not worthwhile for most process engineering calculations. The constants for some commonly encountered gases are listed in Table 4.1, giving heat capacity in kJ kmol^{-1} K^{-1} as a function of temperature in K.

Note that the heat capacity of air at constant pressure varies from 1.004 kJ kg^{-1} K^{-1} (29.070 kJ kmol^{-1} K^{-1}) at 275 K to 1.0106 kJ kg^{-1} K^{-1} (29.267 kJ kmol^{-1} K^{-1}) at 375 K. Other values are listed in steam tables (see section 4.4.4).

TABLE 4.1
Heat Capacity of Gases: Coefficients for Use in
Equation (4.17)

	a	$b \times 10^3$	$c \times 10^5$
Hydrogen	27.143	9.273	−1.380
Water vapour	32.243	1.923	1.055
Oxygen	28.106	−3.680	1.745
Nitrogen	31.150	−13.56	2.679
Carbon dioxide	19.795	73.43	−5.601
Ammonia	27.315	23.83	1.707
Ethylene	3.806	156.5	−8.348
Ethylene oxide	−7.519	222.2	−12.560
Ethanol	9.014	214.0	−8.390

Example 4.9

Calculate the heat capacity at constant pressure of carbon dioxide at 300 K.

Substituting the relevant constants from Table 4.1 into Equation (4.17) gives

$$c_p = 19.795 + (0.07343 \times 300) - 5.601 \times 10^{-5}(300)^2 \, kJ \, kmol^{-1}$$

and

$$c_p = 36.78 \, kJ \, kmol^{-1} \, K^{-1}$$

The molecular weight of carbon dioxide is 44 and therefore the *mass specific* heat capacity at constant pressure is

$$c_p = \frac{36.78}{44} \, kJ \, kg^{-1} \, K^{-1} \quad or \quad c_p = 0.8359 \, kJ \, kg^{-1} \, K^{-1}$$

Heat capacity is an additive property, that is the heat capacity of a mixture of gases can be determined from the heat capacities of the constituents weighted accordingly. Thus, for a mixture of n components,

$$c_{p_{mixture}} = \sum_{i=1}^{i=n} x_i c_{p_i} \tag{4.18}$$

where x_i is the mass fraction of a component with a *mass specific* heat capacity c_{p_i}. If molar specific heat capacities are used then of course they must be weighted with mole fractions.

Example 4.10

Determine the heat capacity of a gas mixture at 40°C containing 2.5 mol% carbon dioxide, 4.0 mol% ethylene and 93.5 mol% nitrogen.

The molar heat capacities at constant pressure of carbon dioxide, ethylene and nitrogen can be obtained from Equation (4.17) using the data of Table 4.1 and taking care to use an absolute temperature of 313 K. Thus for carbon dioxide, ethylene, and nitrogen $c_p = 37.29$, 44.61, and 29.53 kJ kmol^{-1} K^{-1}, respectively. Now from Equation (4.18) the heat capacity of the mixture is

$$c_{p_{mixture}} = (0.025 \times 37.29) + (0.04 \times 44.61) + (0.935 \times 29.53) \, kJ \, kmol^{-1}$$

and

$$c_{p_{mixture}} = 30.33 \, kJ \, kmol^{-1}$$

Of course, if heat capacity varies with temperature then a problem arises as soon as it is necessary to calculate the heat required to raise the temperature of a substance; clearly the heat capacity at the final temperature T_2 will be different to that at the initial temperature T_1. This problem can be overcome in two ways. Over a small temperature range the heat capacity can be evaluated at a mean temperature T_m defined by

$$T_m = \frac{T_2 + T_1}{2} \tag{4.19}$$

Inevitably this approach involves a degree of inaccuracy which in some cases will be unacceptable. The alternative is to calculate a mean heat capacity c_{p_m} by dividing the total enthalpy required to change the temperature from T_1 to T_2 by the temperature difference. Thus

$$c_{p_m} = \frac{\int_{T_1}^{T_2} c_p \, dT}{T_2 - T_1} \tag{4.20}$$

Substituting for c_p from Equation (4.17) gives

$$c_{p_m} = \frac{\int_{T_1}^{T_2} (a + bT + cT^2) \, dT}{T_2 - T_1} \tag{4.21}$$

which on integration yields

$$c_{p_m} = \frac{[aT + (bT^2/2) + (cT^3/3)]_{T_1}^{T_2}}{T_2 - T_1} \tag{4.22}$$

and

$$c_{p_m} = a + \frac{b}{2}(T_2 + T_1) + \frac{c}{3}\left(T_2^2 + T_2 T_1 + T_1^2\right) \tag{4.23}$$

Example 4.11

Determine the *mass specific* mean heat capacity of ethylene over the temperature range 200–400 K and compare this value with that obtained using an arithmetic mean temperature.

From Equation (4.23) the mean molar heat capacity of ethylene is

$$c_{p_m} = 3.806 + \frac{0.1565}{2}(400+200) - \frac{8.348 \times 10^{-5}}{3}\{(400)^2 + (400 \times 200) + (200)^2\}\,\text{kJ kmol}^{-1}$$

and

$$c_{p_m} = 42.96\,\text{kJ kmol}^{-1}$$

Thus, taking the molecular weight of ethylene to be 28, the *mass specific* mean heat capacity is

$$c_{p_m} = \frac{42.96}{28}\,\text{kJ kg}^{-1}, \qquad c_{p_m} = 1.534\,\text{kJ kg}^{-1}$$

From Equation (4.19) the arithmetic mean temperature is

$$T_m = \frac{400 + 200}{2}\,\text{K} \quad \text{or} \quad T_m = 300\,\text{K}$$

which gives a heat capacity, from Equation (4.17), of $43.24\,\text{kJ kmol}^{-1}\,\text{K}^{-1}$ or $1.544\,\text{kJ kg}^{-1}\,\text{K}^{-1}$. Clearly the difference is not great.

The effect of pressure on heat capacity of a gas is negligible below atmospheric pressure, however above atmospheric pressure heat capacity generally increases with pressure.

(b) *Solids* The heat capacities of solids cannot be predicted from fundamental relationships but they are amenable to experimental measurement. However a particularly useful method of estimating heat capacity in the absence of experimental data is Kopp's rule which is based on the fact that heat capacity is an additive property. It may be stated as follows:

> The heat capacity of a solid compound is approximately equal to the sum of the heat capacities of the constituent elements.

Table 4.2 lists the values of heat capacity for commonly encountered elements.

Example 4.12

Use Kopp's rule to estimate the heat capacity of tartaric acid $(C_4H_6O_6)$, in the solid phase, at 20°C.

Tartaric acid contains four carbon atoms each of which (see Table 4.2) contributes $7.53\,\text{kJ kmol}^{-1}\,\text{K}^{-1}$ to the heat capacity of the molecule at 20°C. Adding this to the contributions from hydrogen and oxygen gives

$$c_p = (4 \times 7.53) + (6 \times 9.62) + (6 \times 16.74)\,\text{kJ kmol}^{-1}\,\text{K}^{-1}$$

or

$$c_p = 188.28\,\text{kJ kmol}^{-1}\,\text{K}^{-1}$$

The molecular weight of tartaric acid is 150 and therefore the *mass specific* heat capacity becomes

$$c_p = \frac{188.28}{150}\,\text{kJ kg}^{-1}\,\text{K}^{-1} \quad \text{or} \quad c_p = 1.255\,\text{kJ kg}^{-1}\,\text{K}^{-1}$$

TABLE 4.2
Kopp's Rule: Heat Capacities of Commonly
Encountered Elements

	c_p at 20°C (kJ kmol^{-1} K^{-1})	
	Solids	Liquids
Carbon	7.53	11.72
Hydrogen	9.62	18.00
Silicon	15.90	24.27
Oxygen	16.74	25.10
Fluorine	20.92	29.29
Boron		19.66
Sulphur		30.96
Phosphorus	22.59	30.96
Others	25.94	33.47

(c) *Liquids*   The heat capacities of liquids are generally greater than those of solids and of gases; heat capacity usually increases with temperature and this variation may be expressed in the form

$$c_p = c_{p_o} + aT \tag{4.24}$$

where c_{p_o} is a datum value and a is a temperature coefficient for a specified temperature range. Kopp's rule may be used in the absence of other data but the contributions of individual elements are different to those in the solid phase. They are listed in Table 4.2.

The most commonly encountered liquid in food processing is water. Its heat capacity is listed in steam tables (see section 4.4.4) and is approximately 4.19 kJ kg^{-1} K^{-1} (75.4 kJ kmol^{-1} K^{-1}) in the range 0–100°C. Very few liquids have higher heat capacities and the values for dilute aqueous solutions may be approximated to that for water.

(d) *Heat capacities of foods*   It is still the case that very little data exists for foods. There is no adequate theory to predict thermochemical data and therefore a heavy reliance is placed on experimental measurement which is difficult for multi-component materials with complex structures. The heat capacities for some food stuffs are listed in Table 4.3. Heat capacity is strongly dependant upon moisture content and, in the absence of a strong theoretical framework, many empirical equations have been proposed which are simple functions of the mass fraction of water in the food x_w. Of these one of the most often quoted is that due to Siebel where the units of c_p are kJ kg^{-1} K^{-1}.

$$c_p = 3.349x_w + 0.837 \tag{4.25}$$

Other models have been proposed by Lamb and by Dominguez and are given in Equations (4.26) and (4.27), respectively:

$$c_p = 2.720x_w + 1.470 \tag{4.26}$$

$$c_p = 2.805x_w + 1.382 \tag{4.27}$$

Note that all models converge at 100% moisture content ($x_w = 1$) to ≈4.19 kJ kg^{-1} K^{-1}. At moisture contents close to zero, however, there is some divergence because each

TABLE 4.3
Approximate Heat Capacities of Selected Foods

	Moisture content (%)	c_p (kJ kmol^{-1} K^{-1})	
		Unfrozen	Frozen
Fish	80	3.52–3.73	1.85
Beef, lean	72–74	3.43–3.52	
Beef, fat	51	2.89	
Pork	60	2.85	1.60
Lamb	60–70	2.80–3.20	1.50–2.20
Poultry	74	3.31	1.55
Bread, white	38–44	2.60–2.72	1.40
Bread, wholewheat	42–49	2.68–2.85	1.40
Milk, whole	87.5	3.85	
Milk, skim	91	3.95–4.00	
Ice cream		3.30–3.60	1.90–2.10
Egg yolk	40–50	2.80–3.10	
Egg white	87	3.80–3.85	
Butter	16	2.05–2.14	
Cream cheese	80	2.93	1.88
Margarine	9–15	1.80–2.10	
Castor oil		2.07–2.25	
Soybean oil		1.96–2.06	
Saturated fatty acids (liquid)		2.09–2.42	
Apple juice	88	3.85	
Orange juice	87	3.89	
Apple sauce	83	3.73	
Apple	84	3.70–4.00	1.80–2.10
Banana	75	3.35	
Raspberry	83	3.60–3.73	1.84–1.90
Strawberry	89	3.89–3.94	1.10–2.00
10% w/w aqueous sucrose solution 20°C		3.94	
60% w/w aqueous sucrose solution 20°C		2.76	
Peas	74	3.31	1.75
Carrot	88	3.80–3.94	1.80–1.90
Potato	75–78	3.43–3.52	1.80

model assumes a different non-water content of the food. Overall there is little to choose between them.

Siebel produced a relationship for the heat capacities of frozen foods which is based upon the (lower) heat capacity of frozen water:

$$c_p = 1.256x_w + 0.837 \tag{4.28}$$

A more sophisticated model due to Heldman and Singh takes account of the different heat capacities of the various constituents in food. They suggested

$$c_p = 4.187x_w + 0.837x_a + 1.424x_c + 1.549x_p + 1.675x_f \tag{4.29}$$

where x_a, x_c, x_p, and x_f are the mass fractions of ash, carbohydrate, protein, and fat, respectively. A fuller review of the thermochemical properties of foods is given by Rao and Rizvi.

Example 4.13

Estimate the heat capacity at constant pressure of (a) orange juice which may be assumed to have a composition by mass of 88.3% water, 0.4% ash, 10.4% carbohydrate, 0.7% protein, and 0.2% fat; and (b) rice with a composition of 12.0% water, 0.5% ash, 80.4% carbohydrate, 6.7% protein, and 0.4% fat.

For orange juice, the model of Heldman and Singh [Equation (4.29)] gives

$$c_p = (4.187 \times 0.883) + (0.837 \times 0.004) + (1.424 \times 0.104) + (1.549 \times 0.007)$$
$$+ (1.675 \times 0.002) \, \text{kJ kg}^{-1} \text{K}^{-1}$$

and

$$c_p = 3.863 \, \text{kJ kg}^{-1} \text{K}^{-1}$$

Whereas Siebel's equation [Equation (4.25)] gives the result

$$c_p = (3.349 \times 0.883) + 0.837 \, \text{kJ kg}^{-1} \text{K}^{-1}$$

or

$$c_p = 3.794 \, \text{kJ kg}^{-1} \text{K}^{-1}$$

For rice, Heldman and Singh's relationship gives

$$c_p = (4.187 \times 0.12) + (0.837 \times 0.005) + (1.424 \times 0.804) + (1.549 \times 0.067)$$
$$+ (1.675 \times 0.004) \, \text{kJ kg}^{-1} \text{K}^{-1}$$

and

$$c_p = 1.762 \, \text{kJ kg}^{-1} \text{K}^{-1}$$

whilst Siebel's equation produces a significantly lower value:

$$c_p = (3.349 \times 0.12) + 0.837 \, \text{kJ kg}^{-1} \text{K}^{-1}$$

or

$$c_p = 1.239 \, \text{kJ kg}^{-1} \text{K}^{-1}$$

As these examples indicate, Siebel's equation tends to underestimate the value of heat capacity. This is particularly true where the moisture content of the food is low because Equation (4.25) assumes the non-water content to be ash with a heat capacity of $0.837 \, \text{kJ kg}^{-1} \text{K}^{-1}$ which is somewhat lower than the heat capacity of either protein, carbohydrate or fat.

4.4.2. Latent Heat of Vaporisation

Again water is the most commonly encountered liquid in food processing and latent heats can be obtained from steam tables (section 4.4.4). The latent heats of vaporisation of other materials are listed in many reference works and Rogers and Mayhew give data for common refrigerants. Approximate values can be found using Trouton's rule which states that the ratio of molal latent heat of vaporisation at atmospheric pressure (h_{fg}) to boiling point at atmospheric pressure (T_b) is a constant. Thus

$$\frac{h_{fg}}{T_b} = \text{constant} \tag{4.30}$$

For organic liquids the constant may be assumed to have a value of $88\,\text{kJ}\,\text{kmol}^{-1}\,\text{K}^{-1}$.

4.4.3. Latent Heat of Fusion

Pure water freezes at $0°C$ and the latent heat of fusion of pure water is $343.2\,\text{kJ}\,\text{kg}^{-1}$. This heat is removed during the phase change at a constant temperature of $0°C$. However when foods are frozen the phase change rarely occurs at a fixed and clearly-defined temperature because of freezing point depression. The initial freezing temperature is usually one or two degrees below $0°C$. As the water begins to freeze, the solid content of the food is dissolved in, or associated with, a progressively smaller quantity of remaining liquid water. The freezing point of this water is lowered with the consequence that the entire water content freezes over a range of temperature. Values for the latent heat of fusion of various foods are available but it is important to understand that it is only the water within the food that experiences the phase change. The composition and structure of the food will influence the effective latent heat very considerably but an approximate value can be obtained by multiplying the latent heat of water by the mass fraction of water in the food.

4.4.4. Steam Tables

The enthalpies and heat capacities of water and water vapour are listed as a function of temperature and pressure in what are commonly known as 'steam tables'. Many published versions exist but those compiled by Rogers and Mayhew are particularly easy to read and are referred to here. In addition, data on viscosity, thermal conductivity, density, and other properties are presented. Selected thermodynamic properties of saturated water and dry saturated steam are listed in Appendix D. Because steam tables are such a useful source of information it will repay the student of food engineering to learn how to read them correctly.

As a guide to the use of steam tables consider 1 kg of pure water, initially at $0°C$ and $101.325\,\text{kPa}$, to which heat is added at constant pressure.

(i) Heating to $100°C$

From Appendix D, the enthalpy of saturated liquid water, h_f, at $100°C$ is $419.1\,\text{kJ}\,\text{kg}^{-1}$. Note that the triple point of water ($0.01°C$ and $6.112\,\text{kPa}$) is used as the datum at which enthalpy is defined as zero. In other words, 419.1 kJ represents the change in enthalpy required to raise the temperature of 1 kg of water from $0.01°C$ to $100°C$. Dividing this quantity by the temperature change ($99.99\,\text{K}$) gives a mean heat capacity at constant pressure over this temperature range of approximately $4.19\,\text{kJ}\,\text{kg}^{-1}\,\text{K}^{-1}$. Note also that the pure component vapour pressure of water equals atmospheric pressure at $100°C$, that is, this is the definition of boiling point.

(ii) Change of phase

On complete vaporisation, 1 kg of saturated water produces 1 kg of dry saturated water vapour at a temperature of $100°C$ and at a constant pressure of 101.325 kPa. From Appendix D the enthalpy of 1 kg of saturated water vapour h_g is 2675.8 kJ kg^{-1} and thus the change in enthalpy on vaporisation (the enthalpy, or latent heat, of vaporisation) is $2675.8 - 419.1 = 2256.7$ kJ per kg; the addition of latent heat produces a phase change at constant temperature and pressure. In Rogers and Mayhew, as in many other versions, the latent heat of vaporisation is listed separately. Note that:

$$h_{fg} = h_g - h_f \tag{4.31}$$

(iii) Addition of heat to dry saturated steam

If further heat is added to dry saturated water vapour at $100°C$ and 101.325 kPa its enthalpy will increase without further phase change. Steam raised to a temperature above the saturation temperature is known as superheated steam. The enthalpy of superheated steam is listed, for larger temperature intervals, in Appendix D. For example, at $150°C$ (and still at a pressure of 101.325 kPa) the enthalpy of superheated steam (h_g) is 2777 kJ kg^{-1}.

These changes are illustrated in Figure 4.6 which shows the relationship between enthalpy and temperature for water at 101.325 kPa. Note that the temperature remains constant as latent heat is added or removed. The temperature changes only with the addition or removal of sensible heat. The gradient of the line in zone C is about twice as steep as that in zone A because the heat capacity of water vapour (2.01 kJ kg^{-1} K^{-1}) is rather less than half that of liquid water at $100°C$.

Steam tables are very useful compilations of information but only a limited number of data points can be tabulated. Other values must be obtained by interpolation, that is, by assuming a linear relationship between any two points.

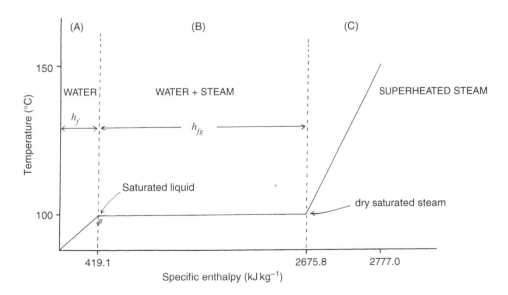

Figure 4.6. Relationship between enthalpy and temperature for water.

Example 4.14

Determine the enthalpy of dry saturated steam at a temperature of 78°C from steam tables using interpolation.

From steam tables the enthalpy of dry saturated steam at 75°C is 2634.7 kJ kg^{-1} and h_g at 80°C is 2643.2 kJ kg^{-1}. Now, assuming that the relationship between enthalpy and temperature is linear between 75°C and 80°C, h_g at 78°C is given by

$$h_g = \left(\frac{78-75}{80-75}\right)(2643.2 - 2.634.7) + h_g \quad \text{at } 75°C$$

that is,

$$h_g = \tfrac{3}{5}(2643.2 - 2634.7) + 2634.7 \, \text{kJ kg}^{-1}, \qquad h_g = 2639.8 \, \text{kJ kg}^{-1}$$

Other tabulated data includes the specific volume of dry saturated vapour v_g (the reciprocal of density), internal energy and entropy. The tabulated values of internal energy, enthalpy, pressure, and specific volume can be used to show the interrelation of thermodynamic properties. Thus for dry saturated steam:

$$h_g = u_g + P v_g \tag{4.32}$$

At a pressure of 0.80 bar, tables give the following values: $h_g = 2665 \, \text{kJ kg}^{-1}$, $u_g = 2498 \, \text{kJ kg}^{-1}$, and $v_g = 2.087 \, \text{m}^3 \, \text{kg}^{-1}$. Now, from Equation (4.32), $h_g = 2498 + (0.80 \times 10^5 \times 2.087)10^{-3} \, \text{kJ kg}^{-1}$ and $h_g = 2664.96 \, \text{kJ kg}^{-1}$.

4.5. ENERGY BALANCES

The remainder of this chapter consists of examples of energy balances over simple processes. Examples 4.15 and 4.16 concern enthalpy changes involving sensible heat and employ Equation (4.15).

Example 4.15

What quantity of heat is required to raise the temperature of 5 kg of tomato pulp from 20°C to 70°C? Assume the mean heat capacity of tomato pulp to be 3.97 kJ kg^{-1} K^{-1}.

The quantity of heat required is the product of mass, heat capacity and the relevant temperature change. Thus

$$Q = 5 \times 3.97(70 - 20) \, \text{kJ}, \qquad Q = 992.5 \, \text{kJ}$$

Example 4.16

Milk (mean heat capacity 3.80 kJ kg^{-1} K^{-1}) flows in a pipe at a rate of 0.19 kg s^{-1} and heat is supplied through the pipe wall. Calculate the rate of heat transfer to the milk if it enters at 280 K and leaves at 318 K.

The rate at which heat must be supplied continuously to a continuous flow of milk is the product of mass flow rate, heat capacity and temperature change. Therefore

$$Q = 0.19 \times 3.80(318 - 280)\,\text{kW} \qquad Q = 27.44\,\text{kW}$$

Liquid foods are usually heated or cooled by exchanging heat with another liquid or with steam. If there are no heat losses in the transfer of heat from stream A to stream B, then

$$m_A c_{p,A}\, \Delta T_A = m_B\, c_{p,B}\, \Delta T_B \tag{4.33}$$

where ΔT_A and ΔT_B are the changes in temperature of the respective streams. Equation (4.33) can be used to determine an unknown flow rate or a single unknown temperature, given that all the other variables are known.

Example 4.17

Orange juice (with a mean heat capacity of $3.8\,\text{kJ}\,\text{kg}^{-1}\,\text{K}^{-1}$) enters a heat exchanger at $12°\text{C}$ with a flow rate of $500\,\text{kg}\,\text{h}^{-1}$. It is heated by water flowing at $0.11\,\text{kg}\,\text{s}^{-1}$ and the water temperature falls from $80°\text{C}$ to $30°\text{C}$. What is the final temperature of the orange juice?

The mean water temperature is $50°\text{C}$ at which, from steam tables, the heat capacity is $4.182\,\text{kJ}\,\text{kg}^{-1}\text{K}^{-1}$. Taking care to convert the juice flow rate to $\text{kg}\,\text{s}^{-1}$, an energy balance may be written as follows

$$\frac{500}{3600} \times 3.80(T - 12) = 0.11 \times 4.182(80 - 30)$$

where T is the final temperature of the orange juice. Note that the left-hand side term represents the enthalpy gained by the juice (kW) and that the right-hand term is the enthalpy loss of the water. Solving for T gives

$$T = 55.6°\text{C}$$

The remaining examples concern energy balances which involve a change a phase [see Equation (4.16)].

Example 4.18

Steam at a pressure of $150\,\text{kPa}$ is condensed at the rate of $0.25\,\text{kg}\,\text{s}^{-1}$. What is the rate of heat transfer to the cooling medium?

The latent heat of vaporisation of water at $150\,\text{kPa}$ is $2226\,\text{kJ}\,\text{kg}^{-1}$. Thus $2226\,\text{kJ}$ is transferred to the cooling medium for every kg of water vapour which is condensed. The rate of heat transfer Q is then the product of the rate of condensation and latent heat.

$$Q = 0.25 \times 2226\,\text{kW} \quad \text{or} \quad Q = 556.5\,\text{kW}$$

Example 4.19

A liquid food containing 12% solids is heated to 120°C under pressure by the direct injection of steam. The food enters at 50°C at a rate of 120 kg min^{-1}. If the heat capacity of the food is 3.40 kJ kg^{-1} K^{-1} and that of the steam-heated product is 4.10 kJ kg^{-1} K^{-1}, determine the necessary steam pressure to ensure that the product contains 10.5% solids.

Firstly a material balance is needed to determine the flow rates of steam S and product X; only then can an enthalpy balance be undertaken. The overall balance (units of kg s^{-1}) is then

$$2 + S = X$$

and the component balance for solids is

$$(0.12 \times 2) + 0 = 0.105X$$

solving these relationships simultaneously yields

$$X = 2.286 \, \text{kg s}^{-1}, \qquad S = 0.286 \, \text{kg s}^{-1}$$

An enthalpy balance can now be written. Because the enthalpies in steam tables are based upon a datum of approximately 0°C, the same datum must be used to calculate enthalpy from a temperature difference and a value of heat capacity. Therefore

$$\left(\frac{120}{60}\right) \times 3.40(50 - 0) + 0.286 \, h_g = 2.286 \times 4.10(120 - 0)$$

from which the enthalpy of the inlet steam is

$$h_g = 2743.7 \, \text{kJ kg}^{-1}$$

At 4 bar $h_g = 2739$ kJ kg^{-1} and at 4.5 bar $h_g = 2744$ kJ kg^{-1} and therefore by interpolation the steam pressure P is

$$P = \frac{(2743.7 - 2739)}{(2744 - 2739)} \times (4.5 - 4.0) + 4.0 \, \text{bar}, \qquad P = 4.47 \, \text{bar}$$

Example 4.20

Calculate the total enthalpy change involved when a 10 kg block of cod, initially at 14°C, is placed in a plate freezer and frozen down to −30°C. The phase change may be assumed to occur at the initial freezing temperature of −2°C and the latent heat of fusion to be 280 kJ kg^{-1}. Use Siebel's equations to estimate heat capacity. Take the water content of cod to be 81%.

Using Siebel's equations for heat capacity gives the following:

$$c_p(\text{unfrozen}) = (3.349 \times 0.81) + 0.837 = 3.55 \, \text{kJ kg}^{-1} \, \text{K}^{-1}$$

and

$$c_p(\text{frozen}) = (1.256 \times 0.81) + 0.837 = 1.85 \, \text{kJ kg}^{-1} \, \text{K}^{-1}$$

The sensible heat removed from the fish then consists of the heat removed above the assumed freezing point $\Delta h_{\text{unfrozen}}$ and the heat removed in decreasing the temperature from $-2°C$ to $-30°C$, Δh_{frozen}. Hence

$$\Delta h_{\text{unfrozen}} = 10 \times 3.55(14 - (-2)) = 568\,\text{kJ}$$

and

$$\Delta h_{\text{unfrozen}} = 10 \times 1.85(-2 - (-30)) = 518\,\text{kJ}$$

The enthalpy change due to the change of phase is simply $10 \times 280 = 2800\,\text{kJ}$. The total enthalpy change therefore is $3886\,\text{kJ}$ of which 72% is due to the change of phase.

Example 4.21

Steam at an absolute pressure of 2 bar is used to blanch $600\,\text{kg}\,\text{h}^{-1}$ of carrots (in an equal mass flow of water) in a continuous process. The carrots enter at $25°C$ and carrots, water, and condensate are all discharged at $95°C$. Assuming that the moisture content of the carrots remains unchanged and that heat losses from the process amount to 50 kW, determine the steam flow rate required. Assume that the mean heat capacity of carrot is $3.75\,\text{kJ}\,\text{kg}^{-1}\,\text{K}^{-1}$.

The total enthalpy of the streams entering the blancher has three contributions: carrot, water, and steam. The enthalpy term for carrot is based on the datum temperature of $0°C$. The enthalpy of the water could be written either in terms of a mean heat capacity and temperature difference above the datum or in terms of the specific enthalpy at $25°C$ obtained from steam tables, as is the case here. The total steam enthalpy is the product of the unknown flow rate S and the specific enthalpy at 2 bar.

Thus the input enthalpy is

$$\left\{ \frac{600}{3600} \times 3.75(25 - 0) \right\} + \left\{ \frac{600}{3600} \times 104.8 \right\} + 2707S\,\text{kW}$$

The total enthalpy of the streams leaving the blancher is determined in the same way. Note, however, that the steam condenses and now has the enthalpy of saturated water at $95°C$. The heat loss term of 50 kW must also be included.

The output enthalpy is, therefore,

$$\left\{ \frac{600}{3600} \times 3.75(95 - 0) \right\} + \left\{ \frac{600}{3600} \times 398 \right\} + 398S + 50\,\text{kW}$$

The input and output terms are equated and the expression solved to give

$$S = 0.0618\,\text{kg}\,\text{s}^{-1}$$

NOMENCLATURE

a Coefficient

A Cross-sectional area

b Coefficient

c Coefficient

c_p Heat capacity at constant pressure

c_{p_m} Mean heat capacity

c_{p_o} Datum value of heat capacity

C Velocity

h Enthalpy

H Humidity

m Mass; mass flow rate

n Molar mass

P Pressure

Q Rate of heat addition

T Temperature

T_b Boiling point

T_m Mean temperature

u Internal energy

W Rate of working

x Distance; mass or mole fraction in liquid phase

X Mass ratio in liquid phase

y Mass or mole fraction in the vapour phase

Y Mass ratio in the vapour phase

Z Height above a datum

GREEK SYMBOLS

Δh Enthalpy difference

ΔT Temperature difference

υ Specific volume

SUBSCRIPTS

a   ash

A   component A

B   component B

c   carbohydrate

f   saturated liquid; fat

fg   phase change liquid to vapour

g   saturated vapour

p   protein

w   water

PROBLEMS

4.1. An aqueous solution contains 100 g of sucrose (molecular weight 342) in 0.8 kg of water. Express the sucrose concentration as a mass fraction, mass percentage, mole fraction, and mole percentage, respectively.

4.2. A solvent used in a process to extract vegetable oil from seeds contains oil at a mass ratio of 0.05. If 100 kg of the used solvent extracts a further 30 kg of oil what is the final mass ratio of oil to pure solvent?

4.3. How much water is removed from moist food of total mass 100 kg if the moisture content (wet weight basis) is reduced from 12.5% to 8%?

4.4. A solution contains 15% sodium chloride and 85% water by mass. Calculate (a) the required evaporation rate to concentrate the solution to 20% NaCl if the feed rate is $800 \, \text{kg} \, \text{h}^{-1}$ and (b) the final composition if the solution is boiled until 70% of the original water is removed.

4.5. Air, with an initial absolute humidity of 0.016 kg water per kg dry air, is to be dehumidified for a drying process by an air conditioning unit which reduces humidity to 0.002 kg per kg. However the required humidity at the drier inlet is 0.004 kg per kg and part of the air therefore must by-pass the conditioning unit. Determine: (a) the percentage of dry air by-passing the unit and (b) the mass of water removed per 100 kg of dry air feed.

4.6. Calculate the heat capacity (in $\text{kJ} \, \text{kmol}^{-1} \, \text{K}^{-1}$) of nitrogen at 10°C.

4.7. Calculate the heat capacity ($\text{kJ} \, \text{kg}^{-1} \, \text{K}^{-1}$) of gaseous ammonia at 0°C.

4.8. A mixture of gases contains, by volume, 30% carbon dioxide, 45% nitrogen, and 25% oxygen. Determine the heat capacity of this mixture at 283 K.

4.9. How do the estimates of the heat capacities of maltose, lactose, and sucrose ($C_{12}H_{22}O_{11}$), in the solid phase at 20°C, calculated from Kopp's rule compare to the literature values of 1.339, 1.201, and $1.251 \, \text{kJ} \, \text{kg}^{-1} \, \text{K}^{-1}$, respectively?

4.10. Use steam tables to determine the following quantities:

(a) The enthalpy of water vapour at 0.50 bar.
(b) The latent heat of water at 70°C.
(c) The vapour pressure of water at 25°C.
(d) The enthalpy of superheated steam at 400°C and 1.5 bar.
(e) The boiling point of water at a pressure of 12,000 Pa.
(f) The boiling point of water under a vacuum of 81.3 kPa.
(g) The density of dry saturated steam at a pressure of 2 bar.
(h) The enthalpy of dry saturated steam at a pressure of 7.3 bar.
(i) The vapour pressure of water at 46.2°C.
(j) The heat capacity of dry saturated steam at 104°C.

4.11. Use the models of Siebel, Lamb, Dominguez, and Heldman and Singh, respectively to calculate the heat capacity at constant pressure of whole milk with a mass composition of 87.4% water, 0.7% ash, 4.9% carbohydrate, 3.5% protein, and 3.5% fat.

4.12. Calculate the enthalpy change when 3 kg of egg white (mean heat capacity 3.81 kJ kg^{-1} K^{-1}) is heated from 10°C to 30°C.

4.13. Milk flowing in a pipe at rate of 540 kg h^{-1} is heated from 10°C to 60°C. If the mean heat capacity of milk is 3.90 kJ kg^{-1} K^{-1}, what flow of heat is required? The heat is provided by hot water which cools from 75°C to 27°C. What flow of water is necessary?

4.14. A cuboid of ice measures 4 m × 3 m × 2 m. Heat is supplied at a rate of 800 kW. The latent heat of fusion of water is 334 kJ kg^{-1} and the density of the ice is 900 kg m^{-3}. How long will it take for the ice to melt completely, assuming that all the heat supplied is used in melting ice?

4.15. A juice is heated from 60°C to 115°C at a rate of 500 kg h^{-1} using steam at an absolute pressure of 2 bar. Condensate leaves the heat exchanger at 115°C. Determine the mass flow rate of steam, assuming that the juice has a mean heat capacity of 3.85 kJ kg^{-1} K^{-1}.

FURTHER READING

G.D. Hayes, *Food Engineering Data Handbook*, Longman, London (1987).

D.R. Heldman and D.B. Lund, *Handbook of Food Engineering*, Marcel Dekker, New York (1992).

O.A. Houghen, K.M. Watson, and R.A. Ragatz, *Chemical Process Principles Part I: Material and Energy Balances*, Wiley, New York (1954).

M.A. Rao and S.S.H. Rizvi, *Engineering Properties of Foods*, Dekker, New York (1995).

G.F.C. Rogers and Y.R. Mayhew, *Thermodynamic and Transport Properties of Fluids, SI units*, Blackwell, Oxford (1995).

R.P. Singh and D.R. Heldman, *Introduction to Food Engineering*, Academic Press, New York (1993).

The Fundamentals of
Rate Processes

5.1. INTRODUCTION

Any analysis of the processing of foods must be based on a thorough understanding of the transfer of heat, of momentum and of mass. It is self-evident that a knowledge of heat transfer to and within foods (both solids and fluids) is vitally important; simply consider the following list of common processing operations, all of which involve the transfer of heat: sterilisation, evaporation, freezing, drying. The theory of heat transfer will be developed in chapter seven. Momentum transfer will be developed into a treatment of fluid flow in chapter six where the concern is the behaviour of both Newtonian and non-Newtonian fluids (amongst the latter may be numbered very many foodstuffs), especially in pipe flow. Mass transfer (chapter eight) is very much under-used in the study of food processing. Many operations, of which drying is a good example, involve both heat and mass transfer and an analysis of both is needed for a full understanding of such processes.

This short chapter is intended to stress the profound similarity that exists between the mechanisms of the transfer of heat, momentum and mass and between the equations which describe them. The mathematical treatment outlined here is very much a simplified one and reduces the complexity of the analysis in two ways. First, only steady-state transfer is considered. That is, at any given point, there is no variation of temperature, velocity or concentration, respectively, with time. Second, only uni-dimensional transfer is considered. In other words temperature, velocity or concentration, respectively, are assumed to vary in only one of the three spatial dimensions and to remain constant in the other two. Whilst it is not expected that the significance of the concepts covered in this chapter will be appreciated fully until chapters six, seven and eight have been read and understood; it is important at this stage to draw attention to the similarity referred to above and to underline the importance of determining transport properties, namely thermal conductivity, viscosity, and diffusivity. The reader is encouraged to return to this chapter at a later point.

5.2. HEAT TRANSFER

Figure 5.1 illustrates the variation in temperature in the z direction through a slab. The rate at which heat is transferred by conduction in the z direction, due to the temperature gradient

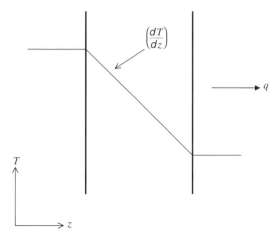

Figure 5.1. Temperature gradient in uni-dimensional conduction.

dT/dz, is given by Fourier's first law of conduction

$$q = -k\frac{dT}{dz} \tag{5.1}$$

where q is the heat flux, that is, the rate of heat transfer per unit cross-sectional area, and k is the thermal conductivity of the material of which the slab is made. Note that the units of heat flux are $W\,m^{-2}$ and the units of thermal conductivity are $W\,m^{-1}\,K^{-1}$. Referring to Figure 5.1, the heat flux is positive in a positive z direction yet the temperature gradient is negative. This explains the minus sign in Equation (5.1). This is simply another way of saying that heat flows from a high temperature to a low temperature in order to reduce the temperature gradient. Equation (5.1) has been written in terms of an ordinary differential because only uni-dimensional conduction is being considered. This relationship should properly be

$$q = -k\frac{\partial T}{\partial z} \tag{5.2}$$

where the partial differential acknowledges that temperature varies in the x and y directions also.

5.3. MOMENTUM TRANSFER

Consider a fluid on a surface (Figure 5.2) to which a shear stress is applied in a direction parallel to that surface. A velocity gradient will develop within the fluid in the z direction, with a velocity u_x at a given distance z from the surface. The relationship between the shear force, at a distance z, and the resulting velocity gradient is then

$$R_z = -\mu\frac{du_x}{dz} \tag{5.3}$$

The shear force R_z, of course, is the same as momentum flux (i.e., momentum per unit time per unit area) because force is proportional to the rate of change of momentum (see section 2.2.3)

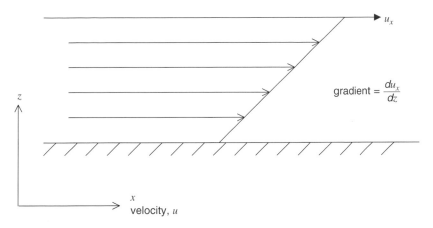

Figure 5.2. Velocity gradient in fluid flow.

and stress is equal to force per unit area. Thus Equation (5.3) has the same general form as Fourier's law and is usually referred to as Newton's law of viscosity or Newton's law of fluid friction. The constant of proportionality in Equation (5.3) is the dynamic viscosity μ. The partial differential equation, indicating the variation of velocity in directions other than the z direction, is then:

$$R_z = -\mu \frac{\partial u_x}{\partial z} \tag{5.4}$$

5.4. MASS TRANSFER

Consider a large room of still air in one corner of which is a stoppered bottle containing a volatile liquid A. If the stopper is removed vapour will be released and eventually will be detected in all parts of the room. The movement of molecules of A is due to a random process known as molecular diffusion and is an example of mass transfer. Rates of mass transfer are expressed frequently in molar quantities and that approach will be adopted here. Thus the molar flux of A (the molar rate of mass transfer of A per unit area) J_A is given by Fick's first law of diffusion:

$$J_A = -D_{AB} \frac{dC_A}{dz} \tag{5.5}$$

for transfer in the z direction. The partial differential form is then

$$J_A = -D_{AB} \frac{\partial C_A}{\partial z} \tag{5.6}$$

In Equation (5.5), C_A is the molar concentration of A (kmol m^{-3}) and D_{AB} is a constant of proportionality relating molar flux with the concentration gradient dC_A/dz. Specifically, D_{AB} is the diffusivity of A in B, where B is the medium through which A diffuses. Diffusivities are often referred to as diffusion coefficients and less commonly as mass diffusivities. Equation (5.5) indicates that mass transfer by diffusion occurs because of the existence of concentration gradients and acts to diminish such gradients.

5.5. TRANSPORT PROPERTIES

5.5.1. Thermal Conductivity

Thermal conductivity is a physical property which is a direct measure of the ability of a material to conduct heat. The SI unit is $W\,m^{-1}\,K^{-1}$. In general the thermal conductivities of solids are greater than those of liquids which in turn are greater than the conductivities of gases. There is some correlation between thermal conductivity and electrical conductivity; metals have higher conductivities than non-metals, crystalline structures have higher conductivities than amorphous materials.

The thermal conductivities of monoatomic gases can be predicted with considerable accuracy by the kinetic theory of gases although the theory for polyatomic gases is less well developed. For gases in general thermal conductivity is proportional to the square root of absolute temperature. There is significant variation of thermal conductivity with pressure only at very high pressures. Liquids are far less amenable to unified molecular theories and, because experimental measurement is made difficult by the presence of convection currents, greater use is made of empirical relationships than for gases. For solids there is generally little difficulty in measuring thermal conductivity directly, although there is a scarcity of data for foods.

5.5.2. Viscosity

Dynamic viscosity is a physical property which is a measure of the ability of a fluid to resist an imposed shear stress. Inspecting Equation (5.3), it is clear that a given shear stress will produce a higher velocity gradient (shear rate) in a low viscosity fluid (e.g., water) than in a higher viscosity material (e.g., treacle). Imagine water and treacle respectively flowing down an inclined plane under the same gravitational force; the water will flow more quickly than the treacle. For these real fluids the layer immediately in contact with the inclined plane is stationary but successive layers of fluid flow with increasing velocity giving rise to the velocity gradient du_x/dz in Equation (5.3). This velocity gradient will not necessarily be linear; for example, in pipe flow the shape of the velocity profile is often parabolic. It is essential to understand that Equation (5.3) is valid only for the case where viscosity is not a function of the velocity gradient (or shear rate), that is, for Newtonian fluids. Although water, milk, many aqueous solutions and vegetable oils (as well as most organic liquids) are Newtonian, very many food fluids are non-Newtonian and in these cases it is inappropriate to refer to a dynamic viscosity.

The SI unit of viscosity is the Pascal second (Pa s). Unfortunately many non-SI units for viscosity are in common use. The student is recommended strongly to avoid these. However it is worth noting that 1 centipoise (cP), the cgs unit of viscosity, is equivalent to 10^{-3} Pa s.

Again, the kinetic theory of gases is able to predict the viscosity of gases with reasonable accuracy. Viscosity is proportional to the square root of absolute temperature and is not a function of pressure below about 10^6 Pa. For Newtonian liquids, viscosity is a strong function of temperature and of concentration.

5.5.3. Diffusivity

The diffusivity D_{AB} is a measure of the rate at which molecules of A are able to pass between molecules of B so as to eliminate the existing concentration gradient. Thus diffusivity is a function *inter alia* of the relative sizes of the two molecules A and B. For a gas, diffusivity is of the order of 10^{-5} m^2 s^{-1} and can be predicted from kinetic theory as a function of pressure, temperature and the molecular volumes of A and B. Diffusivity is proportional to temperature to a power of 1.5 and is inversely proportional to pressure. Diffusivities in the liquid phase are considerably smaller, of the order of 10^{-9} m^2 s^{-1}. Once again it is necessary to rely upon empiricism to predict values; diffusivity is proportional to absolute temperature, inversely proportional to liquid dynamic viscosity and dependant upon solution concentration.

The diffusion of fluids within the pore spaces of a porous solid is of some interest to food processing especially in drying and in solid–liquid extraction. Here it is possible to speak of an 'effective diffusivity' which is an empirical quantity and not a precisely defined physical property. It is very difficult to predict the value of an effective diffusivity but measurements are possible and these indicate a dependence upon temperature, pressure, and the geometry of the pore spaces.

5.6. SIMILARITIES BETWEEN HEAT, MOMENTUM, AND MASS TRANSFER

Clearly there is some similarity between Equations (5.1), (5.3), and (5.5) and this can be developed a little further. The differential quantity in Equation (5.1) represents a temperature gradient in an equation which describes heat transfer. More logically, this relationship should contain a 'heat gradient' term. Introducing the heat capacity at constant pressure c_p and the density of the fluid ρ into the differential produces the term $d(c_p\rho T)/dz$ which now represents 'heat content' (i.e., mass × heat capacity × temperature above a datum) per unit volume of fluid. Note that $c_p\rho T$ has dimensions of ML^{-1}T^{-2} and SI units of Jm^{-3}. Assuming both density and heat capacity to be constant, Equation (5.1) can now be rewritten as

$$q = -\frac{k}{c_p\rho}\frac{d(c_p\rho T)}{dz} \tag{5.7}$$

where the quantity $k/c_p\rho$ is known as thermal diffusivity. Students should satisfy themselves that this quantity has the dimensions L^{2}T^{-1} and thus SI units of m^2 s^{-1}.

Equation (5.3) can be manipulated in a similar way to yield

$$R_z = -\frac{\mu}{\rho}\frac{d(\rho u_x)}{dz} \tag{5.8}$$

which is now more properly a momentum transfer rate equation and where $d(\rho u_x)/dz$ is the momentum gradient, the term ρu_x being momentum (i.e., mass × velocity) per unit volume of fluid. The constant of proportionality μ/ρ is called the kinematic viscosity (or less commonly the kinematic diffusivity) and has dimensions of L^{2}T^{-1}.

The mass transfer relationship, Equation (5.5), Fick's law, does not require to be changed in this way. The gradient already represents molar mass per unit volume per unit length and D_{AB} was described as a diffusivity in section 5.5.3.

The foregoing is summarised in Table 5.1. It will be seen that for each transport equation the rate of transport is given by the product of a diffusivity (thermal, kinematic, and mass diffusivity, respectively) and a gradient term. Each diffusivity has dimensions of L^{2}T^{-1} and units of m^2 s^{-1} in the SI system. However the dimensions of each flux term and each gradient term vary. This simple analogy has been developed to stress the profound similarity between

TABLE 5.1
The Similarity Between Heat, Momentum, and Mass Transfer

Flux	Diffusivity	Gradient
Heat ($W\,m^{-2}$)	Thermal ($m^2\,s^{-1}$)	Heat per unit volume ($J\,m^{-4}$)
Momentum ($kg\,m^{-1}\,s^{-2}$) (= shear stress, Pa)	Kinematic ($m^2\,s^{-1}$)	Momentum per unit volume ($N\,s\,m^{-4}$)
Molar ($kmol\,m^{-2}\,s^{-1}$)	(Mass) diffusivity ($m^2\,s^{-1}$)	Molar mass per unit volume ($kmol\,m^{-4}$) (= Molar concentration)

the three rate equations which form the basis of the study of fluid behaviour in food process engineering.

NOMENCLATURE

C_A	Molar concentration of component A
C_p	Heat capacity at constant pressure
D_{AB}	Diffusivity of A in B
J_A	Molar flux of A (molar rate of mass transfer of A per unit area)
k	Thermal conductivity
q	Heat flux (rate of heat transfer per unit cross-sectional area)
R_z	Shear stress in the z direction
T	Temperature
u_x	Velocity in the x direction at a given distance z from the surface
z	Distance

GREEK SYMBOLS

μ	Dynamic viscosity
ρ	Density

FURTHER READING

C.J. Geankoplis, *Transport Processes and Unit Operations*, Allyn and Bacon (1983).

J.M. Kay and R.M. Nedderman, *Fluid Mechanics and Transfer Processes*, Cambridge University Press, Cambridge (1985).

J.R. Welty, C.E. Wicks, and R.E. Wilson, *Fundamentals of Momentum, Heat, and Mass Transfer*, Wiley, New York (1976).

The Flow of Food Fluids

6.1. INTRODUCTION

This chapter develops the concept of momentum transfer into a treatment of the flow of fluids over surfaces and in pipes and ducts. A later chapter will deal with the behaviour of food particles in a fluid stream. Many foods are of course liquid and the study of fluid flow, or fluid mechanics, is necessary to understand how fluids are transported, how they can be pumped, mixed, and so on. The high viscosity of many liquid foods means that laminar flow is particularly important. However very many foodstuffs are non-Newtonian and later sections of this chapter cover a wide variety of rheological models; these are treated as mathematical descriptions of physical behaviour with the objective of enabling the reader to apply models to experimental data in order to determine whether or not they can be used predictively.

6.2. FUNDAMENTAL PRINCIPLES

6.2.1. Velocity and Flow Rate

As we saw in section 5.3, when a viscous fluid flows under the influence of a shear stress a velocity gradient develops across the flow channel. However it is convenient to use a mean velocity u which is equal to the volumetric flow rate V divided by the cross-sectional area of the conduit A. Thus

$$V = uA \qquad (6.1)$$

The mean velocity is often referred to as the superficial velocity.

Example 6.1

A cylindrical vessel, 2 m in diameter, is to be filled to a depth of 1.5 m. The liquid discharges through a 2.5 cm diameter pipe with a velocity of $3.0 \, \mathrm{m\,s^{-1}}$. How long will it take to fill the vessel? If the outlet valve is accidentally left partially open, and the velocity in the outlet pipe of 2.5 cm diameter is $1 \, \mathrm{m\,s^{-1}}$, determine the new time required to fill the vessel.

The volume of liquid required is the volume of a cylinder 2 m in diameter and 1.5 m deep, that is, the volume is $1.5\pi (2.0)^2/4$ or $4.71 \, \mathrm{m}^2$. The volumetric flow rate through the pipe V is the product of the superficial or mean velocity and the pipe cross-sectional area, therefore

$$V = 3.0 \frac{\pi (0.025)^2}{4} \, \mathrm{m^3\,s^{-1}}, \qquad V = 1.47 \times 10^{-3} \, \mathrm{m^3\,s^{-1}}$$

The time to fill the vessel t is then equal to the volume divided by flow rate. Thus

$$t = \frac{4.71}{1.47 \times 10^{-3}} \, \text{s}, \qquad t = 3200 \, \text{s}$$

Now, because the diameter of the outlet pipe is the same as that of the inlet, the effective net inlet velocity is reduced to $(3.0 - 1.0) = 2.0 \, \text{m s}^{-1}$ and the time to fill the vessel will be greater by a factor of 1.5. Hence

$$t = \tfrac{3}{2} \times 3200 \, \text{s} \ \text{ or } \ t = 4800 \, \text{s}$$

6.2.2. Reynolds' Experiment

Figure 6.1 illustrates Reynolds' classic experiment which demonstrated the nature of fluid flow. A glass tube contains a continuous flow of water and a second very narrow tube enters through the wall and is positioned along the centre line of the larger tube. This second tube is fed with a dye from a reservoir such that the flow is negligible relative to the flow of water. Two broad types of flow are observable: in laminar flow (Figure 6.1a) the trace of dye remains on the centre line indicating that layers of water flow over one another and do not mix. Individual elements or packets of fluid move only in the direction of flow and their components of velocity in other directions are negligible. However under certain conditions turbulent flow is observed and the trace moves from side to side and eventually breaks up with the dye being dispersed throughout the water. In turbulent flow (Figure 6.1b) there is a significant component of velocity in a direction at right angles to the direction of flow and rapid fluctuations in velocity and pressure. Where temperature gradients exist there will be rapid fluctuations in temperature also. Thus turbulent flow promotes mixing, heat transfer and mass transfer but there is a disadvantage to this general improvement. The pressure drop in laminar flow along a given length of pipe is proportional to the flow rate of the fluid whereas in turbulent flow the pressure drop increases more rapidly with flow rate and follows an approximately squared relationship.

These differences are reflected in the velocity profiles illustrated in Figure 6.2. In pipe flow the maximum velocity in the direction of flow is along the centreline of the pipe; at the wall the fluid velocity is zero. For laminar flow the velocity profile adopts a parabolic shape with the mean velocity equal to about one half of the maximum, however when the flow is turbulent a much flatter profile develops and the mean velocity is approximately 82% of the maximum. The region of more slowly moving fluid close to the wall is known as a boundary layer. The edge of this layer is sometimes defined as the point where the velocity

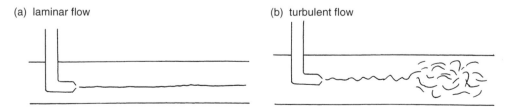

(a) laminar flow   (b) turbulent flow

Figure 6.1. Reynolds experiment: (a) laminar flow; (b) turbulent flow.

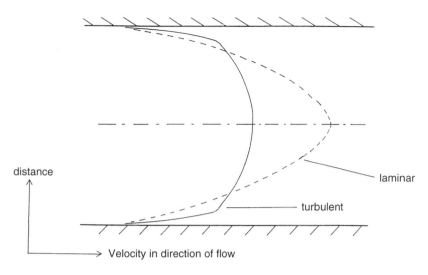

Figure 6.2. Velocity profiles in pipe flow.

is 99% of the maximum velocity. However the structure of a flowing fluid is complex and this definition is not absolute.

Reynolds showed that turbulent flow could be obtained by increasing the mean velocity u, the density ρ and the diameter of the tube d and by decreasing the viscosity μ. Thus the dimensionless group known as the Reynolds number is used to indicate the degree of turbulence.

$$Re = \frac{\rho u L}{\mu} \qquad (6.2)$$

where L is a length characteristic of the geometry. For pipe flow the characteristic length is the pipe diameter d and

$$Re = \frac{\rho u d}{\mu} \qquad (6.3)$$

In pipe flow the transition between laminar and turbulent flow occurs at a Reynolds number of about 2,000. However there is not always a sharp boundary between the two and a transition region is often referred to in the range $2,000 < Re < 4,000$.

Example 6.2

Calculate the Reynolds number for a vegetable oil of viscosity 0.03 Pa s and density $850 \, \text{kg m}^{-3}$ flowing in a 50 mm bore pipe at a mean velocity of $0.75 \, \text{m s}^{-1}$.

Taking care to ensure that all quantities are in SI units, the Reynolds number is

$$Re = \frac{850 \times 0.75 \times 0.050}{0.03} \quad \text{or} \quad Re = 1062.5$$

The flow is therefore laminar.

It is often convenient to express the Reynolds number in terms of a mass flow rate m (which is more easily measurable) than as a function of velocity. Thus the mean velocity is

$$u = \frac{4m}{\rho \pi d^2} \qquad (6.4)$$

and the Reynolds number becomes

$$Re = \frac{4m}{\mu \pi d} \qquad (6.5)$$

Whilst it is clear that the characteristic length should be the diameter of a pipe or sphere, for annular flow the hydraulic mean diameter d_e should be used. This is defined as

$$d_e = \frac{4 \times \text{cross-sectional area of annulus}}{\text{wetted perimeter}} \qquad (6.6)$$

If d_o is the outer diameter of the annular channel (i.e., the inner diameter of the outer pipe) and d_i is the inner diameter of the annular channel (i.e., the outer diameter of the inner pipe), then

$$d_e = \frac{4[(\pi d_o^2/4) - (\pi d_i^2/4)]}{\pi d_o + \pi d_i} \qquad (6.7)$$

which simplifies to

$$d_e = \frac{d_o^2 - d_i^2}{d_o + d_i} \qquad (6.8)$$

and

$$d_e = \frac{(d_o - d_i)(d_o + d_i)}{d_o + d_i} \qquad (6.9)$$

and finally to

$$d_e = d_o - d_i \qquad (6.10)$$

Example 6.3

Calculate the Reynolds number for water at 50°C flowing at a rate of 12 l min^{-1} through an annulus between two concentric pipes. The outer diameter of the inner pipe is 10 mm and the inner diameter of the outer pipe is 20 mm. For water at 50°C: density $= 988$ kg m^{-3} and viscosity $= 5.44 \times 10^{-4}$ Pa s.

The mass flow rate of water m can be found readily from the volumetric flow rate and the density, so that

$$m = \frac{988 \times 12 \times 10^{-3}}{60} \text{ kg s}^{-1} \text{ or } m = 0.1976 \text{ kg s}^{-1}$$

The hydraulic mean diameter [Equation (6.10)] is the difference between the outer diameter of the inner cylinder and the inner diameter of the outer cylinder, that is,

$$d_e = 0.02 - 0.01 \text{ m} \text{ or } d_e = 0.01 \text{ m}$$

Substituting into Equation (6.5) the Reynolds number becomes

$$Re = \frac{4 \times 0.1976}{5.44 \times 10^{-4} \pi 0.01}, \qquad Re = 46249$$

Consequently the flow is clearly turbulent.

6.2.3. Principle of Continuity

The analysis of fluid flow is complex. However considerable understanding can be gained by considering an ideal fluid and applying mass and energy balances, respectively. An ideal, or inviscid, fluid is one which has no viscosity and for which the velocity is uniform across a given cross section of the flow channel. Referring to Figure 6.3 and equating the mass flow rate at points 1 and 2 gives

$$\rho_1 u_1 A_1 = \rho_2 u_2 A_2 + \frac{d(\rho_{\text{av}} V)}{dt} \tag{6.11}$$

where A is the cross section and ρ and u the density and velocity of the fluid, respectively, ρ_{av} being the average density, and V is the volume of the flow channel between points 1 and 2. The differential term represents the accumulation of mass. This result is known as the principle of continuity. At steady state there is no accumulation of mass and Equation (6.11) becomes

$$\rho_1 u_1 A_1 = \rho_2 u_2 A_2 \tag{6.12}$$

A distinction must now be made between liquids and gases. The density of most liquids remains reasonably constant with pressure, certainly within the pressure ranges encountered in most food processes. A liquid thus approximates to an incompressible fluid where $\rho_1 = \rho_2$ and Equation (6.12) can now be written as

$$u_1 A_1 = u_2 A_2 \tag{6.13}$$

Clearly gases are compressible and the gas laws must be used to find the relationship between ρ_1 and ρ_2.

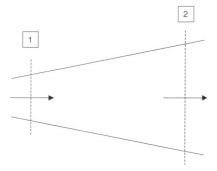

Figure 6.3. Principle of continuity.

Example 6.4

Two pipelines of 20 and 30 mm internal diameter, respectively, merge into a single line of 50 mm diameter. The same liquid flows in each branch at mean velocities of 1.2 and 0.8 m s^{-1}, respectively. Determine the fluid velocity downstream of the junction.

This problem can be solved using the principle of continuity. Expanding Equation (6.13), the sum of the product of velocity and area for the 20 and 30 mm pipelines must equal the product of velocity and area for the merged pipe, in which the velocity is u. Thus, again taking care to use SI units throughout,

$$\frac{\pi}{4}[1.2(0.02)^2 + 0.80(0.03)^2] = u\frac{\pi}{4}(0.05)^2$$

which can be solved to give

$$u = 0.48\,\text{m s}^{-1}$$

6.2.4. Conservation of Energy

The conservation of energy relationship for a flowing fluid is known as Bernoulli's equation. In chapter four the steady-flow energy equation was derived for a general process involving the addition or removal of heat and work to a fluid. If there is no transfer of heat and no work done on or by the fluid then the temperature and the internal energy of the fluid remain constant. Equation (4.10) then becomes

$$\frac{mu_1^2}{2} + mZ_1g + mP_1v_1 = \frac{mu_2^2}{2} + mZ_2g + mP_2v_2 \tag{6.14}$$

Dividing through by the constant mass flow rate m and by the acceleration due to gravity and replacing the specific volume of the fluid v with density ρ gives

$$\frac{u_1^2}{2g} + Z_1 + \frac{P_1}{\rho_1 g} = \frac{u_2^2}{2g} + Z_2 + \frac{P_2}{\rho_2 g} \tag{6.15}$$

which is the usual form of Bernoulli's equation. It should be stressed that the velocity u is the average velocity and that as written above Bernoulli's equation is valid only for the flat velocity profile found in turbulent flow. Each term in Equation (6.15) has units of metres and represents an energy loss in terms of a *head* of the fluid. The terms on each side of the equation are referred to as the velocity head, the potential head, and the pressure head, respectively. For example water flowing along a pipe with a mean velocity of 3 m s^{-1} corresponds to a velocity head of 0.459 m of water. The corresponding pressure is then $0.459 \times 1000 \times 9.81 = 4500$ Pa.

Particularly for higher velocities, a loss term should be included in Equation (6.15) as energy is dissipated between points 1 and 2. This frictional loss is called the head loss due to friction and here is given the symbol h_F. An expression for the frictional loss in straight pipe sections will be derived in section 6.4 but in addition there are standard values which can be used for the losses through valves, bends and other pipe fittings. Bernoulli's equation may now be written as

$$\frac{u_1^2}{2g} + Z_1 + \frac{P_1}{\rho_1 g} = \frac{u_2^2}{2g} + Z_2 + \frac{P_2}{\rho_2 g} + h_F \tag{6.16}$$

Example 6.5

Water flows through a horizontal pipe. At a particular cross-section X the velocity of the water is $1.5 \, \mathrm{m \, s^{-1}}$ and the pressure is 175 kPa. The pipe tapers gradually from 150 mm at X to 75 mm at a section Y. Determine the pressure at Y, assuming that the frictional losses are negligible. What must be the diameter at Y for the pressure there to be reduced to 55 kPa?

This problem concerns a horizontal pipe, there is no change in height above the datum, and therefore $Z_1 = Z_2$. In addition there are no frictional losses and thus Bernoulli's equation reduces to

$$\frac{u_1^2}{2g} + \frac{P_1}{\rho_1 g} = \frac{u_2^2}{2g} + \frac{P_2}{\rho_2 g}$$

Using the principle of continuity [Equation (6.12)], the velocity at cross-section Y, u_2, is

$$u_2 = 1.5 \times \left(\frac{150}{75}\right)^2 \, \mathrm{m \, s^{-1}} \quad \text{or} \quad u_2 = 6.0 \, \mathrm{m \, s^{-1}}$$

Now multiplying through by the acceleration due to gravity and re-arranging Bernoulli's equation explicitly for the pressure at Y gives

$$P_2 = P_1 + \frac{\rho}{2}\left(u_1^2 - u_2^2\right)$$

and therefore

$$P_2 = 1.75 \times 10^5 + \frac{1000}{2}\left(1.5^2 - 6.0^2\right) \, \mathrm{Pa}, \qquad P_2 = 158.1 \, \mathrm{kPa}$$

Now re-arranging Bernoulli's equation to give the velocity at Y yields

$$u_2 = \left[u_1^2 + \frac{2}{\rho}(P_1 - P_2)\right]^{0.5}$$

Hence, for a pressure at Y of 55 kPa,

$$u_2 = \left[(1.5)^2 + \frac{2}{1000}(175 - 55)10^3\right]^{0.5} \, \mathrm{m \, s^{-1}}, \qquad u_2 = 15.6 \, \mathrm{m \, s^{-1}}$$

Again, using the principle of continuity,

$$d_2^2 = \frac{1.5}{15.6}(0.15)^2 \, \mathrm{m}$$

where d_2 is the diameter at Y and thus

$$d_2 = 0.0465 \, \mathrm{m}$$

6.3. LAMINAR FLOW IN A PIPELINE

The most useful relationship concerned with the transport of fluids in pipelines is that between flow rate and the pressure gradient responsible for flow. Many food liquids have a high dynamic viscosity and consequently laminar flow is commonly observed. For the particular case of laminar flow it is possible to derive an exact relationship between flow rate and pressure.

Figure 6.4 represents a cylindrical element of fluid, of general radius r, within a circular pipe of radius r_1, length L and over which the pressure drop due to friction between the fluid and the pipe wall is ΔP. The pressure gradient along the pipe is assumed to be uniform and therefore equal to $\Delta P/L$. If the momentum of the fluid is constant then the net force acting on the fluid must be zero. In other words the shear force acting on the curved surface of the element must equal the difference between the forces acting on the respective ends of the cylinder. This latter force is simply the product of the pressure difference ΔP and the cross-sectional area πr^2. Thus

$$2\pi r L R = \pi r^2 \Delta P \tag{6.17}$$

where R is the shear stress at a radius r and $2\pi r L$ represents the curved surface of the element. Consequently

$$R = \frac{r}{2}\frac{\Delta P}{L} \tag{6.18}$$

but from Equation (5.3) the shear stress equates to the product of viscosity and velocity gradient, in this case du_r/dr where u_r is the velocity in the direction of flow, at a radius r. Hence

$$\frac{r}{2}\frac{\Delta P}{L} = -\mu\frac{du_r}{dr} \tag{6.19}$$

Separating the variables and integrating between $r = r_1$ (the pipe wall), where the fluid is stationary, and $r = r$, where the velocity is u_r, gives

$$\mu \int_0^{u_r} du_r = -\frac{\Delta P}{2L} \int_{r_1}^r r\, dr \tag{6.20}$$

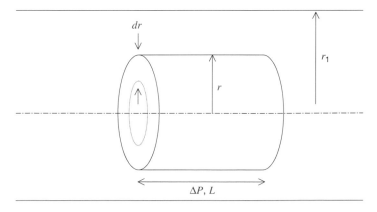

Figure 6.4. Laminar flow in a pipe.

and the result

$$u_r = \frac{\Delta P}{4L\mu}(r_1^2 - r^2) \tag{6.21}$$

Equation (6.21) is the equation of a parabola and when plotted gives the velocity profile shown in Figure 6.2. By inspection it can be seen that the velocity is a maximum when $r = 0$ and zero when $r = r_1$ (at the pipe wall). Imagine now that the cylindrical element in Figure 6.4 is replaced by an annulus of thickness dr and radius r. The volumetric flow rate V is obtained from the product of the velocity [Equation 6.21)] and cross section $2\pi r dr$ and by integrating across the pipe radius. Therefore

$$V = \int_0^{r_1} 2\pi r \frac{\Delta P}{4L\mu}(r_1^2 - r^2)\, dr \tag{6.22}$$

which on integration produces

$$V = \frac{\pi \Delta P}{2L\mu}\frac{r_1^4}{4} \tag{6.23}$$

and in terms of the pipe diameter d the volumetric flow rate is now

$$V = \frac{\pi \Delta P d^4}{128L\mu} \tag{6.24}$$

This result is known as the Hagan–Poiseuille relationship and in terms of a mean velocity over the cross section $\pi d^4/4$ becomes

$$u = \frac{\Delta P d^2}{32L\mu} \tag{6.25}$$

Example 6.6

Calculate the pressure gradient along a 0.025 m diameter tube through which a 55% aqueous sugar solution flows at a mean velocity of $0.6\,\mathrm{m\,s^{-1}}$. The solution has a density of $1,260\,\mathrm{kg\,m^{-3}}$ and a viscosity of 0.032 Pa s.

The Reynolds number in the tube is

$$Re = \frac{1260 \times 0.6 \times 0.025}{32 \times 10^{-3}} \quad \text{or} \quad Re = 590.6$$

and therefore the flow is laminar. The pressure drop along the tube can now be found by using the Hagan–Poiseuille relationship for laminar flow in a tube. Rearranging Equation (6.25) gives the pressure drop per unit length

$$\frac{\Delta P}{L} = \frac{32\mu u}{d^2}$$

Substituting values for mean velocity, viscosity, and tube diameter gives

$$\frac{\Delta P}{L} = \frac{32 \times 0.032 \times 0.6}{(0.025)^2}, \qquad \frac{\Delta P}{L} = 983\,\mathrm{Pa\,m^{-1}}$$

6.4. TURBULENT FLOW IN A PIPELINE

For laminar flow, Equation (6.25) represents an exact solution for the relationship between flow rate and pressure drop. For turbulent flow an exact solution is not possible and it is necessary to introduce a *friction factor*, c_f. We can rewrite Equation (6.18) for the shear stress at the pipe wall as

$$R = \frac{d}{4}\frac{\Delta P}{L} \tag{6.26}$$

and eliminate the pressure gradient from this and from Equation (6.25) to give

$$R = \frac{8u\mu}{d} \tag{6.27}$$

Dividing each side by $\rho u^2/2$ provides the definition of the Fanning friction factor

$$c_f = \frac{R}{\frac{1}{2}\rho u^2}\frac{16}{Re} \tag{6.28}$$

Thus the friction factor allows the pressure losses due to the friction between fluid and pipe wall to be expressed as a function of Reynolds number for a given wall surface. This relationship is best presented as the plot shown in Figure 6.5 which is often referred to as a Moody chart.

Below a Reynolds number of 2,000 flow is laminar and the Hagan–Poiseuille relationship is represented by $c_f = 16/Re$, although the use of a friction factor in these circumstances is unnecessary because Equation (6.25) is exact. As the Reynolds number increases there is a

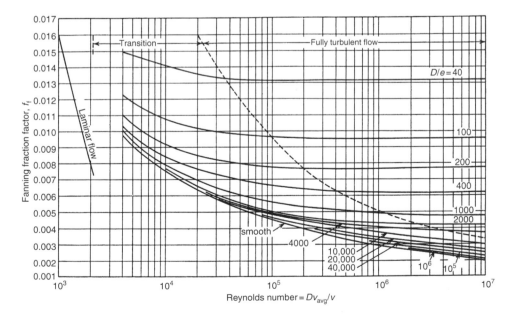

Figure 6.5. Moody chart (from J.R. Welty, C.E. Wicks and R.E. Wilson, Fundamentals of Momentum, Heat and Mass Transfer, John Wiley, 1969, © John Wiley & Sons, Inc., reprinted by permission of John Wiley & Sons, Inc.).

TABLE 6.1
Magnitudes of Surface Roughness

Material	Approximate roughness (μm)
Drawn tubing	1.5
Commercial steel/wrought iron	45
Galvanised iron	150
Stainless steel (hot rolled, mill finish 1)	5
stainless steel (cold rolled, mill finish 2D)	<1
Stainless steel (cold rolled, mill finish 2B)	0.3
Stainless steel (cold rolled, bright annealed, mill finish 2A)	0.1

sudden and significant increase in pressure loss during the development of turbulence. In the transition region it is difficult to predict the friction factor, however for fully developed turbulence a series of curves can be drawn showing c_f as a function of both Reynolds number and the nature of the pipe wall. The latter is quantified in terms of surface roughness, defined as e/d where e is the magnitude of the surface roughness and d is the pipe diameter. At high Reynolds numbers the friction factor becomes independent of Re and the curves in Figure 6.5 become horizontal. The magnitudes of surface roughness for some common materials of construction, including food grade stainless steels, are given in Table 6.1.

Friction factors can be defined in a number of ways and different versions of the Moody chart are available; it is important to be clear which friction factor is being used. The Moody chart reproduced in Figure 6.5 uses the Fanning friction factor whereas the Moody or Darcy friction factor is greater than the Fanning by a factor of 4 so that, for example, the laminar flow curve appears as $c_f = 64/Re$. Also on some charts values of $c_f/2$ rather than c_f are plotted which, for laminar flow, gives a line equal to $8/Re$.

Many expressions have been developed to describe the curves in Figure 6.5. Of these the simplest, for smooth walled pipes, is the Blasius equation

$$c_f = 0.079\, Re^{-0.25} \tag{6.29}$$

which is valid in the range $3,000 < Re < 10^4$. Other equations are overly complex and are unlikely to be required. The numerical examples in this section will be restricted to the use of Figure 6.5 and the Blasius equation.

The pressure drop in a pipeline can now be calculated as follows. Rearranging Equation (6.26) and introducing $\rho u^2/\rho u^2$ gives

$$\Delta P = 4\frac{L}{d}\frac{R}{\rho u^2}\rho u^2 \tag{6.30}$$

and substituting from the definition of friction factor [Equation (6.28)] yields

$$\Delta P = 4\frac{L}{d}\frac{c_f}{2}\rho u^2 \tag{6.31}$$

This pressure loss is more usually expressed as a head loss due to friction

$$h_F = 4\frac{L}{d}\frac{c_f}{2}\frac{u^2}{g}$$ (6.32)

which can now be used in conjunction with the Bernoulli equation [Equation (6.16)].

Example 6.7

A liquid food of density $860\,\mathrm{kg\,m^{-3}}$ and viscosity $0.0012\,\mathrm{Pa\,s}$ flows at $0.5\,\mathrm{l\,s^{-1}}$ from a large header tank through an inclined 20 mm diameter pipe. The pipe is 40 m in length and may be assumed to have a roughness factor of 0.001. Calculate the necessary difference in height between the liquid surface in the tank and the open end of the pipe to give the desired flow rate.

This problem can be solved by applying Bernoulli's relationship in the form of Equation (6.16). Let subscript 1 represent the header tank and subscript 2 the discharge end of the pipe. The pressures at the liquid surface and at the discharge are each atmospheric and therefore $P_1 = P_2$. The liquid level in the tank falls only very slowly because of the large surface area and therefore it is reasonable to approximate the velocity in the tank to zero, that is, $u_1 = 0$. Equation (6.16) now simplifies to

$$Z_1 - Z_2 = \frac{u_2^2}{2g} + h_F$$

where $Z_1 - Z_2$ is the necessary difference in height. The discharge velocity u_2 is

$$u_2 = \frac{0.5 \times 10^{-3} \times 4}{\pi(0.02)^2}\,\mathrm{m\,s^{-1}}, \qquad u_2 = 1.59\,\mathrm{m\,s^{-1}}$$

The Reynolds number is

$$Re = \frac{860 \times 1.59 \times 0.02}{0.0012}, \qquad Re = 22812$$

The Fanning friction factor can be obtained from Figure 6.5. For $Re = 22812$ and $e/d = 0.001$ the friction factor is 0.0067. The head loss due to friction [Equation (6.32)] is now

$$h_F = \frac{4 \times 40 \times 0.0067 \times (1.59)^2}{0.02 \times 2 \times 9.81}\,\mathrm{m}, \qquad h_F = 6.91\,\mathrm{m}$$

Thus the height difference between the tank and the discharge is

$$Z_1 - Z_2 = \frac{(1.59)^2}{2 \times 9.81} + 6.91\,\mathrm{m}, \qquad Z_1 - Z_2 = 7.04\,\mathrm{m}$$

Additional frictional losses due to the presence of pipe bends, fittings and valves must be taken into account if the total pressure drop down a pipeline is to be calculated. Table 6.2 lists the pressure losses which can be expected with some common pipe fittings. These losses are listed both in terms of the equivalent extra pipe length pipe (expressed as a number of pipe diameters) and as a number of velocity heads that is, as multiples of $u^2/2g$.

TABLE 6.2
Frictional Losses in Turbulent Flow through Pipe Fittings and Valves

	No. pipe diameters	No. velocity heads
45° elbow	15	0.35
90° standard elbow	30–40	0.6–0.8
90° large radius elbow	20–23	0.45
90° square elbow	65	1.3
180° bend	75	1.5
T-piece (straight through)	20	0.4
T-piece (branching flow)	60–90	1.2–1.8
Gate valve (fully open)	7	0.15
Gate valve (half open)	200–225	4–4.5
Gate valve (quarter open)	900–1,200	20–24
Globe valve (fully open)	300–350	6.0–7.5

6.5. PRESSURE MEASUREMENT AND FLUID METERING

6.5.1. The Manometer

One of the simplest instruments for the measurement of fluid pressure is the manometer. Figure 6.6 illustrates the use of a U-tube manometer to measure the differential pressure $P_1 - P_2$ as fluid A flows from point 1 to point 2. The manometer is filled with a fluid B which has a density higher than that of A; the height difference h_m is recorded. The pressures in each leg at level Y must be equal and therefore

$$P_1 + \rho_A g(h' + h_m) = P_2 + \rho_A g h' + \rho_B g h_m \qquad (6.33)$$

The differential pressure is then

$$P_1 - P_2 = \rho_A g h' + \rho_B g h_m - \rho_A g h' - \rho_A g h_m \qquad (6.34)$$

or

$$P_1 - P_2 = (\rho_B - \rho_A) g h_m \qquad (6.35)$$

It is often convenient to express this as a head loss h (in metres) of the fluid A, thus equating the pressure difference to the pressure at the base of a column of fluid A gives

$$\rho_A g h = (\rho_B - \rho_A) g h_m \qquad (6.36)$$

and the head loss is then

$$h = \frac{(\rho_B - \rho_A) h_m}{\rho_A} \qquad (6.37)$$

Example 6.8

A U-tube manometer containing mercury (density $13,600 \text{ kg m}^{-3}$) is connected to a pipe through which water flows. The difference in height of mercury between the two legs is 55 mm. What is the pressure drop in the pipe?

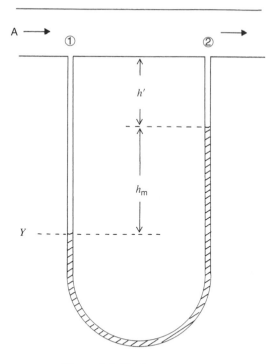

Figure 6.6. U-tube manometer.

From Equation (6.37) the head loss h is

$$h = \frac{(13600 - 1000)9.81 \times 0.055}{13600} \text{ m of water}$$

and

$$h = 0.051 \text{ m of water}$$

This head loss can be expressed as a pressure drop by multiplying by the product of acceleration due to gravity and the density of water. Hence,

$$\Delta P = 1000 \times 9.81 \times 0.051 \text{ Pa}, \qquad \Delta P = 500 \text{ Pa}$$

6.5.2. The Orifice Meter

If a restriction is placed in a pipe through which a fluid is flowing the fluid will accelerate and at the same time the pressure will decrease. It is relatively simple to measure this decrease in pressure which can then be used to determine the velocity, and consequently flow rate, from Bernoulli's equation.

The orifice meter (Figure 6.7) is a disc with a central aperture which is fitted into a pipeline between a pair of flanges. The precise design of the orifice plate, its position and the position of the pressure tappings are specified by British Standards to minimise pressure losses and to give reproducible results. Combining Bernoulli's equation with the principle of continuity

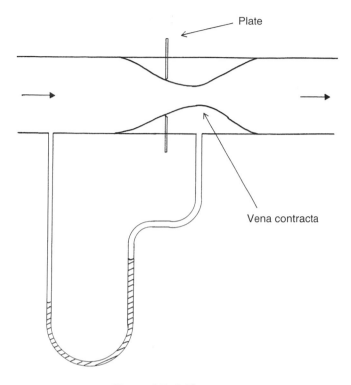

Figure 6.7. Orifice meter.

allows the upstream velocity to be determined. In what follows, subscript 1 refers to a point upstream of the plate and subscript 2 to the point of minimum cross-section or *vena contracta* which exists a short distance downstream of the orifice. The two legs of a manometer should be connected to pressure tappings at these points.

Now, for an incompressible fluid, the density is constant, that is, $\rho_1 = \rho_2 = \rho$, and for a horizontal pipeline $Z_1 = Z_2$. Thus, assuming no frictional losses, Equation (6.16) can be simplified to give

$$\frac{u_1^2}{2} + \frac{P_1}{\rho} = \frac{u_2^2}{2} + \frac{P_2}{\rho} \tag{6.38}$$

From the continuity equation

$$u_2 = u_1 \frac{A_1}{A_2} \tag{6.39}$$

and combining these gives

$$\frac{u_1^2}{2} + \frac{P_1}{\rho} = \frac{u_1^2}{2}\left(\frac{A_1}{A_2}\right)^2 + \frac{P_2}{\rho} \tag{6.40}$$

which can be rearranged to give an expression for the mean velocity at point 1

$$u_1 = \sqrt{\frac{2(P_1 - P_2)}{\rho[(A_1/A_2)^2 - 1]}} \qquad (6.41)$$

The volumetric flow rate V is the product of mean velocity and cross-sectional area and therefore

$$V = A_1 \sqrt{\frac{2(P_1 - P_2)}{\rho[(A_1/A_2)^2 - 1]}} \qquad (6.42)$$

and the mass flow m is then

$$m = A_1\rho \sqrt{\frac{2(P_1 - P_2)}{\rho[(A_1/A_2)^2 - 1]}} \qquad (6.43)$$

In practice energy losses across the orifice plate are significant and therefore it is necessary to introduce a discharge coefficient C_D which is equal to the ratio of the actual flow rate to the flow rate with no energy losses. This coefficient also takes account of the small difference between the cross-section of the *vena contracta* and that of the orifice. Substituting the orifice cross-section A_o for A_2 now gives

$$m = C_D A_1 \rho \sqrt{\frac{2(P_1 - P_2)}{\rho[(A_1/A_o)^2 - 1]}} \qquad (6.44)$$

If the pressure difference is replaced by the term $\rho g h$, where h is the head loss, then the mass flow rate becomes

$$m = C_D A_1 \rho \sqrt{\frac{2gh}{[(A_1/A_o)^2 - 1]}} \qquad (6.45)$$

The discharge coefficient has a value of about 0.62 for an orifice plate. Exact values are specified for designs which conform to British Standards but otherwise individual orifice meters must be calibrated. The orifice plate has the advantages of being inexpensive and simple to use but high pressure losses are sustained.

Example 6.9

An orifice meter, consisting of a 0.1 m diameter orifice in a 0.25 m diameter pipe, has a coefficient of discharge of 0.65. The pipe transports oil with a density of 900 kg m^{-3} and the pressure difference across the orifice plate is measured by a mercury manometer, the leads to the gauge being filled with oil. If the difference in the mercury levels in the manometer is 0.76 m, calculate the flow of oil in the pipe line.

The mass flow rate through the orifice is given by Equation (6.45). The head loss h is

$$h = 0.76 \left(\frac{13600 - 900}{900} \right) \text{ m of oil}$$

or

$$h = 10.724 \text{ m of oil}$$

The cross-sectional area of the pipe, A_1, and the orifice, A_o are, respectively

$$A_1 = \frac{\pi}{4}(0.250)^2\,\mathrm{m}^2 \quad \text{or} \quad A_1 = 0.0491\,\mathrm{m}^2$$

and

$$A_o = \frac{\pi}{4}(0.10)^2\,\mathrm{m}^2 \quad \text{or} \quad A_o = 7.85 \times 10^{-3}\,\mathrm{m}^2$$

Thus the mass flow rate of oil is

$$m = 0.65 \times 0.0491 \times 900\sqrt{\frac{2 \times 9.81 \times 10.724}{[(0.0491/7.85 \times 10^{-3})^2 - 1]}}\,\mathrm{kg\,s^{-1}}$$

or

$$m = 67.5\,\mathrm{kg\,s^{-1}}$$

6.5.3. The Venturi Meter

The venturi meter (Figure 6.8) also presents a restriction to the flow of a fluid in a pipeline but unlike the orifice meter it is characterised by a smooth contraction and expansion of the fluid. The area is a minimum at the throat and there is no vena contracta. Consequently there is a lower pressure loss as the fluid passes through the meter, that is, there is a greater recovery of pressure and venturi meters are used where it is important that pressure is maintained. The discharge coefficient is usually as high as 0.97 or 0.98. However venturi meters have the disadvantage of being more difficult and expensive to construct than orifice plates and they occupy a much greater length of pipeline. Equation (6.45) can now be used for a venturi meter, where A_o becomes the area at the throat and A_1 is the area of the pipeline.

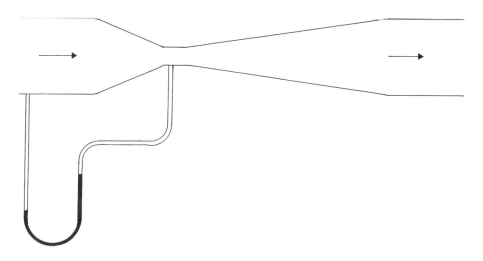

Figure 6.8. Venturi meter.

Example 6.10

A venturi meter used to measure the volumetric flow rate of water has a throat diameter of 75 mm in a pipe diameter of 150 mm. Calculate the flow rate if the difference in level in a mercury U-tube manometer is 178 mm, the mercury being in contact with the water. Assume the coefficient of discharge to be 0.97.

The volumetric flow rate of water is

$$V = C_D A_1 \sqrt{\frac{2gh}{(A_1/A_o)^2 - 1}}$$

where the cross-sectional areas of throat and pipe are, respectively,

$$A_o = \frac{\pi}{4}(0.075)^2 \, \text{m}^2 \quad \text{or} \quad A_o = 4.42 \times 10^{-3} \, \text{m}^2$$

and

$$A_1 = \frac{\pi}{4}(0.150)^2 \, \text{m}^2 \quad \text{or} \quad A_1 = 0.0177 \, \text{m}^2$$

The head loss across the venturi is

$$h = 0.178 \left(\frac{13600 - 1000}{1000} \right) \, \text{m of water}$$

or

$$h = 2.24 \, \text{m of water}$$

Thus the volumetric flow rate of water V is

$$V = 0.97 \times 0.0177 \sqrt{\frac{2 \times 9.81 \times 2.24}{(0.0177/4.42 \times 10^{-3})^2 - 1}} \, \text{m s}^{-1}$$

and

$$V = 0.0294 \, \text{m s}^{-1}$$

6.6. PUMPING OF LIQUIDS

There are two principal types of pump used in food processing applications, the centrifugal pump and the positive displacement pump. However, many other devices are used and these, together with the variants of the principal types, present an immense variety and range of pump from which to select for a given application.

Figure 6.9 represents a liquid being pumped between two tanks, subscript 1 referring to the suction side and subscript 2 to the discharge side of the pump. The liquid, of constant density, flows at a constant velocity through the system and therefore the respective velocity heads can be eliminated from Bernoulli's equation [Equation (6.16)]. The total head Δh which the pump must supply to the liquid is then the difference between the delivery and suction heads, that is,

$$\Delta h = h_2 - h_1 \tag{6.46}$$

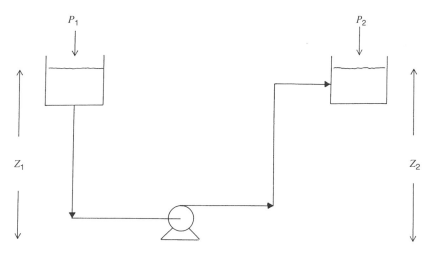

Figure 6.9. A simple pumping system.

where, from Bernoulli,

$$h_2 = Z_2 + \frac{P_2}{\rho g} + h_{F2} \tag{6.47}$$

and

$$h_1 = Z_1 + \frac{P_1}{\rho g} - h_{F1} \tag{6.48}$$

and h_{F1} and h_{F2} are the respective frictional head losses in the pipework. Notice that the suction frictional head loss is negative and that Z_1 would be negative if tank 1 were to be below the level of the pump. Substituting for h_{F1} and h_{F2} from Equation (6.32), and further substituting for velocity in Equation (6.32) in terms of volumetric flow rate V, allows an expression of the form

$$\Delta h = f(V) \tag{6.49}$$

to be developed. This will be necessarily complex in the case of turbulent flow, because of the complex relationship between friction factor and Reynolds number, but for laminar flow Equation (6.49) will take the form of a straight line. Equation (6.49) is the equation of the system characteristic curve.

Example 6.11

Figure 6.10 represents a pumping system for circulating water to a mixing plant one floor above a constant level holding tank. Calculate the total head against which the pump must deliver if the pipe diameter is 0.03 m and the required volumetric flow rate is $72\,l\,min^{-1}$.

The velocity in the pipeline is constant, before and after the pump, and therefore the difference in velocity head in Bernoulli's equation can be omitted. Similarly the difference in the pressure head between the holding tank and the mixing tank is zero (i.e., $P_1 = P_2$). Consequently the total head on each side of the pump is equal to the sum of the potential head and the frictional loss. Thus the delivery head h_2 depends upon the vertical height against

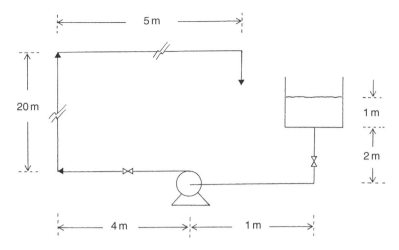

Figure 6.10. Pumping system for Example 6.11.

which the pump must deliver ($Z_2 = 20$ m) and the effective pipe length L_1 which must include frictional losses in valves and elbows:

$$h_2 = 20 + 4\frac{L_2}{d}\frac{c_f}{2}\frac{u^2}{g}$$

For the suction side the frictional loss must be subtracted from the potential head and therefore

$$h_1 = 3 - 4\frac{L_1}{d}\frac{c_f}{2}\frac{u^2}{g}$$

Thus the total head against which the pump must deliver Δh is

$$\Delta h = h_2 - h_1, \qquad \Delta h = 17 + 4\frac{(L_1 + L_2)}{d}\frac{c_f}{2}\frac{u^2}{g}$$

Working in terms of equivalent pipe diameters, and allowing 300 diameters for a globe valve, seven diameters for a gate valve and 40 for each $90°$ bend, for the delivery side $L_2 = 4 + 20 + 5 + 300 + (3 \times 40) = 449$ m and for the suction side $L_1 = 2 + 1 + 7 + 40 = 50$ m.

The mean velocity through the pipework u is the volumetric flow rate divided by the cross-sectional area and

$$u = \frac{72 \times 10^{-3} \times 4}{60\pi (0.03)^2}\,\text{m s}^{-1}, \qquad u = 1.70\,\text{m s}^{-1}$$

Thus the Reynolds number is

$$Re = \frac{1000 \times 1.70 \times 0.03}{10^{-3}} \quad \text{or} \quad Re = 51000$$

Using the Blasius expression to find the friction factor

$$c_f = 0.0079(51000)^{-0.25}, \qquad c_f = 5.26 \times 10^{-4}$$

Hence

$$\Delta h = 17 + 4 \frac{(449 + 50)}{0.03} \frac{5.26 \times 10^{-4}}{2} \frac{(1.70)^2}{9.81} \text{ m}$$

and the pump must deliver against a head of $\Delta h = 22.2$ m.

6.6.1. The Centrifugal Pump

A centrifugal pump (Figure 6.11a) consists of an impeller, with several vanes curved backwards away from the shaft, which rotates at high speed inside a casing. Liquid enters the

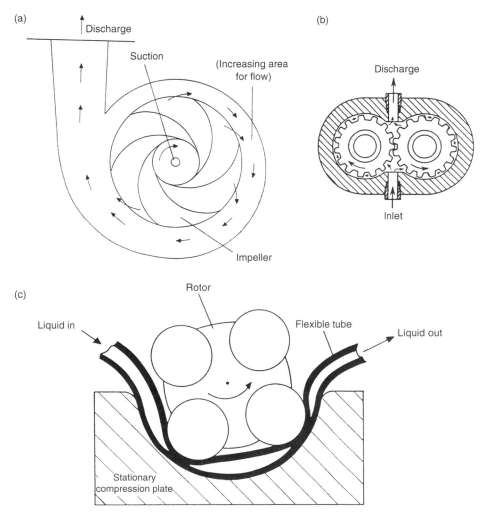

Figure 6.11. Principal pump types: (a) centrifugal pump; (b) gear pump; (c) peristaltic pump ((a) and (c) from J.G. Brennan, J.R. Butters, N.D. Cowell and A.E.V. Lilly, Food Engineering Operations, 3rd ed., Elsevier, 1990, reproduced with permission; (b) from W.L. McCabe and J.C. Smith, Unit Operations of Chemical Engineering, 5th ed., McGraw-Hill, 1993, reproduced with the permission of The McGraw-Hill Companies).

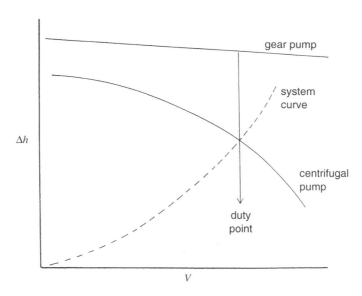

Figure 6.12. Pump and system characteristic curves.

pump axially and is thrown radially outwards at high velocity by the centrifugal effect before being discharged tangentially. The volute, the space between the vanes and the outer casing, increases in cross-sectional area towards the outlet thus reducing the velocity of the liquid, and converting some of the kinetic energy of the liquid into pressure energy as it is discharged.

At increasing volumetric flow rates the total head against which the pump will deliver decreases markedly (Figure 6.12); the maximum head being at zero delivery. The gradient of this curve (the characteristic curve, for a given impeller speed) increases sharply as the viscosity of the liquid increases. A pump may operate at only one point on the pump characteristic curve. This point, the duty point, is found at the intersection of the pump and system characteristic curves, as shown in Figure 6.12. Centrifugal pumps are very widely used but are not self priming and must not be run dry.

6.6.2. Positive Displacement Pumps

Such pumps displace a fixed volume of liquid in a given movement which is then repeated continuously to deliver a flow of liquid. Two broad categories exist: rotary and reciprocating types.

(*a*) *Gear pump*   The gear pump (Figure 6.11b) is perhaps the most common rotary type and consists of two inter-meshing gear wheels inside a casing; one gear is driven and the other idles. The gears rotate such that they move in the direction of flow when close to the casing allowing liquid to be pushed forward in the spaces between consecutive teeth. Liquid is discharged before the teeth mesh together. In a lobe pump the teeth on a wheel are replaced by 2, 3 or 4 lobes which operate in a similar manner. The gear pump is self priming. There is some leakage or slip between the gears and the casing but this decreases as the fluid viscosity increases. The pump is therefore particularly useful for very viscous liquids but cannot be used for suspensions. Gear pumps will deliver against very high pressures and the delivery is more or less constant with pressure. Thus the pump characteristic curve (Figure 6.12) is almost horizontal.

(*b*) *Piston pump* A piston pump is an example of a reciprocating type. Here the liquid is moved by the action of a piston within a cylinder and the delivery then depends upon the length and frequency of the stroke. The movement of the piston is sinusoidal and therefore the delivery is not constant. A more even delivery can be achieved either by using a double-acting pump, in which both sides of the piston are used and delivery occurs during the whole cycle, or by using two pistons in series. In the latter case if each piston is double-acting it is known as a duplex pump and a still more even delivery rate is obtained. The particular advantage of the reciprocating pump is its ability to deliver against very high pressures and at an almost constant output.

(*c*) *Diaphragm pump* An important variant of the piston pump is the diaphragm pump. The piston forces a hydraulic fluid against a diaphragm which is in contact with the liquid to be pumped. Although developed for corrosive liquids this type of device has two advantages for food processing. The lack of contact between the liquid food and any moving parts means that it is inherently hygienic. Second, by means of a relief valve on the hydraulic fluid it is possible to pre-set the maximum pressure exerted on the liquid food and avoid excessive shear which may produce undesirable rheological behaviour (see section 6.8).

(*d*) *Peristaltic pump* Another positive displacement pump of particular use in food applications is the peristaltic pump (Figure 6.11c). An elastic tube looped around three rollers, which are placed on a triangular pitch, is flattened when the disc to which the rollers are attached is rotated. This action has the effect of pushing liquid along the tube in a manner similar to peristalsis in the oesophagus. Flow rate is proportional to the speed of rotation; very small flow rates are possible and the pump is often used as a metering pump. The great advantage of the peristaltic pump is that there is no contact between the liquid and moving parts of the pump and thus it is inherently hygienic. However it can deliver against only very small pressures.

6.6.3. Net Positive Suction Head

There is a danger in operating some pumps (especially centrifugal types) that if the pressure of the liquid at the suction point falls below the pure component vapour pressure of the liquid p' then vaporisation will occur. If this happens inside the pump the formation and collapse of bubbles, known as cavitation, can cause mechanical damage. To avoid cavitation a net positive suction head (*NPSH*) is specified for a given pump. This is the amount by which the suction head must exceed the vapour pressure of the liquid (expressed as a head of the liquid being pumped). Thus net positive suction head is defined as

$$NPSH = h_1 - \frac{p'}{\rho g} \tag{6.50}$$

6.6.4. Hygienic Design

In addition to the flow rate of liquid required and the head against which it is to be pumped, hygiene must also be considered in the selection of pumps for food processing operations. In this respect the hygienic design of pumps follows the general rules and principles of hygienic design: the avoidance of sharp changes in cross-section and dead spaces, ease of dismantling for cleaning, avoidance of screw threads in contact with food, minimising the number of parts and so on.

The pump types mentioned above may be ranked in order of decreasing inherent hygiene, thus:

peristaltic
diaphragm
centrifugal
rotary positive displacement (gear, lobe)
reciprocating

6.7. NON-NEWTONIAN FLOW

6.7.1. Introduction

Very many liquid foods do not obey Newton's law of viscosity [Equation (5.3)]. They do not have a dynamic viscosity as such and the relationships between flow rate and pressure developed in sections 6.3 and 6.4 do not apply. The study of such non-Newtonian behaviour is the provenance of rheology.

The subject of rheology may be thought of as a physical approach to the analysis of mechanical behaviour as opposed to a reliance on subjective judgements based on visual observations, such as the thickness and consistency of liquids for example. A further definition of rheology might then be the science of the deformation of matter. The rheology of liquid foods has a profound influence on a number of processing issues. A proper understanding of rheology is important in calculating pressure drops along pipelines, in pump design, for the correct design and selection of mixers and in ensuring correct rates of heat transfer. It has implications for the flow of foods through nozzles, extruders and packing machinery, and even in the delivery of food to the plate, for example whether cream can be spooned or poured from a container.

Rheological models should be seen as mathematical descriptions of physical behaviour. Sometimes such models are based on an understanding of the structure of the liquid but it is more often the case that attempts have been made to generate a mathematical model of wide applicability and then test whether experimental data from particular foods will fit the model. Thought of in this way rheology becomes a rigorous mathematical description which requires precise and well-defined physical measurements, rather than a way of describing individual materials. Rheology and rheological models can be applied to solid, semi-liquid and soft-solid materials as well as to liquids although theories for solids are not as well advanced as those for liquids and experimental measurements are rather more difficult and unreliable.

6.7.2. Stress, Strain, and Flow

(*a*) *Stress*   Rheological behaviour is defined in terms of the relationship between stress and strain. Stress is simply the force responsible for the deformation of a body divided by the area over which the force acts. It is analogous to pressure and has the same SI unit: the Pascal. Thus when a body is in a state of stress it undergoes deformation. A body may be subject to a

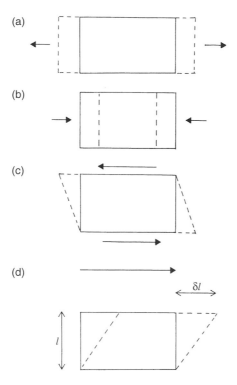

Figure 6.13. Tension, compression, and shear: (a) tension, (b) compression, (c) shear, and (d) simple shear strain.

number of different kinds of force of which the most commonly encountered are compression, tension, and shear. These are illustrated in Figure 6.13. A given material may react in very different ways to each of these forces. In processing, liquids are most usually subjected to shear which, when a liquid flows in a pipeline, produces a velocity gradient, as we saw in section 6.2.2.

The behaviour of solid foods, usually under compressive or tensile forces, is also of interest. There are a variety of empirical tests available to assess the behaviour of solid foods and many of these are product specific; however they are outside the scope of this book.

(*b*) *Simple shear strain*　Strain may be defined as the deformation of a body relative to its original dimensions. Imagine a body undergoing simple shear as in Figure 6.13d. Simple shear strain can be defined as the relative displacement of two parallel layers divided by the separation of those layers. This quantity is then equal to the tangent of the angle γ. Thus

$$\tan \gamma = \frac{\delta l}{l} \tag{6.51}$$

and for small deformations

$$\gamma = \frac{\delta l}{l} \tag{6.52}$$

In rheology the important factor is the rate at which strain is produced. This quantity, designated $\dot{\gamma}$, might best be called the strain rate but is known almost universally, albeit confusingly, as the shear rate. The latter term is used here. Thus the shear rate is the rate of change of γ with time and is defined by

$$\dot{\gamma} = \frac{d\gamma}{dt} \tag{6.53}$$

and has units of s^{-1} or Hz.

Now, for linear flow, where the sheared layer has a velocity u, the displacement δl will occur within a period δt such that

$$\delta l = u \, \delta t \tag{6.54}$$

Thus from Equations (6.52) and (6.53)

$$\dot{\gamma} = \frac{d}{dt} \left(\frac{\delta l}{l} \right) \tag{6.55}$$

and

$$\dot{\gamma} = \frac{u}{l} \tag{6.56}$$

In other words the shear rate is equivalent to the velocity gradient du/dz in Equation (5.3) and the equation of motion for a Newtonian fluid can now be written as

$$R = \mu \dot{\gamma} \tag{6.57}$$

However for non-Newtonian fluids there is no simple linear relationship between shear stress and shear rate and at any particular value of either R or $\dot{\gamma}$ an apparent viscosity μ_a may be defined as

$$\mu_a = \frac{R}{\dot{\gamma}} \tag{6.58}$$

Clearly for a Newtonian fluid the apparent viscosity is equal to the dynamic viscosity.

(c) *Flow*   In many cases it is not a straightforward matter to determine whether a particular food material is a liquid, a solid or in some intermediate state. Consider a fluid, held between two very large parallel plates, to which a shear stress is applied. The stress is removed after the fluid deforms. However if the resultant strain eventually returns to zero then the material has not flowed. On the other hand if the strain does not return to zero then flow has occurred. When flow occurs for a vanishingly small shear stress the material is said to be liquid, if not then it is solid. However over what time scale is the material to be observed? How long must we wait to see if flow occurs? Highly viscous materials may appear to be solid over short periods but are seen to flow (e.g., find their own level in a container) when observed over much longer periods.

6.8. TIME-INDEPENDENT RHEOLOGICAL MODELS

There are many rheological models which can be used to describe the mechanical behaviour of foods. Some have been developed especially for foods, some are complex extensions of existing models. However most models utilise combinations of a relatively small number of basic ideas. These are covered in the following sections broadly in the order of increasing complexity.

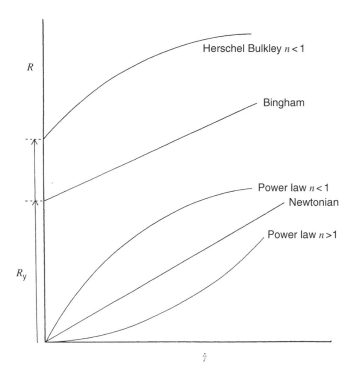

Figure 6.14. Rheograms for non-Newtonian fluids.

6.8.1. Hookean Solids

Many elastic solids obey Hooke's law which states that the ratio of shear stress to shear strain is constant. The constant is known as the shear modulus G. Little more needs to be said about solids here; the rheology of solids will be covered only as it relates to visco-elastic behaviour.

6.8.2. Newtonian Fluids

It is often helpful to visualise rheological models by plotting shear stress against shear rate on a rheogram, or flow curve, as in Figure 6.14. For a Newtonian fluid [Equation (6.57)] there is a linear relationship between shear stress and shear rate and the gradient of the line is equal to the dynamic viscosity. Relatively few food liquids are Newtonian. However the single most important liquid, water, is Newtonian with a dynamic viscosity of 10^{-3} Pa s at a temperature of $20°$C. Consequently many aqueous solutions are also Newtonian. Other materials in this category are milk, organic liquids, most vegetable oils, and some clarified fruit juices.

6.8.3. Bingham Fluids

Many fluids exhibit a yield stress, that is there is stress R_Y below which the fluid will not flow. However when stresses greater than R_Y are applied the fluid behaves as a Newtonian

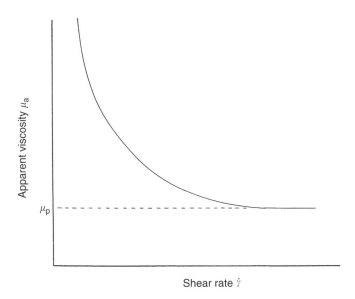

Figure 6.15. Apparent viscosity of a Bingham fluid.

TABLE 6.3
Observed Yield Stresses for Bingham Fluids

	Yield stress (Pa)
Molten milk chocolate	5–200
Strawberry jam	15
Apricot jam	12–30
Strawberry puree	4
Meat extracts	17
High fat cream	3–23
Soft cheeses	70–333
Apple sauce	18–46
Orange juice concentrate	0.9–15

liquid and the rheogram is linear. This behaviour can be represented by

$$R = R_Y + \mu_p \dot{\gamma} \tag{6.59}$$

where μ_p is known as the plastic viscosity. Such fluids are often referred to as Bingham plastics.

Figure 6.15 shows the relationship between apparent viscosity and shear rate. As shear rate increases the apparent viscosity approaches the plastic viscosity which is equal to the gradient of the rheogram, however the apparent viscosity becomes infinite as the shear rate is reduced to zero. Thus when $\dot{\gamma} = 0$ and $R \le R_Y$ a Bingham material behaves rather as a solid. The measured yield stresses for foodstuffs are usually in a range below 100 Pa; Table 6.3 lists some foods for which Bingham behaviour has been observed together with approximate values of their yield stress.

Despite this data there is considerable doubt about whether a yield stress exists. If measurements are made at sufficiently low shear rates then many foodstuffs which otherwise appear to be solid, such as margarine and ice cream, can be made to flow. It is simply that they have a very high apparent viscosity at zero shear. This is a good illustration of the need to specify the range of shear rates over which any particular rheological measurements have been made. Even so, the behaviour which can be characterised by a yield stress has considerable implications for processing. For example, if flow is interrupted it may be difficult to exert a sufficient shear stress to restart flow with the consequence that a Bingham fluid 'sets' at strategic points in the process line such as in valves or pumps.

Example 6.12

Rheological measurements on a new sauce gave the following data:

Shear rate (s^{-1})	Shear stress (Pa)
100	36.0
200	46.5
400	67.0
600	87.6
800	108.0

Which rheological model best fits this data?

From the rheogram (plotted on Cartesian co-ordinates) it is clear that this material is described by the Bingham model with a yield stress of 26 Pa at a shear rate of zero. There is a linear relationship between shear stress and shear rate; the gradient of the line gives the plastic viscosity as $\mu_p = 0.103$ Pa s.

6.8.4. The Power Law

The power law is one of the most useful of all rheological models. It has been found to model the behaviour of a very wide range of foodstuffs yet it is relatively simple, containing only two parameters. However care should be taken not to attempt to apply it over too wide a range of shear rates for any one material. It may be expressed as

$$R = K\dot{\gamma}^n \qquad (6.60)$$

where K is the consistency coefficient and n is a flow behaviour index. On a rheogram the curve passes through the origin, there being no yield stress, but the relationship between shear stress and shear rate is not linear (Figure 6.14). For values of n less than unity the apparent viscosity decreases with increasing shear rate (Figure 6.16). Thus such fluids display *shear-thinning* behaviour and are usually, but not always accurately, referred to as pseudoplastic fluids.

Taking logarithms of each side of Equation (6.60) yields

$$\ln R = \ln K + n \ln \dot{\gamma} \qquad (6.61)$$

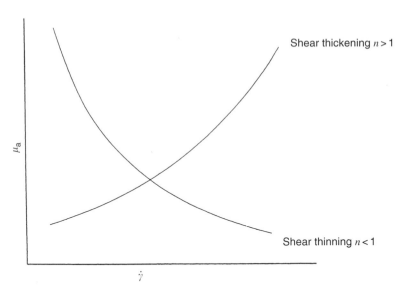

Figure 6.16. Apparent viscosity of power law fluids.

Thus, when shear stress is plotted against shear rate on logarithmic co-ordinates, data which fits the power law will give a straight line of gradient n. Knowing the value of n, and one pair of representative values of R and $\dot{\gamma}$, the consistency coefficient can be obtained from either Equation (6.60) or Equation (6.61).

Example 6.13

A 15% solids peach puree yielded the following rheological information:

Shear rate (s^{-1})	Shear stress (Pa)
100	32.0
300	50.0
500	61.0
1,000	80.0
2,000	105.0
4,000	140.0

Determine the consistency coefficient and flow behaviour index.

Plotting the rheogram on Cartesian co-ordinates gives a curve which suggests shear thinning behaviour. If the data are replotted on logarithmic scales a linear relationship results.

The gradient of the line is 0.4 and thus $n = 0.4$. The point $R = 42\,\text{Pa}$, $\dot{\gamma} = 200\,\text{s}^{-1}$ is representative of the line and therefore

$$K = \frac{42}{(200)^{0.4}}\,\text{Pa}\,\text{s}^{0.4}, \qquad K = 5.04\,\text{Pa}\,\text{s}^{0.4}$$

The consistency coefficient is often erroneously referred to as the apparent viscosity. This is very misleading; it will be seen that K has the dimensions $ML^{-1}T^{n-2}$ and therefore units of Pa s^n. A power law fluid cannot be said to have a viscosity as such; the apparent viscosity when flowing under shear depends upon the rate at which that shear is applied. From Equation (6.58)

$$\mu_a = \frac{K\dot{\gamma}^n}{\dot{\gamma}} \tag{6.62}$$

and therefore

$$\mu_a = K\dot{\gamma}^{n-1} \tag{6.63}$$

Note that μ_a has the dimensions of dynamic viscosity, that is, $ML^{-1}T^{-1}$. Thus for a power law fluid the apparent viscosity is a function of the shear rate. In other words it is essential to determine the shear rate to which the fluid is subjected in order to quantify the viscosity which the material appears to have.

Example 6.14

Calculate the apparent viscosity of yoghurt ($K = 1.2\,\text{Pa s}^{0.6}$, $n = 0.60$) when subjected to a shear rate of $50\,\text{s}^{-1}$.

The apparent viscosity of a power law fluid is given by

$$\mu_a = K\dot{\gamma}^{n-1}$$

and thus, for yoghurt, at the given shear rate,

$$\mu_a = 1.2 \times 50^{(0.6-1)}\,\text{Pa s}, \qquad \mu_a = 0.251\,\text{Pa s}$$

Shear thinning is a very common phenomenon amongst liquid foods and Table 6.4 lists some examples which have been found to exhibit this behaviour. The measured values of consistency coefficient and flow behaviour index vary considerably from food to food and, for a given foodstuff, may vary considerably with composition. It should be appreciated however that the materials listed in Table 6.4 may also display other characteristics and that however well a simple model such as the power law may fit experimental data it does not necessarily offer a complete description.

Shear thinning occurs with dilute suspensions and dilute concentrations of large, often asymmetric, molecules, small particles or other structures. When the fluid is at rest the molecules or particles are entangled in a random order. However on shearing there is a disentanglement and an orientation of molecules in the direction of shear which is proportional to the shear rate. Thus the extent of shear affects the structure and appearance of the fluid and may well influence its processing. Mixing is a good example: shear thinning fluids will adopt a relatively low apparent viscosity in high shear regions close to the impeller in a mixing vessel. As a result good mixing may be confined to this region whilst the bulk of the liquid, which experiences lower shear rates and therefore has a higher apparent viscosity, remains poorly mixed.

TABLE 6.4

Consistency Coefficients and Flow Behaviour Indices
for Power Law Fluids

	K	n
Skim milk concentrate	0.02–4.28	0.72–0.92
Yoghurt	0.3–1.4	0.55–0.65
Egg white	0.12–0.19	0.56–0.6
Apple sauce	7–37	0.17–0.4
Various jams	5–25	0.55–0.7
Apricot puree	5–300	0.27–0.53
10% solids peach puree	4.5	0.34
Guava	2.6–11	0.38–0.68
Tomato paste	7–83	0.09–0.36
Tomato puree	204–406	0.22–0.34

When n is greater than unity the fluid is said to be *shear-thickening* or dilatant and both Figures 6.14 and 6.16 show that shear-thickening behaviour is the reverse of pseudoplasticity. This phenomenon is much less common than pseudoplasticity but is shown by some starch suspensions, honey and guar gum. Shear thickening occurs with very concentrated suspensions so that there is greater interaction between molecules and greater entanglement as shear increases. Some concentrated solutions expand or dilate upon shearing and are also very often shear thickening. Hence the use of the term dilatant to describe shear thickening behaviour. However shear thickening power law fluids are not necessarily dilatant and the terms are not interchangeable.

6.8.5. Laminar Flow of Power Law Fluids

The Hagan–Poiseuille equation developed in section 6.3, which describes the laminar flow of Newtonian fluids, is clearly inadequate for non-Newtonian fluids. Concentrating on the power law because of its widespread applicability, this section will examine the relationship between flow rate and pressure drop in laminar flow within circular cross-section pipes. Equation (6.21) described the parabolic shape of the velocity profile across a pipe diameter. Combining this with the Hagan–Poiseuille equation [Equation (6.25)] gives

$$u_r = 2u \left(1 - \frac{r^2}{r_1^2} \right) \tag{6.64}$$

which is simply a restatement of the velocity profile in terms of the velocity u_r at radius r and the mean velocity u. The shear rate at the pipe wall is then the value of the velocity gradient du_r/dr at $r = r_1$. Differentiating Equation (6.64)

$$\frac{du_r}{dr} = 0 - \frac{4ur}{r_1^2} \tag{6.65}$$

and at $r = r_1$ this gives the wall shear rate as

$$\frac{du_r}{dr} = -\frac{4u}{r_1} \tag{6.66}$$

or in terms of pipe diameter

$$\dot{\gamma}_{wall} = -\frac{8u}{d} \tag{6.67}$$

The quantity $8u/d$ is known as the flow characteristic and is only equal to the shear rate at the wall for a Newtonian fluid. For a power law fluid it is possible to show that the shear rate is given by

$$\dot{\gamma} = \left(\frac{3n+1}{4n}\right)\left(\frac{8u}{d}\right) \tag{6.68}$$

This is usually known as the Rabinowitsch–Mooney equation. Its derivation is complicated and beyond the scope of this book. From inspection it will be seen that the correction factor in the Rabinowitsch–Mooney equation has a value greater than unity for shear-thinning fluids (i.e., for $n > 1$) implying that the wall shear rate for pseudoplastics is always greater than that for Newtonian fluids. The shear stress at the wall is given by Equation (6.26) and therefore for a power law fluid

$$\frac{\Delta P d}{4L} = K\left(\frac{3n+1}{4n}\right)^{n}\left(\frac{8u}{d}\right)^{n} \tag{6.69}$$

It is sometimes convenient to write the final term of Equation (6.69) as a function of volumetric flow rate rather than velocity, that is,

$$\frac{\Delta P d}{4L} = K\left(\frac{3n+1}{4n}\right)^{n}\left(\frac{32V}{\pi d^{3}}\right)^{n} \tag{6.70}$$

This relationship can now be used to estimate the pressure drop, along a pipeline of given length and diameter, for a power law fluid in laminar flow. If the pressure drop is measured at different flow rates in a series of pipelines of different diameter the data can be plotted as $\log(\Delta P d/4L)$ against $\log(32V/\pi d^{3})$ to give a straight line of gradient n. The consistency coefficient can be determined as before from a point on the line.

Example 6.15

The pressure drop required to transport mayonnaise, a power law fluid, through a series of pipelines of different diameter was measured together with volumetric flow rate and the following data were generated:

Diameter (mm)	Volumetric flow rate (l min^{-1})	Pressure gradient (kPa m^{-1})
20	10.4	28.95
20	20.7	35.64
30	21.2	17.33
30	42.4	20.45
50	58.9	8.55
65	99.5	6.08

Determine the consistency coefficient and flow behaviour index and calculate the pressure drop which must be developed to pump mayonnaise along 15 m of a 40 mm diameter pipeline at a rate of 120 l min^{-1}.

The shear stress, from Equation (6.70), is given by $R = \Delta P d/4L$. For the first data set, the pipe diameter d is 0.020 m and $\Delta P/L$, the pressure gradient, is 2.895×10^4 Pa m^{-1} and therefore

$$R = \frac{2.895 \times 10^4 \times 0.020}{4} \text{ Pa}, \qquad R = 144.8 \text{ Pa}$$

The shear rate $\dot{\gamma}$ can be determined from

$$\dot{\gamma} = \frac{32V}{\pi d^3}$$

Taking care to express the volumetric flow rate V in m^3 s^{-1}, this gives

$$\dot{\gamma} = \frac{32 \times 10.4 \times 10^{-3}}{60\pi (0.020)^3} \text{ s}^{-1}, \qquad \dot{\gamma} = 220.7 \text{ s}^{-1}$$

The following table can now be constructed:

Shear stress (Pa)	Shear rate (s^{-1})
144.8	220.7
178.2	439.3
130.0	133.3
153.4	266.6
106.9	80.0
98.8	61.5

Plotting $\log R$ against $\log \dot{\gamma}$ gives a straight line whose gradient is equal to the flow behaviour index and $n = 0.31$. Taking a pair of representative values, for example $R = 106.9$ Pa and $\dot{\gamma} = 80.0$ s^{-1}, allows the consistency coefficient K to be found. Thus

$$K = \frac{106.9}{(80)^{0.31}} \text{ Pa s}^{0.31}, \qquad K = 27.5 \text{ Pa s}^{0.31}$$

The required pressure drop can now be found from Equation (6.70), where the volumetric flow rate V is 2.0×10^{-3} m^3 s^{-1}, $L = 15$ m and $d = 0.04$ m. Hence

$$\Delta P = \frac{4 \times 15}{0.04} 27.5 \left(\frac{(3 \times 0.31) + 1}{4 \times 0.31} \right)^{0.31} \left(\frac{32 \times 2.0 \times 10^{-3}}{\pi (0.04)^3} \right)^{0.31} \text{ Pa}$$

and

$$\Delta P = 2.82 \times 10^5 \text{ Pa}$$

Using the Rabinowitsch–Mooney equation the Reynolds number for a power law fluid now becomes

$$Re = \frac{\rho u^{2-n} d^n}{(K/8)(6n + 2/n)^n} \qquad \qquad (6.71)$$

6.8.6. Other Time-independent Models

There are many other models available to describe time-independent rheological behaviour, many of which have been used successfully with food materials. Most of these models combine in different ways the concepts which have been used so far, that is yield stress, consistency coefficient, and a flow behaviour index which indicates either shear thinning or shear thickening.

(a) *Herschel–Bulkley model* The Herschel–Bulkley model describes materials which combine power law and Bingham behaviour. Above the yield stress the rheogram is non-linear and may display either shear-thinning or shear-thickening (Figure 6.14). The model can be expressed mathematically as

$$R = R_Y + K\dot{\gamma}^n \tag{6.72}$$

This relationship reduces (i) to the power law when there is no yield stress and $R_Y = 0$ and (ii) to the Bingham model when $n = 1$.

(b) *Ellis model* The Ellis equation is usually written explicitly in terms of shear rate, thus

$$\dot{\gamma} = C_1 R + C_2 R^N \tag{6.73}$$

The parameter N, which is *not* the same as the flow behaviour index, always has a value greater than unity. At low shear stresses the second term therefore becomes far less significant and the model reduces to

$$\dot{\gamma} = C_1 R \tag{6.74}$$

This is now a linear relationship between R and $\dot{\gamma}$ and is equivalent to the Newtonian case with the constant C_1 equal to the reciprocal of the dynamic viscosity. At high shear stresses the first term in Equation (6.73) becomes negligible and it reduces to

$$\dot{\gamma} = C_2 R^N \tag{6.75}$$

which is an expression for the power law where the consistency coefficient is a function of C_2 and the flow behaviour index is the reciprocal of N. Because N is always greater than unity, the Ellis model covers only shear thinning behaviour.

(c) *Casson model* The Casson equation is based on the theory of two-phase suspensions and combines a yield stress with pseudoplasticity and was derived originally for suspensions. It has found particular application to complex food fluids such as molten chocolate.

$$R^{0.5} = R_Y^{0.5} + C_3 \dot{\gamma}^{0.5} \tag{6.76}$$

(d) *Power series* The power series is an extension of the power law. It is a polynomial capable of making very slight corrections to fit experimental data. As with all polynomials it should be used with care and should not be used in cases where the plot is almost, but not quite, linear. In other words there is a tendency to fit over complicated equations to data which require a simple straight line equation

$$R = k_1 \dot{\gamma} + k_2 \dot{\gamma}^3 + k_3 \dot{\gamma}^5 \tag{6.77}$$

6.9. TIME-DEPENDENT RHEOLOGICAL MODELS

The foregoing models all assume that the behaviour of the sample is unaffected by time. However the rheological behaviour of some foods is influenced by sample history. For example, the length of time for which a material has been subject to shear may determine the consistency of the fluid. The simplest forms of time-dependent behaviour are thixotropy and rheopexy; these are described in this section. Visco-elasticity is also a time-dependent phenomenon but is dealt with separately in section 6.10.

Thixotropy is characterised by a decrease in apparent viscosity with prolonged shearing at a fixed shear rate. This is illustrated in Figure 6.17; eventually a minimum value of the apparent viscosity is reached at a given constant shear rate. This observed behaviour should not be confused with the classic shear-thinning phenomenon described by the power law, although the two are closely linked. In Figure 6.17 the apparent viscosity falls with time, not with shear rate. The structure of thixotropic foods are similar to those which display only time-independent pseudoplasticity. At a given shear rate there is a finite time required to reach the shear stress at which there exists a structural equilibrium, that is, the point where the rate of aggregation of large molecules, particles or other structures is equal to the rate of breakdown. In other words there is finite time needed to reach a constant level of orientation or disorientation.

In thixotropic foods these processes are very slow compared to time-independent behaviour and on removal of shear the material thickens only slowly. However the original apparent viscosity will be recovered if the material is left to stand. Thus it is possible to view thixotropy and pseudoplasticity as essentially the same phenomenon, but observed over very different time scales.

Figure 6.18 shows the rheogram which is obtained for thixotropic foods. Again the rheogram is a plot of shear stress against shear rate but here time is implicit as the shear rate is

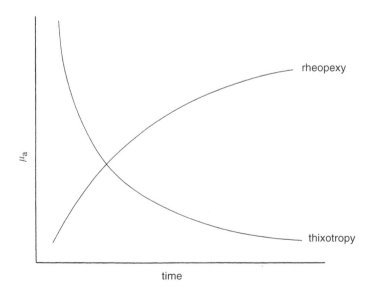

Figure 6.17. Thixotropy and rheopexy: variation in apparent viscosity with time at a fixed shear rate.

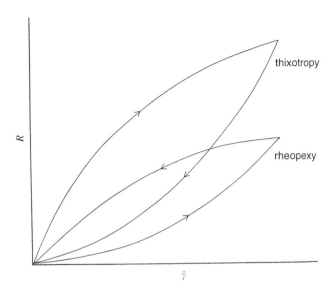

Figure 6.18. Rheogram for thixotropic and rheopectic fluids.

first increased and then reduced as indicated by the arrows. As the shear rate is increased, the shear stress will increase but begins to decline as apparent viscosity decreases with time, in other words as the inherent shear-thinning behaviour makes itself apparent. If the shear rate is now decreased (the lower curve), the shear stress continues to fall because the material thickens only slowly. After allowing sufficient time the material resumes its original structure and returns to its original apparent viscosity. Thus thixotropic behaviour is reversible. Some examples of foods which display thixotropic behaviour are: sweetened condensed milk, soft cheeses, tomato ketchup, and egg white. Rheopexy is the exact reverse of thixotropy. The apparent viscosity increases with time at a given shear rate and hysteresis is also observed in the rheogram. As with shear thickening/shear thinning, rheopectic foods are far less common than thixotropic foods.

6.10. VISCO-ELASTICITY

6.10.1. Introduction

Very many food liquids display both classic viscous flow behaviour and elastic behaviour, that is, the ability to store and recover shear energy. Such materials are known as visco-elastic fluids. Visco-elasticity is associated with a gel structure in liquids in which long chain molecules are stretched and then release their elastic energy. Shear produces not only the usual slippage between layers which results in viscous flow, but also a rotation of large molecules or other structures. This rotation gives rise to normal stresses within the fluid (i.e., stresses perpendicular to the direction of shear). Such normal stresses produce in turn a pressure in the fluid which, for example, leads to the Weissenberg effect; in a Newtonian liquid a free vortex is formed at the liquid surface on stirring but visco-elastic fluids will instead climb the stirrer shaft. Examples of foods which show visco-elasticity include cream, mayonnaise, gums, starches, ice cream, emulsions, dough, cheese, and food gels.

Whether a visco-elastic fluid displays viscous or elastic behaviour usually depends upon the time over which the material is sheared. The Deborah* number De is a dimensionless measure of time which can be used to relate a time constant τ (characteristic of the fluid) to a time t which is characteristic of the process to which the material is subject. Thus

$$De = \frac{\tau}{t} \qquad (6.78)$$

The characteristic process time for flow in a pipe can be approximated by

$$t = \frac{d}{u} \qquad (6.79)$$

where d is the pipe diameter and u is the mean velocity.

6.10.2. Mechanical Analogues

Rheological behaviour, and visco-elasticity in particular, can be modelled by using mechanical analogues. These provide a particularly useful technique with which to visualise the behaviour of complex fluids.

(*a*) *Hookean solid*   A perfect Hookean solid can be modelled by an ideal spring. The force applied to the spring, that is, the weight added, represents shear stress and the extension of the spring is equivalent to shear strain. Thus the spring simulates the behaviour of an ideal elastic material (Figure 6.19a); deformation is instant and the return to zero strain is instant when the applied force (or shear) is removed.

(*b*) *Newtonian liquid*   In Figure 6.19b a Newtonian liquid is represented by a piston in a cylinder (sometimes referred to as a dashpot). The movement of the piston simulates the steady flow of a viscous Newtonian fluid. Extension of the piston is steady with time, representing a constant dynamic viscosity. The strain does *not* return to zero when the applied force is removed; in other words viscous flow cannot be reversed.

(*c*) *Yield stress*   A yield stress can be represented by a St. Venant element or frictional slider. This is a block held in, and then drawn through, a spring clip (Figure 6.19c). Thus the block is rigid below the limiting friction between the block and the clip, which is equivalent to the yield stress exhibited by Bingham fluids. Above this limiting friction the block moves and consequently models viscous flow above the yield stress.

Although not visco-elastic, the use of mechanical analogues can be demonstrated readily by modelling a Bingham fluid. The mechanical analogue for a Bingham fluid has Newtonian and St.Venant elements in parallel, fitted to a Hookean element in series (Figure 6.20). The horizontal bars remain parallel. Now, at low stress only the spring will extend (i.e., the material behaves as a Hookean solid). However, when the applied force (shear stress) exceeds the limiting friction (yield stress) the spring clip will slip and deformation occurs with the limit of extension depending upon the Newtonian element. In other words beyond the yield stress viscous flow occurs and the extent of flow depends upon the plastic viscosity. When the applied force is removed, or falls below the limiting friction, there is some (or perhaps

---

*The Deborah number is so-called after the Jewish prophetess who claimed [Judges 5:5] that 'the mountains melted before God.'

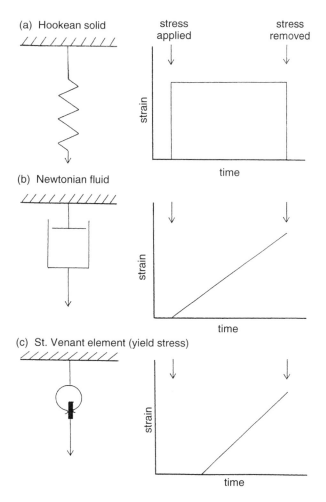

Figure 6.19. Mechanical analogues for rheological behaviour.

complete) recovery of strain in the spring but there will be no recovery in the Newtonian element which is held in place by the spring clip. Once a viscous liquid has flowed it cannot return.

(*d*) *Maxwell model*   The Maxwell model consists of a Newtonian and Hookean element in series (Figure 6.21) and represents the ideal visco-elastic food fluid because the slightest stress causes the material to flow, that is, there is an instant strain (the extension of the spring) followed by the slow elongation of the Newtonian element which simulates viscous flow. The total strain is then the sum of the strain due to each element; Figure 6.21 illustrates the total strain as a function of time. When the stress is removed the ideal spring recovers strain but the Newtonian element does not and thus the strain remains constant with time after removal of the stress.

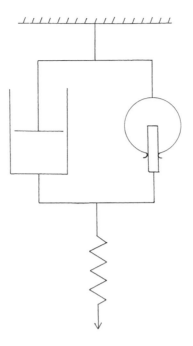

Figure 6.20. Bingham model.

The Maxwell model is represented mathematically by

$$R = \frac{\mu}{G}\left(\frac{dR}{dt}\right) + \mu\dot{\gamma} \tag{6.80}$$

where the quantity μ/G is the characteristic relaxation time τ. If a Maxwell fluid is subjected to slow viscous flow where the rate of change of stress dR/dt is small, Equation (6.80) reduces to the Newtonian case with $\tau = 0$. However when the fluid is stressed suddenly, and dR/dt is large, elastic behaviour is dominant and τ approaches infinity. Putting this another way, viscous flow occurs when the Deborah number is small, that is, deformation is slow and the fluid has time to relax. Elastic behaviour occurs at high Deborah numbers, when the process time is short compared to τ, and the fluid has insufficient time to relax.

The relaxation of a Maxwell fluid is an exponential function and the shear stress at any time t is given by

$$R = R_0 \exp\left[-\left(\frac{t - t_0}{\tau}\right)\right] \tag{6.81}$$

where R_0 is the initial shear stress at time t_0. The relaxation time τ is then the time taken for the imposed shear stress to decay to a fraction e^{-1} of its original value. Cream is an example of a foodstuff whose rheological behaviour is well described by the Maxwell model.

(e) *Kelvin–Voigt model* The Kelvin–Voigt model can be used to represent a more solid-like visco-elastic material. It consists of a Newtonian and Hookean element in parallel (Figure 6.22). When a stress is applied both elements extend together, with the same strain, and the total stress equals the sum of the stresses for each element. On removal of

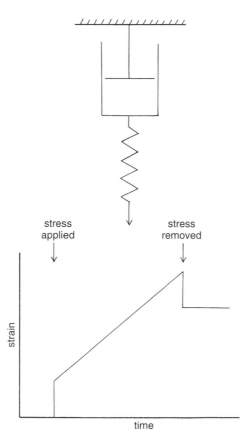

Figure 6.21. Maxwell model.

the stress the spring pulls the Newtonian element back and hence there is delayed elasticity or delayed recovery with complete recovery taking an infinite time. The model can be described by

$$R = G\dot{\gamma} + \mu\dot{\gamma} \qquad (6.82)$$

The recovery of strain is exponential and the strain at any time t is then

$$\dot{\gamma} = \frac{R_0}{G}\left[1 - \exp\left(\frac{-t}{\tau}\right)\right] \qquad (6.83)$$

where τ is the retardation time and

$$\tau = \frac{\mu}{G} \qquad (6.84)$$

The Maxwell and Kelvin–Voigt models are the simplest available for visco-elastic fluids. Many other combinations of these basic ideas have been proposed for more complex foods. For example the Burgers model, used *inter alia* for milk gels, consists of Maxwell and Kelvin–Voigt elements in series.

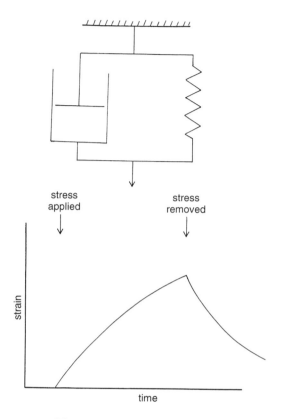

Figure 6.22. Kelvin–Voigt model.

Example 6.16

The following data were obtained from the analysis of two fluids when a constant stress was applied from time = 0.

Time	Strain (%)	
(arbitrary units)	Fluid A	Fluid B
0	7.5	0
1	10.5	10.5
2	13.5	16.5
4	19.0	26.0
6	24.5	31.0
8	30.0	19.0
8.1	22.5	—
10	22.5	16.0
12	22.5	15.5

Propose mechanical analogues for each fluid.

From the strain time curves it is apparent that fluid A approximates to the Kelvin–Voigt model and that fluid B is a Maxwell fluid.

6.11. RHEOLOGICAL MEASUREMENTS

6.11.1. Measurement of Dynamic Viscosity

The dynamic viscosity of Newtonian liquids can be determined relatively easily by measuring a related characteristic or variable. The falling sphere method is applicable to Newtonian fluids only and requires the measurement of the terminal falling velocity u_t of a single particle of density ρ_s and diameter d falling in a fluid of unknown viscosity but of known density ρ. The derivation and validity of this relationship is covered in chapter thirteen. The dynamic viscosity is then

$$\mu = \frac{d^2 g (\rho_s - \rho)}{18 \, u_t} \tag{6.85}$$

This method suffers from two very significant practical difficulties: first, the liquid must be sufficiently transparent to be able to observe the falling sphere and time its descent (more sophisticated methods of timing cannot be justified) and second, a relatively large volume of liquid is required.

A more practical method is the use of a capillary or tube viscometer, the general principle of which is the establishment of the relationship between pressure drop and flow rate for laminar flow in a tube. The flow may be gravitational or pressure driven but the most commonly used instrument is the Ostwald capillary viscometer (Figure 6.23). This relies on the Hagan–Poiseille relationship and consists of a thin capillary tube below a glass bulb. The tube is mounted vertically in a constant temperature bath and the time t, for a constant volume of fluid to drop from A to B, is recorded. The viscometer is used comparatively after calibration with a liquid of known density ρ_2 and viscosity μ_2, usually water, because the diameter of the capillary cannot be known with any precision. Now the velocity is inversely proportional to t, the pressure drop is the hydrostatic head, that is, $\Delta P = \rho g L$, and therefore

$$\mu \propto \rho t \tag{6.86}$$

where the constant of proportionality includes the volume of liquid and the diameter of the capillary. Hence

$$\mu_1 = \mu_2 \frac{t_2}{t_1} \frac{\rho_2}{\rho_1} \tag{6.87}$$

6.11.2. Rheological Measurements for Non-Newtonian Fluids

Although the tube viscometer can be adapted to shear fluids at different rates, and measurements can be made on a series of different diameter pipes at varying flow rates (see Example 6.15), the rheological characteristics of non-Newtonian fluids are best determined by shearing the fluid in a well-defined geometry or configuration. There are two principal types of rotational rheometer: the cone and plate and the concentric cylinder.

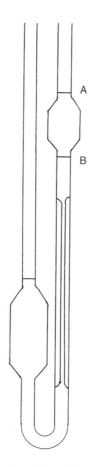

Figure 6.23. Ostwald capillary viscometer.

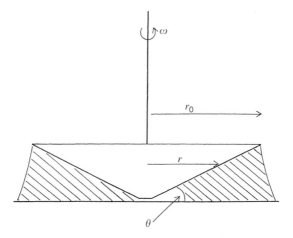

Figure 6.24. Cone and plate rheometer.

(a) *Cone and plate rheometer* The cone and plate rheometer has an inverted cone of very shallow angle, 4° or less, which rotates above a stationary plate (Figure 6.24). The cone is truncated slightly to prevent contact with the plate and the fluid to be sheared is placed in the gap between cone and plate. At the surface of the plate the fluid velocity is zero whilst at a radius r the cone has a local velocity ϖr if the angular velocity is ϖ radians per second. The separation between cone and plate is everywhere $r \sin \theta$ and therefore the shear rate is

$$\dot{\gamma} = \frac{\varpi r}{r \sin \theta} \tag{6.88}$$

or

$$\dot{\gamma} = \frac{\varpi}{\sin \theta} \tag{6.89}$$

Thus the fluid is sheared at a uniform rate and consequently the shear stress is uniform also. This means that the cone and plate is greatly to be preferred for non-Newtonian fluids.

The shear stress is determined by measuring the torque needed to maintain the shear rate $\dot{\gamma}$. The area of an annular element of the plate, of width dr and radius r, is $2\pi r\, dr$ and the torque τ_r acting upon this element is then the product of area, the shear stress R and the distance at which the force acts, that is, the radius r. Thus

$$\tau_r = 2\pi r^2 R \, dr \tag{6.90}$$

and the total torque is then obtained by integrating across the entire plate:

$$\tau = \int_0^{r_o} 2\pi r^2 R \, dr \tag{6.91}$$

where r_o is the outer radius of the plate, yielding

$$\tau = \frac{2\pi r_o^3 R}{3} \tag{6.92}$$

from which the shear stress is calculated. It is assumed in measuring the shear stress and shear rate that the fluid is subject to simple shear. In particular it is important to avoid the phenomenon of secondary flow in which radial circulation patterns are established within the fluid. This occurs particularly when the cone angle is greater than about 4°. In practice the cone angle is usually 2° or less.

(b) *Concentric cylinder rheometer* The fluid of interest is sheared in the narrow gap between two concentric, or coaxial, cylinders (Figure 6.25). Either cylinder may be driven; there is less chance of secondary flow developing if the outer cylinder rotates, although it is easier to arrange for the inner one to rotate. Again the torque required to rotate the inner cylinder is measured. The torque per unit length of cylinder is then

$$\tau_i = 2\pi r_i R_i r_i \tag{6.93}$$

where R_i is the shear stress at the inner cylinder surface. At equilibrium this must be equal and opposite to the torque at the outer surface and therefore

$$2\pi r_i R_i r_i = 2\pi r_o R_o r_o \tag{6.94}$$

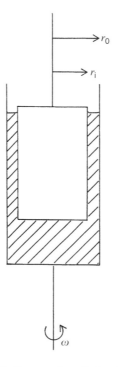

Figure 6.25. Concentric cylinder rheometer.

and

$$\frac{R_{o}}{R_{i}} = \left(\frac{r_{i}}{r_{o}}\right)^{2}$$ (6.95)

Clearly, from Equation (6.95), shear stress is a function of radius and therefore the gap must be small to ensure as far as possible that the shear rate is uniform throughout the fluid. For example, if a 10% variation in shear stress is thought to be acceptable then $r_{i}/r_{o} = 1.049$ and, for a cylinder diameter of 50 mm, the gap must be no greater than 1.22 mm. Similarly the shear rate varies across the gap and an average value must be used which is obtained from

$$\dot{\gamma} = \frac{r_{a}\varpi}{r_{o} - r_{i}}$$ (6.96)

where r_{a} is the arithmetic mean radius.

(c) *Infinite fluid viscometer*   An infinite fluid viscometer consists of a rotating disc on a thin spindle which is immersed in a bath of fluid. Now, because the fluid is not sheared in a precise geometry, an estimate of the average shear rate to which the fluid is subjected is required; this is assumed to be a simple function of the speed of rotation of the disc. However, because fluid close to the disc is sheared at a higher rate than fluid some distance away, this approximation means that such devices cannot be used to make precision measurements. The torsion on the spindle is measured to give the shear stress and a calibration chart provides an instant read out of apparent viscosity. Despite its limitations, the infinite fluid viscometer is portable, easy to use and useful for on-site measurements.

NOMENCLATURE

A	Cross-sectional area
A_o	Cross-sectional area of orifice or venturi throat
c_f	Friction factor
C_D	Discharge coefficient
C_1	Parameter in Ellis model
C_2	Parameter in Ellis model
C_3	Parameter in Casson model
d	Diameter
d_e	Hydraulic mean diameter
d_i	Inner diameter of annulus
d_o	Outer diameter of annulus
De	Deborah number
e	Magnitude of surface roughness
g	Acceleration due to gravity
G	Shear modulus
h	Head loss
h_F	Head loss due to friction
h_m	Manometer height difference
h'	Height
k_1, k_2, k_3	Coefficients in power series
K	Consistency coefficient
l	Separation of layers in simple shear
L	Length; characteristic length
m	Mass flow rate
n	Flow behaviour index
N	Parameter in Ellis model
$NPSH$	Net positive suction head
p'	Pure component vapour pressure
P	Pressure
r	Radius
r_a	Arithmetic mean radius
R	Shear stress
R_0	Initial shear stress
R_Y	Yield stress
Re	Reynolds number
t	Time
u	Mean velocity
u_r	Velocity at a radius r
u_t	Terminal falling velocity
V	Volumetric flow rate
Z	Height above datum

GREEK SYMBOLS

γ Displacement angle in simple shear
$\dot{\gamma}$ Shear rate
$\dot{\gamma}_{\text{wall}}$ Wall shear stress
ΔP Pressure drop
θ Cone angle
μ Viscosity
μ_{a} Apparent viscosity
μ_{p} Plastic viscosity
υ Specific volume
ρ Density
ρ_{av} Average density
τ Torque; time constant
ϖ Angular velocity

SUBSCRIPTS

A   Fluid A
B   Fluid B
i   Inner
o   Outer
s   Solid

PROBLEMS

6.1. Water discharges at a flow rate of $0.004 \, \text{m}^3 \, \text{s}^{-1}$ through an opening 50 mm in diameter. Calculate the velocity of discharge.

6.2. Determine the mass flow rate of a liquid, density $950 \, \text{kg} \, \text{m}^{-3}$, flowing in a 25 mm diameter pipe at a mean velocity of $1.5 \, \text{m} \, \text{s}^{-1}$.

6.3. Calculate the Reynolds number in each of the following cases and state whether the flow is laminar or turbulent: (a) fluid density $1{,}000 \, \text{kg} \, \text{m}^{-3}$, fluid viscosity $1.2 \times 10^{-3} \, \text{Pa} \, \text{s}$, flow rate $0.5 \, \text{l} \, \text{s}^{-1}$, pipe diameter 15 mm; (b) fluid density $800 \, \text{kg} \, \text{m}^{-3}$, fluid viscosity $60 \times 10^{-3} \, \text{Pa} \, \text{s}$, mean velocity $2 \, \text{m} \, \text{s}^{-1}$, pipe diameter 20 mm.

6.4. A simple concentric tube heat exchanger is used to cool a roux sauce from 70°C to 10°C. The inner tube has internal and external diameters of 50 and 55 mm, respectively and the outer tube has an internal diameter of 110 mm. The water flows in the annulus at a velocity of $2 \, \text{m} \, \text{s}^{-1}$ and sauce flows in the inner tube at a rate of $2 \, \text{kg} \, \text{s}^{-1}$. Determine the Reynolds number for each fluid. Data for roux sauce at 40°C: density $= 1{,}042 \, \text{kg} \, \text{m}^{-3}$, viscosity $= 0.190 \, \text{Pa} \, \text{s}$. Take the properties of water at 5°C.

6.5. Oil of density $850 \, \text{kg} \, \text{m}^{-3}$ flows in a pipe at $2.5 \, \text{m} \, \text{s}^{-1}$. Express this as a velocity head in metres of oil. What is the corresponding pressure in kPa?

6.6. What pressure must be developed to pump $4,500 \, \text{kg h}^{-1}$ of vegetable oil through a 0.05 m diameter pipeline if the density and viscosity are $850 \, \text{kg m}^{-3}$ and 0.03 Pa s, respectively?

6.7. Determine the manometer reading on a mercury manometer corresponding to a head of 6 m of oil (density $850 \, \text{kg m}^{-3}$).

6.8. What deflection will be recorded on a mercury manometer, connected to a pipe carrying water, if the pressure drop being measured is 50 kPa?

6.9. A venturi meter with a throat diameter of 50 mm measures the flow of water in a 75 mm diameter pipe. A mercury manometer is used to measure the difference in head between the throat and the entrance to the meter. What should be the manometer deflection for a water flow rate of $600 \, \text{l min}^{-1}$. Take the discharge coefficient to be 0.97.

6.10. A cone and plate rheometer was used to measure shear stress as a function of shear rate for quark, a soft cheese, at $10°C$. Assuming that quark is a power law fluid, determine the consistency coefficient and flow behaviour index.

R (Pa)	$\dot{\gamma} \, (\text{s}^{-1})$
3,450	20
4,000	40
4,600	80
6,000	300
6,600	500

6.11. A fruit puree gave the following data when subjected to rheological measurement:

R (Pa)	$\dot{\gamma} \, (\text{s}^{-1})$
170	20
270	100
320	200
420	500
540	1,000
850	5,000

Determine which model best fits the data and the find the values of the parameter in the model.

6.12. Before installing a line to transport tomato concentrate between two stages of a process, the following data was obtained from pilot plant trials in which pressure drop was measured as a function of flow rate for different pipe diameters:

Diameter (mm)	Volumetric flow rate (1 min^{-1})	Pressure gradient (kPa m^{-1})
25	184.0	17.6
25	92.0	14.1
50	441.0	5.92
50	147.3	4.10
75	124.3	1.63

Determine the pressure drop required to transfer tomato concentrate along 25 m of 50 mm diameter pipe at a rate of 3.6 m^3 h^{-1}.

6.13. Estimate the apparent viscosity of tomato concentrate during transfer in the pipeline specified in Problem 6.12.

6.14. Tomato puree is assumed to be a shear thinning fluid with a consistency coefficient of 330 Pa s^n and a flow behaviour index of 0.27. It is pumped at a rate of 120 1 min^{-1} through a horizontal pipeline over a distance of 5 m. If the pump is limited to a delivery pressure of 300 kPa, what is the minimum pipe diameter which can be used?

6.15. A constant stress was applied from time $t = 0$ to a new food material. The resultant strain, as a function of time, is given in the table:

Time (arbitrary units)	Strain (%)
0	7.5
0.1	8.5
0.6	13.0
1.7	19.5
3.0	24.5
5.0	29.2
5.9	32.0
6.1	25.0
6.2	23.2
7.0	19.6
9.1	16.6

Suggest a suitable model to describe the rheological behaviour of this material.

FURTHER READING

H. A. Barnes, J. F. Hutton, and K. Walters, *An Introduction to Rheology*, Elsevier, Amsterdam (1989).

J. F. Douglas, J. M. Gasiorek, and J. A. Swaffield, *Fluid Mechanics*, Longman Scientific & Technical, New York (1995).

P. J. Fryer, D. L. Pyle, and C. D. Rielly (eds.), *Chemical Engineering for the Food Industry*, Chapman and Hall, London (1997).

J. M. Kay and R. M. Nedderman, *Fluid Mechanics and Transfer Processes*, Cambridge University Press, Cambridge (1985).

W. Madill, *Fluid Mechanics*, Macdonald and Evans, (1983).

J. H. Prentice, *Measurements in the Rheology of Foodstuffs*, Elsevier Applied Science, Amsterdam (1984).

P. Sherman, *Industrial Rheology*, Academic Press, New York (1970).

R. W. Whorlow, *Rheological Techniques*, Ellis Horwood, Chichester (1980).

Heat Processing of Foods

7.1. INTRODUCTION

Very many food processing operations involve the transfer of heat: cooking, roasting, drying, evaporation, sterilisation (either of bulk liquids or packaged foods), chilling and freezing are utilised to preserve food or to prepare it directly for eating. Thus the student of food engineering needs a thorough understanding of the mechanisms of the transfer of heat together with a knowledge of heat exchange equipment. Many applications of heat transfer to food processing require a knowledge of unsteady-state theory; this is particularly true of freezing, for example. However, it is undoubtedly easier to grasp the principles of heat transfer by studying steady-state processes first and, although steady-state is simply a special case of the general unsteady-state theory, the approach adopted here is to study the former first. More complex problems will be introduced only when steady-state heat exchange has been covered in detail.

Heat may be transferred by one or more of the three mechanisms of conduction, convection and radiation. Most industrial heat transfer operations involve a combination of these but it is often the case that one mechanism is dominant. Conduction is associated with heat transfer in solids because it describes the transfer of vibrational energy from one molecule to another, and is, therefore, more significant in solids where molecules are arranged more closely together than in the freer structures of liquids or gases. Convection is associated with liquids and gases because heat is transferred due to the bulk movement of a fluid. In both conduction and convection the rate at which heat is transferred is proportional to the temperature difference between the heat source and the heat sink. Radiative heat transfer is significantly different to either conduction or convection. Thermal radiation is part of the electromagnetic spectrum and, therefore, can be transferred through a vacuum. It is especially significant at high temperatures because of its dependence on temperature to the fourth power. Microwave radiation is also part of the electromagnetic spectrum but heats food by a different mechanism; it is treated separately in section 7.8.

7.2. CONDUCTION

7.2.1. Steady-state Conduction in a Uniform Slab

The heat flux due to conduction in the x direction through a uniform homogeneous slab of material (Figure 7.1) is given by Fourier's first law of conduction

$$q = -k \frac{dT}{dx} \tag{7.1}$$

131

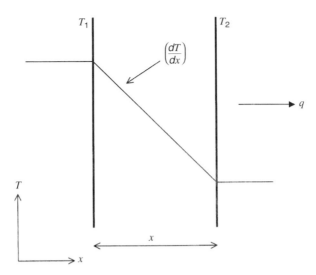

Figure 7.1. Steady-state conduction through a uniform homogeneous slab.

and is related to the rate of heat transfer Q through the cross-sectional area of the slab A by

$$q = \frac{Q}{A} \tag{7.2}$$

Combining Equations (7.1) and (7.2) and integrating between the relevant limits gives

$$Q \int_0^x dx = -kA \int_{T_1}^{T_2} dT \tag{7.3}$$

where T_1 and T_2 are the surface temperatures of the respective faces of the slab and x is its thickness in the x direction. Thus the rate of heat transfer is given by

$$Q = \frac{kA(T_1 - T_2)}{x} \tag{7.4}$$

which can be rearranged in the form

$$Q = \frac{A\Delta T}{x/k} \tag{7.5}$$

Equation (7.5) shows that for uni-dimensional, steady-state conduction the rate of heat transfer Q is proportional to a driving force or potential (i.e., the temperature difference across the slab ΔT), to the area through which heat is transferred and is inversely proportional to the quantity x/k or thermal resistance. Thermal resistance is a measure of the difficulty with which heat is conducted through a solid and depends upon the thermal conductivity of the material and a dimension in the direction of heat flow, in this case the slab thickness. Note that thermal resistance has units of $m^2\,K\,W^{-1}$. Table 7.1 lists the thermal conductivities of a range of food and non-food materials.

TABLE 7.1

Thermal Conductivities of Food and Non-food Materials

	Material	Temperature (°C)	Thermal conductivity (W m^{-1} K^{-1})
Metals	Copper	0	388
		100	377
	Aluminium	0	202
		100	205
	Carbon steel	100	45
	316 stainless steel		15
Construction materials	Wood		0.11–0.21
	Concrete		0.8–1.0
	Brick		0.70
	Soda glass		0.52–0.76
Insulation	Polystyrene		0.025–0.040
	Polyurethane		0.025
	Glass wool		0.042
	Mineral wool		0.037–0.043
	Cork		0.060
Packaging	Corrugated cardboard		0.064
	Paper		0.13
Gases	Air	−23	0.0223
		2	0.0243
		52	0.0282
		102	0.0319
	Oxygen	0	0.0246
		65	0.0299
	Nitrogen	0	0.0239
		65	0.0287
	Carbon dioxide	0	0.0145
		65	0.0190
Foods	Apple juice		0.55
	Apple sauce	23	0.692
	Beef (lean)	−10	1.35
	Butter	4	0.197
	Fish (fresh)	0	0.43–0.54
	Fish (frozen)	−10	1.2
	Honey	2–70	0.50–0.62
	Ice	−25	0.45
		0	2.25
	Lamb	5	0.41–0.48
	Margarine		0.234
	Milk (whole)	20	0.50–0.53
	Milk (skim)	2	0.538
	Milk (condensed 50% water)	26	0.329
		78	0.364
	Orange juice		0.55
	Peanut oil	4	0.168
	Pork (fresh)	2	0.44–0.54
	Pork (frozen)	−15	1.11
	Salmon	−29	1.30
		4	0.502
	Water	0	0.569
		20	0.603
		50	0.643
		100	0.681
	Water vapour	100	2.48×10^{-5}

Example 7.1

In an experiment to measure the thermal conductivity of meat, beef was formed into a square-section block 5 cm × 5 cm and 1 cm thick. The edges of the block were insulated and heat was supplied continuously to one face of the block at a rate of 0.80 W. The temperatures of each face were measured with thermocouples and found to be 28.5°C and 23.3°C, respectively. What is the thermal conductivity of beef?

The integrated form of Fourier's law, Equation (7.4), can be rearranged explicitly in terms of thermal conductivity:

$$k = \frac{Qx}{A(T_1 - T_2)}$$

Thus, taking care to convert both the area and the slab thickness into SI units,

$$k = \frac{0.80 \times 0.01}{(0.05)^2(28.5 - 23.3)} \, \text{W m}^{-1}\,\text{K}^{-1}, \qquad k = 0.615\,\text{W m}^{-1}\,\text{K}^{-1}$$

Note that the temperature difference across the block of 5.2°C is equal to a temperature difference of 5.2 K.

7.2.2. Conduction in a Composite Slab

If now the slab is a composite of materials of both different thickness and different thermal conductivity, as in Figure 7.2, Equation (7.4) cannot be used for the whole slab; there is no single value of either x or k and thermal conductivities cannot be added together. However,

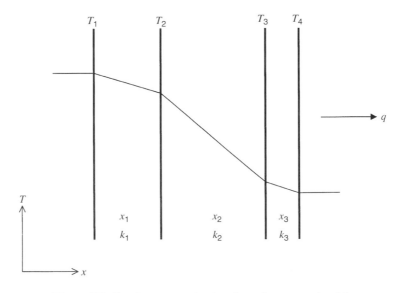

Figure 7.2. Steady-state conduction through a composite slab.

applying Equation (7.5) to each layer in turn gives the respective temperature differences as

$$\left.\begin{aligned}
\Delta T_1 &= \frac{Q_1}{A}\frac{x_1}{k_1} \\
\Delta T_2 &= \frac{Q_2}{A}\frac{x_2}{k_2} \\
\Delta T_3 &= \frac{Q_3}{A}\frac{x_3}{k_3}
\end{aligned}\right\} \tag{7.6}$$

where Q_1, Q_2, and Q_3 are the rates of heat transfer in each layer, respectively. Now

$$\left.\begin{aligned}
\Delta T_1 &= T_1 - T_2 \\
\Delta T_2 &= T_2 - T_3 \\
\Delta T_3 &= T_3 - T_4
\end{aligned}\right\} \tag{7.7}$$

and the overall temperature difference across the composite slab is

$$\Delta T = T_1 - T_4 \tag{7.8}$$

and

$$\Delta T = \Delta T_1 + \Delta T_2 + \Delta T_3 \tag{7.9}$$

Further, there is no generation or accumulation of heat at the interfaces between each layer so that

$$Q_1 = Q_2 = Q_3 = Q \tag{7.10}$$

Thus, substituting from Equations (7.6) and (7.10) into Equation (7.9), the overall temperature difference becomes

$$\Delta T = \frac{Q}{A}\left(\frac{x_1}{k_1} + \frac{x_2}{k_2} + \frac{x_3}{k_3}\right) \tag{7.11}$$

and therefore

$$Q = \frac{A\,\Delta T}{(x_1/k_1 + x_2/k_2 + x_3/k_3)} \tag{7.12}$$

The rate of heat transfer through a composite slab is now proportional to area and to the overall temperature difference across the whole slab and is inversely proportional *to the sum of the thermal resistances of each layer*. This is an extremely important point: thermal resistance is an additive property whereas thermal conductivity is not. A general version of Equation (7.12) is now

$$Q = \frac{\text{area} \times \sum \text{temperature difference}}{\sum \text{thermal resistance}} \tag{7.13}$$

and this relationship can be applied to any steady-state problem involving either conduction or convection or combinations of the two.

Example 7.2

It is proposed to build a cold store having an outer wall of concrete (100 mm thick) and an inner wall of wood (10 mm thick), with the space in between (100 mm) filled with polyurethane foam. If the inner wall temperature is 5°C and the outer wall is maintained at the ambient air temperature of 20°C, calculate the rate of heat penetration.

The wall area is not specified and therefore only the rate of heat penetration per unit area can be determined. Hence, from Equation (7.12),

$$\frac{Q}{A} = \frac{\Delta T}{(x_1/k_1 + x_2/k_2 + x_3/k_3)}$$

The overall temperature difference across the composite wall is the difference between $20°C$ and $5°C$. Let the subscripts 1, 2, and 3 represent concrete, insulation, and wood, respectively. Thus, taking thermal conductivity data from Table 7.1,

$$\frac{Q}{A} = \frac{(20 - 5)}{(0.10/0.80 + 0.10/0.025 + 0.01/0.17)} \text{ W m}^{-2}$$

and

$$\frac{Q}{A} = \frac{(20 - 5)}{(0.125 + 4.0 + 0.0588)} \text{ W m}^{-2}$$

or

$$\frac{Q}{A} = 3.59 \text{ W m}^{-2}$$

Note the magnitudes of the thermal resistances. That of the insulation is significantly greater than those of the structural materials.

Example 7.3

For the cold store wall in Example 7.2, calculate the temperature at the interfaces of the three layers.

The equation for a uniform slab can be applied to each layer in turn. The heat flux is known and therefore it is possible to find one unknown temperature. Alternatively the problem can be solved by recognising that temperature difference is proportional to thermal resistance. The total thermal resistance of the composite wall is $4.184 \text{ m}^2 \text{ K W}^{-1}$. Thus the temperature difference across the concrete layer is $\frac{0.125}{4.184} (20 - 5) = 0.45 \text{ K}$ and the temperature at the concrete/insulation interface is $(20 - 0.45) = 19.55°C$. Similarly the temperature drop across the layer of wood is $\frac{0.0588}{4.184} (20 - 5) = 0.21 \text{ K}$ and the temperature between wood and insulation is $5.21°C$. Consequently the greatest temperature change (14.34 K) is that across the insulation, which has the largest thermal resistance.

7.2.3. Radial Conduction

Many problems in food engineering involve the conduction of heat in a radial direction through the walls of pipes and tubes (Figure 7.3). If a hot liquid is conveyed along a pipe heat will be lost from the liquid to the surroundings, assuming the surroundings to be at a lower temperature, even if the pipe is insulated. In some heat exchangers, in which heat is deliberately transferred from one fluid to another, the fluids are carried in tubular conduits.

Fourier's law now takes the form

$$q = -k \frac{dT}{dr} \tag{7.14}$$

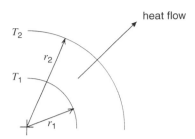

Figure 7.3. Steady-state conduction through a thick-walled cylinder.

where the temperature varies with radius r but is constant along the length of the pipe L. The rate of heat transfer is then given by

$$Q = -k A \frac{dT}{dr}$$ (7.15)

where the area through which heat is conducted is the curved surface area of the pipe

$$A = 2\pi r L$$ (7.16)

Care should be taken not to confuse this with the cross-sectional area of the pipe $\pi d^2/4$. Thus, on integration and substitution, Equation (7.15) becomes

$$Q \int_{r_1}^{r_2} \frac{dr}{r} = -2\pi L k \int_{T_1}^{T_2} dT$$ (7.17)

where T_1 is the temperature at a radius r_1 and T_2 is the temperature at a radius r_2. Evaluating Equation (7.17) gives

$$Q \ln \left(\frac{r_2}{r_1} \right) = -2\pi L k (T_2 - T_1)$$ (7.18)

and the rate of heat transfer becomes

$$Q = \frac{2\pi L k (T_1 - T_2)}{\ln(r_2/r_1)}$$ (7.19)

Equation (7.19) can now be rearranged into the general form of Equation (7.13) by multiplying through by r_2/k, hence

$$Q = \frac{2\pi r_2 L (T_1 - T_2)}{(r_2/k) \ln(r_2/r_1)}$$ (7.20)

where

$$\frac{r_2}{k} \ln \left(\frac{r_2}{r_1} \right)$$

is the thermal resistance of the cylinder wall and $2\pi r_2 L$ is the (external) surface area through which heat is transferred. It is possible to base the rate of heat transfer on the internal surface area but it is conventional to use the external area. The thermal resistance is inversely proportional to thermal conductivity and is a function of the inner and outer radii of the cylinder which define the distance in the r direction through which heat is transferred. As before this has units of $m^2\,K\,W^{-1}$. It is essential to realise that the temperatures in Equation (7.20) are *surface* temperatures and *not* the temperatures of the fluids (the hot food fluid and the air surrounding the pipe).

Example 7.4

Water flows through an uninsulated 0.05 m diameter pipe which has a wall thickness of 0.01 m. The thermal conductivity of the pipe wall is $50\,W\,m^{-1}\,K^{-1}$ and the inside and outside surface temperatures of the pipe are $70°C$ and $69.5°C$, respectively. Calculate the radial heat loss per metre length.

Adapting Equation (7.20), the rate of heat transfer in a radial direction per unit length of pipe is

$$\frac{Q}{L} = \frac{2\pi r_2 (T_1 - T_2)}{(r_2/k)\ln(r_2/r_1)}$$

For convenience, the external radius may be cancelled from this expression and therefore

$$\frac{Q}{L} = \frac{2\pi (70 - 69.5)}{(1/50)\ln(0.035/0.025)}\,W\,m^{-1}, \qquad \frac{Q}{L} = 466.8\,W\,m^{-1}$$

(*a*) *Simplified expressions for radial conduction*   Equation (7.20) may be put in a slightly different form by making use of the logarithmic mean radius which is defined by

$$r_{lm} = \frac{r_2 - r_1}{\ln(r_2/r_1)} \tag{7.21}$$

and thus

$$Q = \frac{2\pi r_{lm} L k (T_1 - T_2)}{r_2 - r_1} \tag{7.22}$$

This expression can be simplified for thin-walled cylinders, where $r_2/r_1 \approx 1$, by replacing the logarithmic mean radius with the arithmetic mean radius r_a such that

$$Q = \frac{2\pi r_a L k (T_1 - T_2)}{r_2 - r_1} \tag{7.23}$$

where the arithmetic mean radius is defined by

$$r_a = \frac{r_1 + r_2}{2} \tag{7.24}$$

(*b*) *Radial temperature distribution*   From Equation (7.13) it can be seen that the temperature difference across a body is proportional to the thermal resistance. Therefore, for the case of radial conduction, for a general radius r at which the temperature is $T(r)$,

$$\frac{T_1 - T(r)}{T_1 - T_2} = \frac{\ln(r/r_1)}{\ln(r_2/r_1)} \tag{7.25}$$

This can be rearranged to give the temperature $T(r)$ as a function of radius r and the boundary conditions T_1, T_2, r_1, and r_2.

$$T(r) = T_1 - \left(\frac{(T_1 - T_2) \ln(r/r_1)}{\ln(r_2/r_1)} \right) \tag{7.26}$$

7.2.4. Conduction in a Composite Cylinder

In order to reduce heat loss from a pipe conveying a hot food fluid a layer of insulation material of low thermal conductivity may be placed around the pipe (Figure 7.4). If the insulation has a thermal conductivity k_B and an outer radius r_3 then the rate of heat transfer through it is

$$Q = \frac{2\pi r_3 L (T_2 - T_3)}{(r_3/k_B) \ln(r_3/r_2)} \tag{7.27}$$

where the external surface area of the insulation is $2\pi r_3 L$. As before the overall temperature difference is

$$\Delta T = (T_1 - T_2) + (T_2 - T_3) \tag{7.28}$$

Substituting for $(T_1 - T_2)$ and $(T_2 - T_3)$ gives

$$\Delta T = Q \left[\frac{(r_2/k_A) \ln(r_2/r_1)}{2\pi r_2 L} + \frac{(r_3/k_B) \ln(r_3/r_2)}{2\pi r_3 L} \right] \tag{7.29}$$

or

$$\Delta T = Q \left[\frac{(r_3 r_2/k_A) \ln(r_2/r_1) + (r_2 r_3/k_B) \ln(r_3/r_2)}{2\pi r_2 r_3 L} \right] \tag{7.30}$$

Rearranging and cancelling r_3 results in the following expression for the rate of heat transfer (or heat loss) through the combined resistance of the pipe (thermal conductivity k_A) and insulation:

$$Q = \frac{2\pi r_2 L \, \Delta T}{(r_2/k_A) \ln(r_2/r_1) + (r_2/k_B) \ln(r_3/r_2)} \tag{7.31}$$

Note that Equation (7.31) is based on the external surface area of the pipe $2\pi r_2 L$.

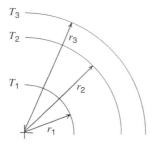

Figure 7.4. Steady-state conduction through a composite cylinder.

Example 7.5

The pipe in Example 7.4 is insulated with a 0.03 m thick layer of material with a thermal conductivity of 0.20 W m^{-1} K^{-1} which reduces the external surface temperature of the insulation to 25°C. Determine the reduction in heat loss per meter length of pipe.

The addition of a layer insulation greatly increases the total thermal resistance. The radial heat loss is now reduced to

$$\frac{Q}{L} = \frac{2\pi 0.035(70 - 25)}{(0.035/50)\ln(0.035/0.025) + (0.035/0.20)\ln(0.065/0.035)} \text{ W m}^{-1}$$

and

$$\frac{Q}{L} = 91.15 \text{ W m}^{-1}$$

7.2.5. Conduction through a Spherical Shell

The rate of heat transfer by conduction through a spherical shell is derived here for completeness although it is rather less useful in food processing than the relationships for cylindrical bodies. Convection usually dominates heat transfer problems involving food in spherical forms, for example, the drying of milk droplets in a spray drier or the freezing of peas in a fluidised bed.

Assuming that Figure 7.3 now represents the cross section through a spherical shell then T_1 is the temperature of the inner face of the shell, at a radius r_1, and T_2 is the temperature of the outer face at a radius r_2. The area in Equation (7.15) now becomes

$$A = 4\pi r^2 \tag{7.32}$$

and therefore

$$Q \int_{r_1}^{r_2} \frac{dr}{r^2} = -4\pi k \int_{T_1}^{T_2} dT \tag{7.33}$$

Evaluating the integral results in

$$Q \left(\frac{1}{r_1} - \frac{1}{r_2} \right) = -4\pi k(T_2 - T_1) \tag{7.34}$$

and thus the rate of heat transfer by conduction through the shell becomes

$$Q = \frac{4\pi k(T_2 - T_1)}{(1/r_1 - 1/r_2)} \tag{7.35}$$

7.3. CONVECTION

7.3.1. Film Heat Transfer Coefficient

Imagine a container of liquid to which heat is supplied as indicated in Figure 7.5. Further, imagine a small element or packet of liquid, perhaps only a few molecules thick, in contact with

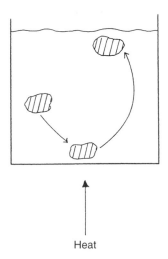

Heat

Figure 7.5. Establishment of a free convection current.

the heated surface of the container. Heat is transferred by conduction through the container wall and initially by conduction to the liquid packet. The temperature of the liquid increases and consequently it experiences thermal expansion and a decrease in density. Thus the packet of liquid rises (because of buoyancy forces) and as it moves away from the hot container wall it is replaced by cooler liquid which will in turn be heated and will rise through the body of liquid. The circulation pattern which is established in this way is known as a convection current and is responsible for the transfer of heat throughout the mass of a fluid (either liquid or gas). The rate at which fluid is circulated, and thus the rate at which heat is transferred, can be increased by deliberately agitating the fluid, for example by the addition of an impeller. A current which is established solely because of thermal expansion is called free or natural convection, whilst deliberately enhanced circulation is called forced convection.

In pipe flow forced convection can be introduced by ensuring that the flow is turbulent. However, as we have seen in chapter six, there is a region of more slowly moving and often laminar fluid adjacent to the wall of the pipe which may be termed the boundary layer. Beyond this layer the fluid has a much greater velocity and may well be turbulent. We may assume that heat is transferred by conduction through this layer and it follows that the layer contains almost all of the resistance to heat transfer, first, because fluids have relatively low thermal conductivities and second, because of the rapid heat transfer from the edge of the boundary layer into the bulk of the fluid. Thus it is valid to write for this layer

$$Q = \frac{A \, \Delta T}{x/k} \tag{7.5}$$

but, because x (the boundary layer thickness) can neither be predicted nor measured easily, the thermal resistance cannot be determined and x/k is replaced with the term $1/h$, where h is a film heat transfer coefficient. Equation (7.5) now becomes

$$Q = hA \, \Delta T \tag{7.36}$$

where

$$h = \frac{k}{x} \tag{7.37}$$

Heat transfer coefficients are thus empirical measures of the ability to transfer heat in particular circumstances and are used to describe heat transfer within the entire fluid both across any boundary layer *and* across turbulent or higher velocity regions. Values of the heat transfer coefficient must be determined in order to solve any realistic industrial heat transfer problem and much effort in process engineering research has been, and is, directed to this end. Increasing the velocity of the fluid has the effect of promoting turbulence (Reynolds' experiment) and thus of reducing the thickness of the boundary layer. Therefore, as x/k can be approximated to $1/h$, the heat transfer coefficient increases with increasing fluid velocity. Values of h depend also upon the physical properties of the fluid and upon geometry. For example, there is a difference between flow in a circular cross-section pipe and flow in an annulus with consequent differences in the rate at which heat is transferred even though materials and temperatures may be the same in each case. By inspecting Equation (7.36) it is clear that the units of a film heat transfer coefficient must be $\mathrm{W\,m^{-2}\,K^{-1}}$.

7.3.2. Simultaneous Convection and Conduction

The solution to most heat transfer problems of industrial interest involves the addition of both conductive and convective thermal resistances. Consider the plane wall in Figure 7.6, on either side of which is a fluid. The inner fluid is at a higher temperature T_i than the outer fluid and therefore heat flows outwards. If the inner surface of the wall is at a temperature T_1 then the rate of heat transfer from the inner fluid to the wall *by convection* is

$$Q = h_i A (T_i - T_1) \tag{7.38}$$

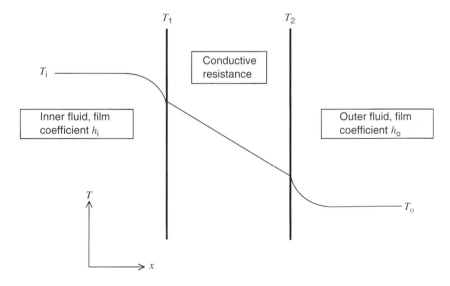

Figure 7.6. Convection and conduction: plane wall of cross-sectional area A.

The film heat transfer coefficient h_i is a measure of the ease with which heat is transferred across the inner fluid. The value of h_i will be greater if the fluid is in laminar flow rather than being stagnant and greater still if the flow is turbulent. Heat must now be transferred *by conduction* through the wall and thus

$$Q = \frac{A(T_1 - T_2)}{x/k} \tag{7.39}$$

However convection is again responsible for transfer from the outer wall surface to the outer fluid and thus

$$Q = h_o A(T_2 - T_o) \tag{7.40}$$

Thermal resistance is additive, as it was for conduction alone, and consequently

$$Q = \frac{A(T_i - T_o)}{(1/h_o + x/k + 1/h_i)} \tag{7.41}$$

where the denominator represents the sum of two convective resistances and the wall resistance. Equation (7.41) again conforms to the pattern of Equation (7.13) where the overall temperature difference is now the difference in temperature *between the fluids*, that is

$$\Delta T = T_i - T_o \tag{7.42}$$

The prediction of heat transfer coefficients will be considered after studying convection in radial geometries. For now students should familiarise themselves as quickly as possible with the magnitudes of film heat transfer coefficients for different circumstances. Table 7.2 gives some generalised values.

One or two points should be noted about the data. First, values of heat transfer coefficients over wide ranges are quoted. It is not possible to be more precise without specifying either the physical properties of the materials involved or the exact geometry. Second, the values

TABLE 7.2
Typical Values of Film Heat Transfer
Coefficient

	h (W m^{-2}K^{-1})
No change of state	
Water	300–11000
Gases (free convection)	1–25
Gases (forced convection)	20–300
Oils	60–700
Evaporation	
Water	2000–12000
Organic solvents	600–2000
Ammonia	1100–2300
Condensation	
Steam (film-wise)	6000–20000
Organic vapours	900–2800
Ammonia	3000–6000

for coefficients in heat transfer operations where a fluid evaporates or condenses at a surface tend to be very much higher than where a change of state is not involved. Then, for the latter case, convection within a liquid is much more rapid than within a gas. The widely varying values for water in Table 7.2 cover the range from slow laminar flow or virtually stagnant conditions through to highly developed turbulence. Note also that for gases the magnitude of forced convection currents is reflected in the values of the film heat transfer coefficient.

Example 7.6

A refrigerator door of area $0.6\,\text{m}^2$ consists of 25 mm of lagging on top of a thin metal sheet. The film heat transfer coefficients inside and outside the refrigerator are 10.0 and $15.0\,\text{W}\,\text{m}^{-2}\,\text{K}^{-1}$, respectively and the thermal conductivity of the lagging is $0.25\,\text{W}\,\text{m}^{-1}\,\text{K}^{-1}$. If the working temperature of the refrigerator is $0°\text{C}$ and ambient temperature is $20°\text{C}$, determine the heat flow through the refrigerator door and the temperature of the inside surface of the door.

The thermal resistance of the air film on the inside of the refrigerator is equal to the reciprocal of the relevant heat transfer coefficient, that is, $0.10\,\text{m}^2\,\text{K}\,\text{W}^{-1}$. Similarly for the air film on the outside, the thermal resistance is $1/15 = 0.0667\,\text{m}^2\,\text{K}\,\text{W}^{-1}$. The thin metal sheet will present very little resistance to the flow of heat and therefore only the insulation need be considered. The conductive resistance is thus $0.025/0.25 = 0.10\,\text{m}^2\,\text{K}\,\text{W}^{-1}$. The overall temperature difference is that between the air on the inside and the air outside the refrigerator, that is, 20 K. Hence the rate of heat transfer through the door is

$$Q = \frac{0.6 \times (20 - 0)}{(0.10 + 0.10 + 0.0667)}\,\text{W}, \qquad Q = 45\,\text{W}$$

Adapting Equation (7.38), the rate of heat transfer *from* the refrigerator door *to* the inner air by convection is

$$45 = 10 \times 0.60 \times (T_1 - 0)$$

and therefore the surface temperature T_1 is $7.5°\text{C}$.

7.3.3. Radial Convection

The analysis of convection can now be extended to radial geometries. Consider again the problem of conveying a hot food fluid in a circular cross-section pipe (see section 7.2.3). The rate of heat transfer from the hot fluid (temperature T_i) inside the pipe to the inner pipe surface (temperature T_1) is

$$Q = h_i 2\pi r_1 L(T_i - T_1) \tag{7.43}$$

where $2\pi r_1 L$ is the inner surface area and h_i is the relevant film heat transfer coefficient. Heat is conducted through the pipe wall at a rate given by Equation (7.20) and thence into the fluid surrounding the pipe (temperature T_o) by convection for which

$$Q = h_o 2\pi r_2 L(T_2 - T_o) \tag{7.44}$$

The overall temperature difference can be summed as in previous cases to give

$$\Delta T = (T_i - T_1) + (T_1 - T_2) + (T_2 - T_o) \tag{7.45}$$

and therefore

$$\Delta T = \left[\frac{1}{h_i 2\pi r_1 L} + \frac{r_2 \ln(r_2/r_1)}{k \, 2\pi r_2 L} + \frac{1}{h_o 2\pi r_2 L} \right] \tag{7.46}$$

Rearranging and multiplying through by the external pipe surface area $2\pi r_2 L$ yields

$$Q = \frac{2\pi r_2 L \Delta T}{(r_2/(r_1 h_i) + (r_2 \ln(r_2/r_1))/k + 1/h_o)} \tag{7.47}$$

Once more this follows the form of Equation (7.13) but the exact form of the thermal resistance term is more complex. The thermal resistance at the outer surface is simply the reciprocal of the film heat transfer coefficient h_o. However the thermal resistance at the inner surface is multiplied by the ratio of the external to internal surface area, that is, r_2/r_1, because the external and not the internal area is the chosen basis for Equation (7.47).

Example 7.7

Steam at 100°C condenses on the outside of an alloy tube (thermal conductivity 180 W m^{-1} K^{-1} through which water flows at a velocity such that the tube-side film heat transfer coefficient is 4,000 W m^{-2} K^{-1}. The film heat transfer coefficient for condensing steam may be assumed to be 10,000 W m^{-2} K^{-1}. The tube is 5 m long, has an external diameter of 25 mm and a wall thickness of 1 mm. If the mean temperature of the water is 15°C, calculate the rate of heat transfer to the water.

This problem may be solved simply by substituting the relevant quantities into Equation (7.47). However it is instructive to consider the magnitudes of the three thermal resistances involved. The external tube radius r_2 is 0.0125 m and the internal radius r_1 is 0.0115 m. Thus the conductive resistance is

$$\frac{r_2}{k} \ln \left(\frac{r_2}{r_1} \right) = \frac{0.0125}{180} \ln \left(\frac{0.0125}{0.0115} \right) \mathrm{m^2 \, K \, W^{-1}}$$

or

$$\frac{r_2}{k} \ln \left(\frac{r_2}{r_1} \right) = 5.79 \times 10^{-6} \, \mathrm{m^2 \, K \, W^{-1}}$$

The tube-side thermal resistance is

$$\frac{r_2}{r_1 h_i} = \frac{0.0125}{0.0115 \times 4000} \, \mathrm{m^2 \, K \, W^{-1}}$$

or

$$\frac{r_2}{r_1 h_i} = 2.72 \times 10^{-4} \, \mathrm{m^2 \, K \, W^{-1}}$$

The film resistance on the outside of the tube is simply the reciprocal of the film coefficient and equals $10^{-4} \, \mathrm{m^2 \, K \, W^{-1}}$.

The total thermal resistance is therefore $3.78 \times 10^{-4} \, \mathrm{m^2 \, K \, W^{-1}}$ and substituting for the temperature driving force and the external tube surface area $(2\pi r_2 L)$ gives the rate of transfer of heat to the water as

$$Q = \frac{2\pi 0.0125 \times 5(100 - 15)}{3.78 \times 10^{-4}} \, \mathrm{W} \quad \text{or} \quad Q = 88.42 \, \mathrm{kW}$$

Film heat transfer coefficients for condensing vapours are high and therefore the tube-side resistance is greater, by a factor of almost three, than that presented by the condensing steam. In a condenser, as in any heat exchanger, the conductive resistance should be minimised and this is achieved by using thin-walled tubes of high thermal conductivity. Thus in Example 7.7 the conductive resistance is by far the smallest of the three and it could justifiably be neglected in the calculation.

7.3.4. Critical Thickness of Insulation

The purpose of placing low thermal conductivity insulation against hot surfaces is to reduce heat loss. However this may not always happen. As the thickness of insulation around a pipe increases, the surface area increases too and thus the rate of heat loss should increase. Let R be the thermal resistance per unit area and per unit length of pipe. The resistance of the insulation layer then becomes

$$R_{\text{insulation}} = \frac{\ln(r/r_{\text{p}})}{2\pi k} \tag{7.48}$$

where r is the outer radius of the insulation and r_{p} is its inner radius (or the outer radius of the pipe).

The thermal resistance of the air surrounding the insulation will depend upon the heat transfer coefficient at the outer surface of the insulation and thus

$$R_{\text{air}} = \frac{1}{2\pi rh} \tag{7.49}$$

Resistance is clearly dependant upon radius and for maximum heat loss

$$\frac{dR}{dr} = 0 \tag{7.50}$$

Substituting from Equations (7.48) and (7.49) produces

$$0 = \frac{1}{2\pi k} \frac{d}{dr}\left(\ln\left(\frac{r}{r_{\text{p}}}\right)\right) + \frac{1}{2\pi h} \frac{d}{dr}\left(\frac{1}{r}\right) \tag{7.51}$$

and the result on differentiating is

$$0 = \frac{1}{2\pi kr} - \frac{1}{2\pi hr^2} \tag{7.52}$$

which simplifies to

$$\frac{1}{k} = \frac{1}{rh} \tag{7.53}$$

If now the radius of the insulation for maximum heat loss is r_{critical} then, from Equation (7.53),

$$r_{\text{critical}} = \frac{k}{h} \tag{7.54}$$

where k is the thermal conductivity of the insulation. It now follows that if the outer radius is smaller than r_{critical} heat loss from the pipe will continue to increase with the addition of insulation until $r = r_{\text{critical}}$. If r is greater than r_{critical} then heat loss will decrease as extra insulation is added. Consequently the critical radius should be as small as possible so that adding insulation does not have the opposite effect from that intended.

7.3.5. Correlations for Film Heat Transfer Coefficients

Unless a film heat transfer coefficient for a given processing operation is to be measured directly, use must be made of empirical dimensionless correlations. These generally take the form

$$Nu = f(Re, Pr, Gr) \tag{7.55}$$

The Nusselt number Nu contains the heat transfer coefficient and is defined as

$$Nu = \frac{hL}{k} \tag{7.56}$$

where L is a linear dimension which is characteristic of the heat transfer geometry, for example the diameter of a pipe or a sphere. The Nusselt number is a kind of dimensionless heat transfer coefficient and represents the ratio of the actual rate of heat transfer to that due to conduction alone. The Reynolds number Re was encountered in chapter six and is included in Equation (7.55) because of the effect of turbulence on the thickness of the boundary layer and consequently upon the film heat transfer coefficient itself. The Prandtl number defined by

$$Pr = \frac{c_p \mu}{k} \tag{7.57}$$

is solely a function of the physical properties of the fluid and is equal to the ratio of kinematic viscosity (kinematic diffusivity) μ/ρ to thermal diffusivity $k/\rho c_p$. It may be thought of as a measure of the relative ability of a fluid to transfer momentum and heat. The Grashof number Gr is defined by

$$Gr = \frac{\beta g \, \Delta T L^3 \rho^2}{\mu^2} \tag{7.58}$$

and is relevant only to cases where natural convection, and therefore buoyancy, is important. Thus under most circumstances Equation (7.55) reduces to

$$Nu = f(Re, Pr) \tag{7.59}$$

This general relationship can be obtained by dimensional analysis and the derivation is set out in detail in Appendix C. However, the precise form of the relationship must now be determined by experiment and often takes the form of Equation (7.60),

$$Nu = C \, Re^m \, Pr^n \tag{7.60}$$

where C is a constant and m and n are indices which depend upon geometry and the nature of the heat transfer application. For pipe flow the Dittus–Boelter equation is commonly used. Thus

$$Nu = 0.023 \, Re^{0.8} \, Pr^n \tag{7.61}$$

where $n = 0.4$ if the fluid is being heated and $n = 0.3$ if being cooled. This relationship is valid over the range $Re > 10^4$ and $0.7 < Pr < 160$. The physical properties of the fluid are evaluated at the mean bulk temperature, that is, the arithmetic mean of the inlet and outlet temperatures. Note that the characteristic length in the Reynolds and Nusselt numbers is the *internal* pipe diameter [although the external radius or diameter will still define the rate of heat transfer in Equation (7.47)]. For annular flow the hydraulic mean diameter, defined by

Equation (6.10), should be used. For very viscous liquids there is a marked difference between the viscosity in the bulk μ and that at the heat transfer surface μ_w. This is taken into account in the Sieder–Tate equation which includes a viscosity correction term:

$$Nu = 0.023 \, Re^{0.8} \, Pr^{0.33} \left(\frac{\mu}{\mu_w}\right)^{0.14} \tag{7.62}$$

Example 7.8

Water flowing at a velocity of $1 \, \text{m s}^{-1}$ in a 0.025 m diameter tube is heated to a mean bulk temperature of 30°C. Determine the film heat transfer coefficient.

The properties of water at 30°C are listed in steam tables as follows: $\rho = 995.6 \, \text{kg m}^{-3}$, $\mu = 7.97 \times 10^{-4} \, \text{Pa s}$, $c_p = 4179 \, \text{J kg}^{-1} \, \text{K}^{-1}$ and $k = 0.618 \, \text{W m}^{-1} \, \text{K}^{-1}$. Thus the Reynolds number is

$$Re = \frac{995.6 \times 1.0 \times 0.025}{7.97 \times 10^{-4}} \quad \text{or} \quad Re = 31230$$

and the Prandtl number is

$$Pr = \frac{4179 \times 7.97 \times 10^{-4}}{0.618} \quad \text{or} \quad Pr = 5.39$$

Note that in steam tables the Prandtl number is often listed separately as a function of temperature. Using the Dittus–Boelter equation, with an exponent of 0.4 on the Prandtl number, the film heat transfer coefficient is now

$$h = \frac{0.618}{0.025} \times 0.023 (31230)^{0.8} (5.39)^{0.4} \, \text{W m}^{-2} \, \text{K}^{-1}$$

and

$$h = 4396 \, \text{W m}^{-2} \, \text{K}^{-1}$$

Example 7.9

An aqueous sucrose solution flowing at a velocity of $1 \, \text{m s}^{-1}$ in a 0.025 m diameter tube is heated to a mean bulk temperature of 30°C. The tube wall is at 70°C. The properties of the solution at 30°C are: $\rho = 1282 \, \text{kg m}^{-3}$, $\mu = 0.0928 \, \text{Pa s}$, $c_p = 3800 \, \text{J kg}^{-1} \, \text{K}^{-1}$ and $k = 0.434 \, \text{W m}^{-1} \, \text{K}^{-1}$. At 70°C $\mu = 0.0388 \, \text{Pa s}$. Use the Sieder–Tate equation to calculate the film heat transfer coefficient.

The Reynolds number is

$$Re = \frac{1282 \times 1.0 \times 0.025}{0.0928}$$

or

$$Re = 345.4$$

and the Prandtl number, significantly larger for a viscous fluid, is

$$Pr = \frac{3800 \times 0.0928}{0.434} \quad \text{or} \quad Pr = 812.5$$

Using the viscosity correction term in the Sieder–Tate equation, the film heat transfer coefficient is

$$h = \frac{0.434}{0.025} \times 0.023(345.4)^{0.8}(812.5)^{0.33} \left(\frac{0.0928}{0.0388}\right)^{0.14} \text{W m}^{-2}\text{k}^{-1}$$

and

$$h = 442 \text{ W m}^{-2}\text{K}^{-1}$$

As would be expected the heat transfer coefficient for a viscous liquid in laminar flow is considerably smaller than that for water in turbulent flow (as in Example 7.8).

7.3.6. Overall Heat Transfer Coefficient

Thermal resistance is a measure of the difficulty with which heat is transferred and is inversely proportional to the film heat transfer coefficient. It is possible to rewrite Equation (7.13) replacing the sum of thermal resistance term with the reciprocal of an overall heat transfer coefficient, U. Thus

$$Q = UA\,\Delta T \tag{7.63}$$

The overall heat transfer coefficient includes both conductive and convective resistances and for the case of heat transfer to or from a fluid in a circular cross-section pipe is given by

$$\frac{1}{U} = \text{resistance of fluid i} + \text{resistance of pipe wall} + \text{resistance of fluid o} \tag{7.64}$$

which, from Equation (7.47), becomes

$$\frac{1}{U} = \frac{r_2}{r_1 h_i} + \frac{r_2 \ln(r_2/r_1)}{k} + \frac{1}{h_o} \tag{7.65}$$

when the surface area for heat transfer is, as is conventional, defined by

$$A = 2\pi r_2 L \tag{7.66}$$

and the overall temperature driving force by

$$\Delta T = T_i - T_o \tag{7.67}$$

Equation (7.63) is a more convenient expression than Equation (7.13) for the rate of heat transfer at steady-state. However, the exact form of overall heat transfer coefficient will depend upon the surface area for heat transfer. If, in this example, the internal surface area is used then

$$A = 2\pi r_1 L \tag{7.68}$$

and Equation (7.65) becomes

$$\frac{1}{U} = \frac{1}{h_i} + \frac{r_1 \ln(r_2/r_1)}{k} + \frac{r_1}{r_2 h_o} \tag{7.69}$$

Some generalised values of overall heat transfer coefficients are given in Table 7.3. Clearly these values reflect the values of the film heat transfer coefficients quoted in Table 7.2.

TABLE 7.3

Typical Values of Overall Heat Transfer Coefficient for a Shell and Tube Heat Exchanger

	Hot fluid	Cold fluid	U (W m^{-2} K^{-1})
No change of state	Water	Water	900–1700
	Water	Brine	600–1200
	Organics	Water	300–900
	Gases	Water	20–300
	Light oils	Water	400–900
	Steam	Water	1500–4000
	Steam	Light oils	300–900
Condensation	Steam	Water	2000–4000
	Organics	Water	600–1200
Evaporation	Water	Refrigerant	400–900

Example 7.10

Calculate the overall heat transfer coefficient for the condenser in Example 7.7.

The overall heat transfer coefficient is the reciprocal of the sum of thermal resistance. Thus

$$U = \frac{1}{(2.72 \times 10^{-4}) + (5.79 \times 10^{-6}) + (10^{-4})} \text{ W m}^{-2}\text{K}^{-1}$$

and

$$U = 2647 \text{ W m}^{-2}\text{K}^{-1}$$

(a) *Simplification of overall heat transfer coefficient*   The definition of overall heat transfer coefficient can be simplified by substituting for the logarithmic mean radius from Equation (7.21) to give

$$\frac{1}{U} = \frac{r_2}{r_1 h_i} + \frac{r_2(r_2 - r_1)}{k r_{lm}} + \frac{1}{h_o} \tag{7.70}$$

or

$$\frac{1}{U} = \frac{r_2}{r_1 h_i} + \frac{x r_2}{k r_{lm}} + \frac{1}{h_o} \tag{7.71}$$

where x is the wall thickness. However for the case of thin-walled tubes of high thermal conductivity, where $r_2/r_1 \approx 1$, the conductive resistance is negligible and Equation (7.65) reduces to

$$\frac{1}{U} = \frac{1}{h_i} + \frac{1}{h_o} \tag{7.72}$$

(b) *Fouling resistances*   Many fluids which pass through process equipment deposit layers on heat transfer surfaces, especially in heat exchangers, which then present an extra thermal resistance to the transfer of heat. This is a very serious problem in a heat exchanger because the performance of the exchanger is likely to deteriorate significantly over time. If the surfaces are not cleaned the capacity will fall with time or the required temperatures may not be reached. This heat transfer problem is in addition to the consequences for hygiene. A fouling resistance, or fouling factor, may be defined in terms of the difference between the overall heat transfer coefficient obtained with no deposit, U_{clean}, and the coefficient when the surface has been fouled, U_{fouled}. Thus

$$R_f = \frac{1}{U_{fouled}} - \frac{1}{U_{clean}} \tag{7.73}$$

Fouling factors can therefore be added to the definition of overall heat transfer coefficient in Equation (7.65) to give

$$\frac{1}{U} = \frac{r_2}{r_1 h_i} + \frac{r_2 \ln(r_2/r_1)}{k} + \frac{1}{h_o} + R_f \tag{7.74}$$

Fouling can arise by a number of mechanisms. Precipitation fouling or scaling occurs when deposits form on heating surfaces because the limit of solubility has been reached. Examples include the precipitation of calcium carbonate from hard water and calcium salt deposition in milk processing and in sugar refining. Chemical reactions activated or enhanced by the transfer of heat may produce compounds which then form deposits on heat transfer surfaces. A prime example of this is the denaturation of proteins during the heat processing of milk, especially in evaporation. Other fouling mechanisms include the deposition of layers of particles, that is, sedimentation, from suspensions and the solidification of melts, either from emulsions (e.g., fat from milk) or of homogeneous molten foods. In biological fouling micro-organisms are deposited, and subsequently grow, on heating surfaces. Often fouling deposits are laid down in regions where the local fluid velocity is low. Thus fouling can be reduced, or perhaps prevented, by increased shear over the fouling layer, that is by maintaining adequate fluid velocities. Other solutions to the problem include minimising temperature driving forces and ensuring that heat transfer surfaces are completely wetted. Dry regions will cause localised hot spots which result in the burning on of deposits.

Values of fouling resistances for food deposits are difficult to find. However the magnitude of the fouling effect can be judged from data quoted by Heldmann and Lund: a 60% reduction in the overall heat transfer coefficient for sugar evaporation from about 4,200 to 1,700 W m^{-2} K^{-1} and a decrease of 80% in the overall heat transfer coefficient for the evaporation of whey. Data for some commonly encountered fouling resistances are given in Table 7.4.

TABLE 7.4
Fouling Resistances

Fluid	R_f (m^2 K W^{-1})
Boiler feedwater	2×10^{-4}
Mineral oils	7×10^{-4}–9×10^{-4}
Alcohol vapour	9×10^{-5}
Steam	9×10^{-5}
Industrial air	4×10^{-4}
Refrigerant	2×10^{-4}
Grease	1×10^{-4}

7.4. HEAT EXCHANGERS

A heat exchanger is a device in which heat is transferred deliberately from one fluid stream to another; either of the fluids may be a liquid or a gas. In food processing the purpose is to heat or cool a liquid food in bulk. In pasteurisation and the bulk sterilisation of liquids the food is heated to a specific temperature and therefore the rate of heat transfer must be controlled carefully; the heating fluid may be steam or hot water. Alternatively, the purpose may be to exchange heat between two food streams one of which is to be heated and the other cooled. Each may require further heating or cooling (e.g., with steam or chilled water) to reach the desired temperature but the overall energy input can be reduced by using what would otherwise be waste heat.

7.4.1. Types of Industrial Heat Exchanger

In this section the major kinds of industrial heat exchanger are described briefly. The purpose is to give sufficient understanding of the principles of construction that students are able to undertake simple preliminary design calculations (e.g., sizing of the heat transfer surface area and determining the number of plates or tubes required) with some confidence. For a more detailed treatment of the very many kinds of exchanger, and their advantages and disadvantages for particular applications, students should consult the specialised texts by Kern and others.

(a) *Double pipe*   The simplest type of exchanger is the double pipe (Figure 7.7) which consists of two concentric tubes with one fluid passing along the centre tube and the second fluid flowing in the annular space created between the tubes. Such exchangers are limited to a relatively small heat transfer area resulting in a long pipe which must then be doubled back on itself to fit conveniently into a process line. Although the capital costs are low the disadvantages of double pipe exchangers (including relatively low operating pressures) are such that they are rarely used except in the form of scraped surface exchangers.

(b) *Shell and tube*   A similar principle, however, is used in the shell and tube heat exchanger (Figure 7.8). Here a series of small diameter tubes, of the order of 0.01 or 0.02 m, are placed in a bundle within a cylindrical shell. The largest exchangers may be up to 1 m in diameter, several metres long and contain several hundred tubes giving a heat transfer surface area of the order of 10^2 m^2. The respective ends of the tubes open into chambers at each end of the shell and thus one fluid passes through the tubes and exchanges heat, through the tube walls, with the second fluid which is circulated through the shell and over the external surface of the tube bundle. The shell-side fluid is usually forced to change direction abruptly by means

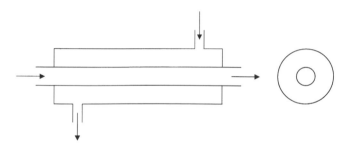

Figure 7.7. Double-pipe heat exchanger.

of a series of transverse baffles. This, together with the close spacing of the tubes (often on a triangular pitch), results in a tortuous flow path and greater turbulence. The tube-side fluid is kept turbulent by maintaining a minimum velocity of 1 or $2 \, m \, s^{-1}$. The great advantage of the shell and tube arrangement over the concentric tube exchanger is that a very high surface area can be packed into a relatively small volume. The tubular construction also allows considerable pressures to be used (tens of atmospheres) although this is unlikely to be necessary in food processing applications. However, shell and tube units are used extensively as condensers and evaporators and in the bulk sterilisation of liquids.

(*c*) *Multiple pass exchangers*   The exchanger in Figure 7.8 has a single tube-side pass and a single shell-side pass. In order to shorten the overall exchanger length the tubes may be arranged within the shell so that half carry the fluid in one direction and the same fluid then passes back down the length of the exchanger in the opposite direction using the other half of the tube bundle. The exchanger would then be said to have two tube-side passes, as in Figure 7.9. Such an arrangement has the added advantage that tube-side velocities are doubled, for the same flow rate, thus increasing the heat transfer coefficient. The number of shell-side passes can also be increased by placing longitudinal baffles in the shell with a consequent increase in the shell-side coefficient. A 2–4 shell and tube heat exchanger would then have two shell passes and four tube passes. Note that there is always an even number of tube passes. The improved heat transfer characteristics for multi-pass heat exchangers is offset, however, by the more complex and costly construction and by higher pressure drops for each fluid.

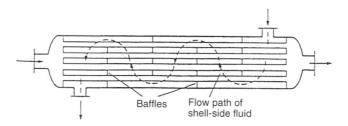

Baffles   Flow path of
shell-side fluid

Figure 7.8. Shell and tube heat exchanger (from P.J. Fryer, D.L. Pyle and C.D. Rielly (eds.), Chemical Engineering for the Food Industry, Chapman and Hall, 1997, with permission).

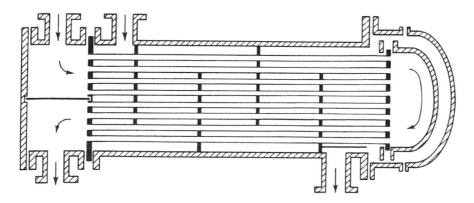

Figure 7.9. 1-2 shell and tube heat exchanger (from McCabe and J.C. Smith, Unit Operations of Chemical Engineering, 5th ed., McGraw-Hill, 1993, reproduced with the permission of The McGraw-Hill Companies).

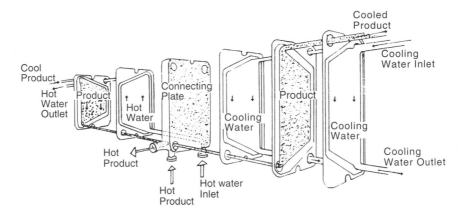

Figure 7.10. Plate heat exchanger (from R.T. Toledo, Fundamentals of Food Process Engineering, 2nd ed., Chapman and Hall, 1994, with permission).

(*d*) *Plate heat exchangers*   A plate heat exchanger (Figure 7.10) consists of a series of corrugated pressed metal plates which are clamped together in a frame so as to allow liquid to flow in the small gap between them. The corrugated faces of the plates consist of troughs at right angles to the direction of flow and thus a channel of constantly changing direction and cross section is formed. The plates are held 2–3 mm apart and leakage is prevented by a gasket which runs around the edge of each plate. The developed area of a plate is up to 30% or 40% greater than the projected area and therefore a high heat transfer surface area is obtained within a small plant volume. Additional heat transfer area is obtained simply by adding extra plates. Because turbulence can be obtained at low fluid velocities, high heat transfer coefficients are experienced which in turn allows small temperature differences between the fluids to be used. This is a particular advantage for food materials; close temperature control is possible with a smaller chance of thermal damage to the food. Overall heat transfer coefficients are about two to three times higher than for a shell and tube heat exchanger.

Plate heat exchangers have a number of other advantages; they are particularly suitable for high viscosity liquids and are very easy to dismantle and clean. However, perhaps the major advantage is the ability to arrange a wide variety of flow patterns within a single exchanger. Several different fluids can flow through separate sections of the same heat exchanger, for example allowing heating, cooling and heat recovery to take place in a single unit (see Example 7.16).

(*e*) *Scraped surface heat exchangers*   This consists of a double-pipe exchanger with a central rotating shaft inside the inner pipe. Scraper blades are attached to the shaft and remove any material which builds up on the inner pipe wall. The second fluid flows in the annular space as in a conventional exchanger. The continual scraping of the heat transfer surface ensures that higher heat transfer coefficients are obtained with highly viscous fluids.

7.4.2. Sizing of Heat Exchangers

Consider a 1–1 shell and tube heat exchanger. The exchanger may be operated in either co-current mode, in which both fluids flow in the same direction, or in counter-current mode, where the fluids flow in opposite directions. In co-current flow (Figure 7.11) the 'hot' fluid enters at a temperature of T_{h_i} and cools by exchanging heat through the tube walls with the

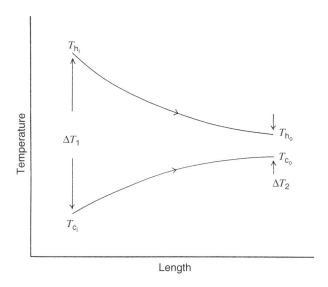

Figure 7.11. Logarithmic mean temperature difference: co-current flow.

'cold' fluid which rises in temperature from T_{c_i} to T_{c_o} approaching the outlet temperature of the hot fluid T_{h_o}. Thus the temperature difference between the two streams at the inlet of the exchanger is very much larger than at the outlet. In the analysis which follows it is assumed that heat losses from the exchanger are minimal, a reasonable assumption if the external surfaces are insulated. Consequently the rate at which heat is lost by the 'hot' fluid must equal the rate at which heat is gained by the 'cold' fluid, which must in turn equal the rate at which heat is transferred through the tube walls. Thus the energy balance may be written as

$$Q = m_c c_{p_c} (T_{c_o} - T_{c_i}) = m_h c_{p_h} (T_{h_i} - T_{h_o}) \qquad (7.75)$$

where m is the mass flow rate and the subscripts h and c refer to the 'hot' and 'cold' streams, respectively, and the rate of heat transfer is given by a relationship of the form of Equation (7.63).

The overall heat transfer coefficient can be calculated from Equation (7.74) taking into account any known fouling resistance. The area A will normally be the total external tube area which for n tubes is

$$A = n 2 \pi r_2 L \qquad (7.76)$$

or

$$A = n \pi \, dL \qquad (7.77)$$

where d is the external tube diameter. The temperature difference between the two fluids in Equation (7.63), ΔT, presents rather more of a difficulty. Referring again to Figure 7.11 it can be seen that the temperature difference changes with position in the exchanger, from ΔT_1 at the inlet to ΔT_2 at the outlet. Clearly some kind of mean temperature difference must be used but this can be neither a simple arithmetic average nor the value half way along the length of the heat exchanger. It can be shown that Equation (7.63) should be replaced by

$$Q = U A \Delta T_{lm} \qquad (7.78)$$

where ΔT_{lm} is the logarithmic mean temperature difference defined by

$$\Delta T_{lm} = \frac{\Delta T_1 - \Delta T_2}{\ln(\Delta T_1/\Delta T_2)} \tag{7.79}$$

if the following assumptions are valid:

i   the overall heat transfer coefficient is constant throughout the exchanger,
ii   heat exchange takes place only between the two fluids, that is, there are no heat losses and the heat balance in Equation (7.75) is satisfied,
iii   the temperatures of each stream are constant across a given cross-section of the exchanger, and
iv   the heat capacities of each stream are constant (although of course heat capacity is a function of temperature).

Equation (7.79) is derived in Appendix E. Note that for the case where the temperature difference is constant at all points in the exchanger then

$$\Delta T_1 = \Delta T_2 = \Delta T_{lm} \tag{7.80}$$

Counter-current flow (Figure 7.12) has a distinct advantage over co-current flow. There is less variation in temperature difference between the two fluids and the exit 'cold' temperature may be higher than the outlet 'hot' temperature, that is, it is possible that $T_{c_o} > T_{h_o}$. As the following example shows, for the same inlet and outlet temperatures, counter-current operation gives a higher logarithmic mean temperature difference than co-current operation. In other words a higher proportion of the heat content of the 'hot' fluid can be transferred in counter-current mode resulting in a smaller heat transfer surface area.

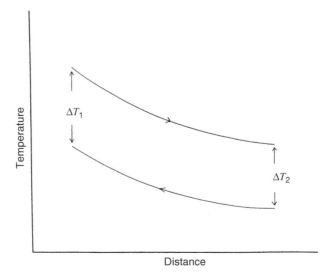

Figure 7.12. Logarithmic mean temperature difference: counter-current flow.

Example 7.11

Calculate the logarithmic mean temperature difference, for both co-current and counter-current flow, in a heat exchanger where one stream rises from 20°C to 70°C and the other falls from to 95°C to 80°C.

In co-current flow the temperature difference at one end of the exchanger will be that between the two inlet temperatures, that is, $\Delta T_1 = (95 - 20) = 75$ K. At the other end the difference between the two outlet temperatures is $\Delta T_2 = (80 - 70) = 10$ K. Therefore using the definition of logarithmic mean temperature difference in Equation (7.79) gives

$$\Delta T_{lm} = \frac{75 - 10}{\ln(75/10)} \text{ K}, \qquad \Delta T_{lm} = 32.2 \text{ K}$$

The selection of ΔT_1 and ΔT_2 is quite arbitrary; reversing the values so that $\Delta T_1 = 10$ K and $\Delta T_2 = 75$ K gives the same result for ΔT_{lm}. For counter-current operation, the direction of flow of one stream is reversed so that $\Delta T_1 = (80 - 20) = 60$ K and $\Delta T_2 = (95 - 70) = 25$ K. Hence

$$\Delta T_{lm} = \frac{60 - 25}{\ln(60/25)} \text{ K}, \qquad \Delta T_{lm} = 40 \text{ K}$$

Consequently the temperature driving force in counter-current flow is greater than that for co-current flow. If now the quantity of heat transferred is equal in each case and the overall heat transfer coefficient is the same then counter-current operation implies that a smaller heat transfer surface area is required.

Example 7.12

Orange juice, flowing at $3,600 \text{ kg h}^{-1}$, is to be pasteurised by heating it from 10°C to 80°C in a simple shell and tube exchanger. Water enters at 90°C and flows counter-currently to the orange juice, leaving at 34°C. The heat exchanger consists of tubes 1.50 m in length and 0.026 m external diameter. If the overall heat transfer coefficient (based on the external area of the tubes) is $1,700 \text{ W m}^{-2} \text{ K}^{-1}$, determine the necessary mass flow rate of water and the number of tubes required. Assume the mean heat capacity of orange juice and water to be 3.80 and $4.18 \text{ kJ kg}^{-1} \text{ K}^{-1}$, respectively.

The rate of heat loss Q from the water must equal the rate of heat gain by the orange juice, assuming no heat loss from the exchanger. Thus for the orange juice, expressing the mass flow rate in kg s^{-1},

$$Q = \frac{3600}{3600} \times 3.80(80 - 10) \text{ kW}, \qquad Q = 266 \text{ kW}$$

For water, if the mass flow rate is W,

$$Q = W \times 4.18(90 - 34)$$

Thus equating expressions for Q gives

$$266 = W \times 4.18(90 - 34), \qquad W = 1.14 \text{ kg s}^{-1}$$

The heat exchanger operates in counter-current mode and the respective temperature differences are $\Delta T_1 = (90 - 80) = 10\,\text{K}$ and $\Delta T_2 = (34 - 10) = 24\,\text{K}$. The logarithmic mean temperature difference is then

$$\Delta T_{lm} = \frac{24 - 10}{\ln(24/10)}\,\text{K} \ \text{ or } \ \Delta T_{lm} = 16\,\text{K}$$

From the rate equation, Equation (7.78), the tube surface area is

$$A = \frac{Q}{U\,\Delta T_{lm}}, \qquad A = \frac{266}{1.7 \times 16}\,\text{m}^2$$

or

$$A = 9.78\,\text{m}^2$$

Care should be taken that the units of Q and U are consistent: W and $\text{W}\,\text{m}^{-2}\,\text{K}^{-1}$ or kW and $\text{kW}\,\text{m}^{-2}\,\text{K}^{-1}$. The total external surface area of the tubes is given by Equation (7.77) and therefore the number of tubes is

$$n = \frac{9.78}{\pi \times 0.026 \times 1.50} \ \text{ or } \ n = 79.8$$

Clearly the number of tubes must be an integer and it is therefore sensible to round up the answer to 80 tubes.

Example 7.13

Water is chilled by brine in a counter-current heat exchanger. If the flow rate of the brine is $1.8\,\text{kg}\,\text{s}^{-1}$ and that of the water is $1.05\,\text{kg}\,\text{s}^{-1}$, estimate the temperature to which the water is cooled if the brine enters at $-8°\text{C}$ and leaves at $10°\text{C}$. The water enters the heat exchanger at $32°\text{C}$. If the heat transfer surface area of the exchanger is $5.50\,\text{m}^2$, determine the overall heat transfer coefficient. Take the mean heat capacity of brine to be $3.38\,\text{kJ}\,\text{kg}^{-1}\,\text{K}^{-1}$.

The mean bulk temperature of the water is unknown but it is reasonable to assume a mean heat capacity for water of $4.18\,\text{kJ}\,\text{kg}^{-1}\,\text{K}^{-1}$. For the brine, the rate of heat gain is

$$Q = 1.80 \times 3.38(10 - (-8))\,\text{kW} \ \text{ or } \ Q = 109.5\,\text{kW}$$

The rate of heat loss from the water is equal to the heat gain of the brine and

$$109.5 = 1.05 \times 4.18(32 - T)$$

where T is the outlet temperature of the water. Thus, $T = 7.05°\text{C}$ and the logarithmic mean temperature difference between the brine and the water becomes

$$\Delta T_{lm} = \frac{(32 - 10) - (7.05 - (-8))}{\ln(22/15.05)}\,\text{K} \ \text{ or } \ \Delta T_{lm} = 18.3\,\text{K}$$

Rearranging Equation (7.78) explicitly in terms of the overall heat transfer coefficient gives

$$U = \frac{Q}{A\,\Delta T_{lm}}$$

Taking care to express Q in W,

$$U = \frac{109.5 \times 10^3}{5.5 \times 18.3} \, \text{W m}^{-2} \text{K}^{-1}$$

and therefore

$$U = 1088 \, \text{W m}^{-2}\text{K}^{-1}$$

Example 7.14

Milk is to be cooled from 90°C to 15°C at a rate of 9.72 t h^{-1}. A single-pass shell and tube heat exchanger with the following specification is available:

Number of tubes	250
Tube length	1.0 m
Internal tube diameter	23.5 mm
External tube diameter	25 mm
Thermal conductivity of tube wall	15 W m^{-1} K^{-1}

Chilled water is supplied at 4°C and is normally discharged at 10°C. From experience it is assumed that the film heat transfer coefficients for the tube-side (water) and the shell-side will be 2,500 and 3,000 W m^{-2} K^{-1}, respectively. Is the exchanger adequate? After some time of operation the tubes become scaled and a fouling factor of 2×10^{-4} m^2 K W^{-1} can be assumed. What is the new overall heat transfer coefficient and, assuming that the inlet and outlet temperatures remain unchanged, what is the reduction in capacity of the exchanger? Take the mean heat capacity of milk to be 3.90 kJ kg^{-1} K^{-1}.

The rate of heat transfer which is required in the exchanger is equal to the product of the mass flow rate, heat capacity and temperature change of the milk. Hence

$$Q_{\text{required}} = \left(\frac{9720}{3600}\right) 3.90(90 - 15) \, \text{kW} \quad \text{and} \quad Q_{\text{required}} = 789.8 \, \text{kW}$$

On the other hand the actual rate of heat transfer achievable is given by the rate equation [Equation (7.78)] and depends upon the overall heat transfer coefficient, the area and the logarithmic mean temperature difference. Thus, before the tubes become scaled, noting that $r_2 = 0.0125$ m and $r_1 = 0.01175$ m, the overall heat transfer coefficient is obtained from

$$\frac{1}{U} = \frac{1}{3000} + \frac{0.0125}{15} \ln\left(\frac{0.0125}{0.01175}\right) + \frac{0.0125}{0.01175 \times 2500} \, \text{W m}^{-2} \text{K}^{-1}$$

and therefore

$$U = 1234 \, \text{W m}^{-2} \text{K}^{-1}$$

The heat transfer area available, the external surface area of the tubes, is $A = 250\pi \times 0.025 \times 1.0$ m^2 and therefore $A = 19.63$ m^2. The logarithmic mean temperature difference for counter-current flow is

$$\Delta T_{\text{lm}} = \frac{(90 - 10) - (15 - 4)}{\ln(80/11)} \, \text{K} \quad \text{or} \quad \Delta T_{\text{lm}} = 34.8 \, \text{K}$$

Thus

$$Q_{\text{actual}} = 1234 \times 19.63 \times 34.8 \, \text{kW} \quad \text{or} \quad Q_{\text{actual}} = 843 \, \text{kW}$$

However for co-current flow ΔT_{lm} is reduced to 28 K and the actual rate of heat transfer (for the same transfer coefficient and area) becomes only 678.3 kW. Consequently the heat exchanger is capable of supplying heat at the required rate of 789.8 kW when operated counter-currently but not when operated co-currently. An energy balance for milk now gives the capacity of the exchanger and the mass flow rate of milk is

$$m = \frac{Q_{\text{actual}}}{c_{\text{p}} \Delta T_{\text{milk}}}, \quad m = \frac{843}{3.90(90 - 15)} \, \text{kg s}^{-1}$$

or

$$m = 2.88 \, \text{kg s}^{-1}$$

This is equal to $10.38 \, \text{t h}^{-1}$.

After a scale is deposited on the tubes the new overall heat transfer coefficient is given by Equation (7.73)

$$\frac{1}{U_{\text{fouled}}} = \left(\frac{1}{1234} \right) + 2 \times 10^4$$

hence

$$U_{\text{fouled}} = 989.7 \, \text{W m}^{-2} \, \text{K}^{-1}$$

Consequently the actual rate of heat transfer is reduced to

$$Q_{\text{actual}} = 989.7 \times 19.63 \times 34.8 \, \text{kW} \quad \text{or} \quad Q_{\text{actual}} = 676.1 \, \text{kW}$$

with a resultant reduction in the capacity of the exchanger to $2.31 \, \text{kg s}^{-1}$ (or $8.32 \, \text{t h}^{-1}$), a decrease of 14.4%.

Example 7.15

A counter-current tubular heat exchanger containing 200 tubes is to be used to cool a liquid food product from 80°C to 35°C at flow rate of 7.6 kg s^{-1}. The tubes have external and internal diameters of 11 and 10 mm, respectively, and a thermal conductivity of 17 W m^{-1} K^{-1}. Water enters the shell at 10°C at a mass flow rate of 22 kg s^{-1}; the shell side heat transfer coefficient is 2,500 W m^{-2} K^{-1}. Calculate the length of the heat exchanger.

Data for liquid food:

Thermal conductivity	$0.5 \, \text{W m}^{-1} \, \text{K}^{-1}$
Viscosity	$2 \times 10^{-3} \, \text{Pa s}$
Mean heat capacity	$3.68 \, \text{kJ kg}^{-1} \, \text{K}^{-1}$
Density	$1,100 \, \text{kg m}^{-3}$

An energy balance on the food stream gives

$$Q = 7.60 \times 3.68 \times (80 - 35)\,\text{kW} \quad \text{or} \quad Q = 1258.6\,\text{kW}$$

Now, for the water,

$$1258.6 = 22.0 \times 4.183 \times (T - 10)$$

and therefore the outlet water temperature T is $T = 23.7°\text{C}$.

The logarithmic mean temperature difference for counter-current flow becomes

$$\Delta T_{\text{lm}} = \frac{(80 - 23.7) - (35 - 10)}{\ln((80 - 23.7)/(35 - 10))}\,\text{K}$$

or

$$\Delta T_{\text{lm}} = 38.6\,\text{K}$$

Substituting from Equation (6.5) for the Reynolds number into the Dittus–Boelter equation, taking care to divide the mass flow rate by the number of tubes, and using an exponent of 0.3 for the Prandtl number, the tube-side film heat transfer coefficient is now

$$h_{\text{i}} = \frac{0.50}{0.010} \times 0.023 \left(\frac{4 \times 7.6}{200 \times 2 \times 10^{-3}\pi 0.010}\right)^{0.8} \left(\frac{3680 \times 2 \times 10^{-3}}{0.50}\right)^{0.3}\,\text{W m}^{-2}\,\text{K}^{-1}$$

and

$$h_{\text{i}} = 1312\,\text{W m}^{-2}\,\text{K}^{-1}$$

The overall coefficient is obtained from

$$\frac{1}{U} = \frac{1}{2500} + \frac{0.0055}{17}\ln\left(\frac{0.0055}{0.0050}\right) + \frac{0.0055}{0.0050 \times 1312}\,\text{W m}^{-2}\,\text{K}^{-1}$$

and therefore

$$U = 788\,\text{W m}^{-2}\,\text{K}^{-1}$$

The rate equation gives the heat transfer surface area as

$$A = \frac{1258.6 \times 10^3}{788 \times 38.6}\,\text{m}^2 \quad \text{or} \quad A = 41.42\,\text{m}^2$$

The area is that of the external surface of the 200 tubes and from Equation (7.77) the length of the tubes is

$$L = \frac{41.42}{200\pi 0.011}\,\text{m}$$

and therefore $L = 6.0\,\text{m}$.

For multi-pass heat exchangers the flow will be co-current in some passes and counter-current in others and therefore there is uncertainty about the true value of the logarithmic mean temperature difference. A graphical method of solving this problem is available which employs a correction factor. Equation (7.78) is replaced by

$$Q = FUA\Delta T_{\text{lm}} \tag{7.81}$$

where F is a function of two dimensionless temperature differences

$$X = \frac{\theta_2 - \theta_1}{T_1 - \theta_1} \qquad (7.82)$$

and

$$Y = \frac{T_1 - T_2}{\theta_2 - \theta_1} \qquad (7.83)$$

which can be determined from a series of charts for different heat exchanger geometries. T refers to shell-side temperatures and θ refers to tube-side temperatures.

Equation (7.78) can be applied equally to plate heat exchangers.

Logarithmic mean temperature differences are determined in exactly the same way as for tubular exchangers and the total area is simply the product of the developed area of each plate and the number of plates.

Example 7.16

An ice cream mix is to be pasteurised at a rate of $3,600\,\mathrm{kg\,h^{-1}}$ in a plate heat exchanger. The mix enters the regenerator section of the exchanger at $25°\mathrm{C}$ and leaves the heating section at $80°\mathrm{C}$ to enter a holding tube before returning to the regenerator and the cooling section. Hot water enters at $90°\mathrm{C}$ and flows counter-currently to the mix in the heating section at a rate of $1.5\,\mathrm{kg\,s^{-1}}$. If 80% of the total heat required to raise the mix to $80°\mathrm{C}$ is supplied in the regenerator, calculate the number of plates needed in the regeneration and heating sections, respectively.

Heat transfer surface per plate	$0.80\,\mathrm{m^2}$
Overall heat transfer coefficient (regenerator)	$2,500\,\mathrm{W\,m^{-2}\,K^{-1}}$
Overall heat transfer coefficient (heater)	$2,700\,\mathrm{W\,m^{-2}\,K^{-1}}$
Mean heat capacity of ice cream mix	$4.0\,\mathrm{kJ\,kg^{-1}\,K^{-1}}$

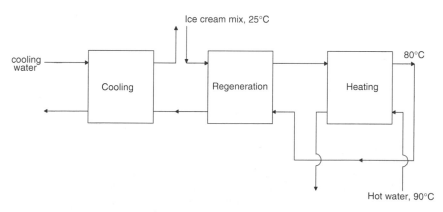

The total rate of heat transfer required to raise the temperature of the ice cream mix to $80°\mathrm{C}$ is

$$Q = \frac{3600}{3600} \times 4.0 \times (80 - 25)\,\mathrm{kW} \quad \text{or} \quad Q = 220\,\mathrm{kW}$$

Of this, 80% is to be supplied from the pasteurised mix returning from the heating section and giving up its heat to the cold mix, that is, $Q_{regenerator} = 176\,kW$. Now, because the two streams in the regenerator are in fact the same stream (with a constant flow rate and a very nearly constant mean heat capacity) the temperature change across the regenerator is simply 80% of the total temperature change, that is, $\Delta T_{regenerator} = 0.80(80 - 25) = 44\,K$. Consequently the temperature at the inlet to the heater is $25 + 44 = 69°C$ and the temperature at the outlet of the regenerator is $80 - 44 = 36°C$. Hence $\Delta T_1 = (36 - 25) = 11\,K$ and $\Delta T_2 = (80 - 69) = 11\,K$. Clearly the logarithmic mean temperature difference must also be 11 K. From the rate equation the total plate area in the regeneration section is

$$A = \frac{176 \times 10^3}{2500 \times 11}\,m^2, \qquad A = 6.4\,m^2$$

The number of plates needed is thus $6.4/0.8 = 8$ plates. The remaining $44\,kW$ must be transferred in the heating section and an energy balance for the hot water allows the outlet water temperature T to be determined:

$$44 = 1.5 \times 4.18 \times (90 - T)$$

from which $T = 83°C$. Thus the logarithmic mean temperature difference between the mix and the water is

$$\Delta T_{lm} = \frac{(90 - 80) - (83 - 69)}{\ln(10/14)}\,K \quad \text{or} \quad \Delta T_{lm} = 11.9\,K$$

The area in the heating section is then

$$A = \frac{44 \times 10^3}{2700 \times 11.9}\,m^2 \quad \text{and} \quad A = 1.37\,m^2$$

The number of plates needed is thus $1.37/0.8$ which when rounded up gives two plates.

7.5. BOILING AND CONDENSATION

7.5.1. Boiling Heat Transfer

Heat transfer to boiling liquids is particularly important in processing operations such as evaporation in which the boiling of liquids takes place either at submerged surfaces or on the inside of vertical tubes, as in a climbing film evaporator. Vapour bubbles form during boiling and for this to occur the liquid must be slightly superheated near the bubble. However, bubble formation is much easier on curved surfaces and on irregularities which may be present on a surface. In these circumstances only a small degree of superheat of the order of 0.5 K is needed. The nature of the surface has a significant effect on the shape of the bubbles which are formed and upon the resultant heat transfer coefficient. This is illustrated in Figure 7.13 for

  i   a non-wettable surface, for example with a thin layer of oil,

  ii   a partially wettable surface, which is the most common occurrence, and

  iii a clean surface which is entirely wetted, for example, when the liquid has detergent properties, and from which bubbles leave when very small giving the minimum area of contact.

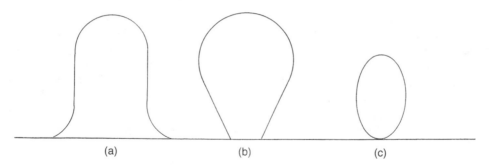

Figure 7.13. Bubble shapes in boiling liquids: (a) non-wettable surface, (b) partially wettable surface, (c) clean surface.

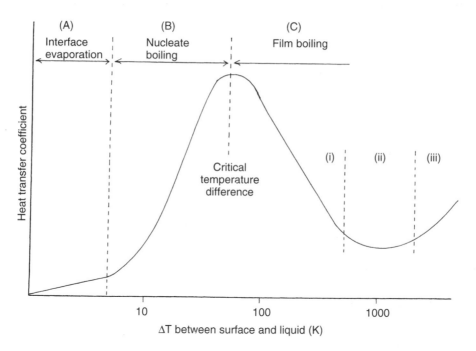

Figure 7.14. Boiling heat transfer: relationship between heat transfer coefficient and temperature difference.

The latter case will result in the maximum rate of heat transfer. Now the heat flux q changes dramatically as a function of the temperature difference between the surface and the boiling liquid ΔT, rising to a peak value and falling away sharply. This is reflected in the strong dependence between the heat transfer coefficient h and the temperature difference shown in Figure 7.14, where the heat transfer coefficient is defined by

$$q = h\,\Delta T \tag{7.84}$$

The curve may be divided into a number of zones. During *interface evaporation* (A), at small temperature differences, the bubbles of vapour leave the surface because of natural convection and therefore there is very little agitation of the liquid. The relationship between the heat transfer coefficient and temperature difference is linear and can be expressed as

$$q = \text{constant} \, \Delta T^{1.25} \tag{7.85}$$

with the heat transfer coefficient correlated by

$$Nu = 0.61(Pr \, Gr)^{0.25} \tag{7.86}$$

As the temperature difference increases the bubbles of vapour form much more rapidly and nucleate freely at many more centres. This means that small bubbles are produced and the majority of the surface is in contact with liquid. The rising bubbles are now responsible for significant agitation of the liquid and consequently the heat transfer coefficient increases dramatically, peaking at the critical temperature difference. Boiling is vigorous and heat transfer rates are high. This is known as *nucleate boiling* (B). The third major zone is called *film boiling* (C). Eventually a film of vapour is formed over the surface because the bubbles form so rapidly and liquid cannot flow into the surface. As the film forms, the heat flux and the heat transfer coefficient begin to fall rapidly with increasing temperature difference. The peak heat flux and the maximum heat transfer coefficient occur just before this happens. Referring to Figure 7.14, the film boiling zone may be sub-divided as follows:

i   A region where h falls rapidly and the vapour film is unstable.
ii A region where a stable film forms above the heat transfer surface and heat is transferred by conduction through the film which contains all of the resistance to heat transfer. The nature of the surface has little or no effect at this point.
iii A region where, as ΔT increases further, radiation becomes significant and both the flux and the heat transfer coefficient begin to increase again.

Ideally, equipment should be operated in the nucleate boiling zone, just below the critical temperature difference in Figure 7.14. At the beginning of nucleate boiling heat transfer coefficients are in the range 5,000 to 12,000 W m^{-2} K^{-1} for water at atmospheric pressure. However the peak flux may be as high as 400 to 1,200 kW m^{-2} with the heat transfer coefficient reaching 55,000 W m^{-2} K^{-1}. Care must be taken with equipment which is not designed for film boiling; the heat transfer rates and surface coefficients are so low, and ΔT so high, that there is a grave danger of overheating and damaging the walls of the heater.

Kutateladze's correlation for the heat transfer coefficient in nucleate boiling is commonly used. This takes the form

$$\left(\frac{h}{k}\right) \Psi^{0.5} = 0.0007 \left[\frac{q_{max}}{\alpha h_{fg} \rho_V} \frac{P}{\sigma} \Psi\right]^{0.7} Pr^{-0.35} \tag{7.87}$$

where the group Ψ is defined by

$$\Psi = \frac{\sigma}{g(\rho_L - \rho_V)} \tag{7.88}$$

P is the absolute pressure, σ and k are the surface tension and thermal conductivity of the liquid, respectively. The Prandtl number Pr and thermal diffusivity α are evaluated for the liquid. The peak flux q_{max} is obtained from

$$q_{max} = 0.16 h_{fg} \rho_V \left(\frac{\sigma g (\rho_L - \rho_V)}{\rho_V^2} \right)^{0.25} \qquad (7.89)$$

where the subscripts L and V refer to the liquid and vapour, respectively.

Example 7.17

Use Kutateladze's correlation to calculate the heat transfer coefficient at a stainless steel surface on which water is boiled at atmospheric pressure. Hence determine the temperature difference between the surface and the water which is required to maintain a heat flux of $2{,}000 \, \text{kW m}^{-2}$. The surface tension of water at $100°C$ is $0.0589 \, \text{N m}^{-1}$.

The following properties of saturated water at $100°C$ are obtained from steam tables: $\mu = 2.79 \times 10^{-4} \, \text{Pa s}$; $\rho_L = 957.9 \, \text{kg m}^{-3}$; $c_p = 4219 \, \text{J kg}^{-1} \text{K}^{-1}$; $\rho_V = 0.598 \, \text{kg m}^{-3}$; $k = 0.681 \, \text{W m}^{-1} \text{K}^{-1}$; $h_{fg} = 2.2567 \times 10^6 \, \text{J kg}^{-1}$.

The peak heat flux is therefore

$$q_{max} = 0.16 \times 2.2567 \times 10^6 \times 0.598 \left(\frac{0.0589 \times 9.81 (957.9 - 0.598)}{(0.598)^2} \right)^{0.25} \, \text{W m}^{-2}$$

or

$$q_{max} = 1.354 \times 10^6 \, \text{W m}^{-2}$$

and the group Ψ is

$$\Psi = \frac{0.0589}{9.81 (957.9 - 0.598)} \, \text{m}^2 \quad \text{or} \quad \Psi = 6.27 \times 10^{-6} \, \text{m}^2$$

The Prandtl number and thermal diffusivity of the condensate are, respectively,

$$Pr = \frac{4219 \times 2.79 \times 10^{-4}}{0.681} \quad \text{or} \quad Pr = 1.728$$

and

$$\alpha = \frac{0.681}{4219 \times 957.9} \quad \text{or} \quad \alpha = 1.658 \times 10^{-7} \, \text{m}^2 \, \text{s}^{-1}$$

Substituting these quantities into Equation (7.87) gives

$$\left(\frac{h}{0.598} \right) (6.27 \times 10^{-6})^{0.5} = 0.0007 \left[\frac{1.354 \times 10^6}{1.685 \times 10^{-7} \times 2.2567 \times 10^6 \times 0.598} \right.$$

$$\left. \times \frac{1.01325 \times 10^5 \times 6.27 \times 10^{-6}}{0.0589} \right]^{0.7} (1.728)^{-0.35}$$

from which the film heat transfer coefficient is

$$h = 40{,}311 \, \text{W m}^{-2} \, \text{K}^{-1}$$

Consequently the required temperature difference must be

$$\Delta T = \frac{2 \times 10^6}{40{,}311} \, \text{K}$$

and therefore

$$\Delta T = 49.6 \, \text{K}$$

7.5.2. Condensation

The heat transfer coefficients experienced when a liquid is vaporised or when a vapour is condensed are considerably greater than for heat transfer to a liquid or gas without a phase change (see Table 7.2). However, it is rather more difficult to measure heat transfer coefficients under these conditions and the correlations used to predict film coefficients are inevitably more complex.

(a) *Mechanisms of condensation*   A vapour which condenses on a cold surface will do so in one of two distinct ways. In filmwise condensation the vapour forms a liquid film which is amenable to mathematical analysis. However if the condensed vapour does not fully wet the surface then droplets of liquid will form which then grow and fall from the surface under their own weight. This is called dropwise condensation. With this second mechanism fresh surface is continually exposed and, because the dominant thermal resistance lies in the liquid film, heat transfer coefficients in dropwise condensation are between four and eight times higher than in filmwise condensation. It is important to realise that a given vapour will condense only in either one or other of these two distinct mechanisms. There is no change between filmwise and dropwise, or vice versa, under different process conditions. The only pure vapour definitely known to give dropwise condensation is water vapour (steam). In addition, some mixed vapours, especially immiscible mixtures, are also known to condense dropwise.

(b) *Presence of non-condensable gases*   A non-condensable gas is effectively a super-heated vapour which does not attain the saturation temperature as the vapour is condensed, an example being traces of air in steam. Consequently, the vapour must diffuse through the gas which tends to concentrate in a layer adjacent to the condensate film (e.g., the steam must diffuse through the air). There is then a temperature gradient across the gas layer in addition to that across the condensate film. This will affect the rate of condensation and the film heat transfer coefficient may be reduced very considerably. For example, the presence of 1% air in steam may reduce the heat transfer coefficient by up to 50%. Therefore it is important to fully vent steam chests in evaporators or sterilising retorts in order to maximise heat transfer coefficients and thermal efficiencies.

(c) *Nusselt theory*   The film heat transfer coefficient for condensation can be predicted from the Nusselt theory which considers a liquid film of condensate flowing down an inclined plane. The algebraic derivation is lengthy and beyond the scope of this chapter but the assumptions behind the theory may be listed as follows:

    i   the film flows under gravity and flow is laminar,

    ii   the film thickness varies with distance down the inclined surface,

    iii the surface is smooth and clean, with the surface temperature remaining constant, and

    iv heat is transferred by conduction through the film.

Taking the film heat transfer coefficient as thermal conductivity divided by film thickness the Nusselt theory, for a vertical surface, gives the mean film coefficient h_m to be

$$h_m = 0.943 \left[\frac{\rho^2 k^3 g h_{fg}}{\mu L \, \Delta T} \right]^{0.25} \tag{7.90}$$

and, for a horizontal tube

$$h_m = 0.725 \left[\frac{\rho^2 k^3 g h_{fg}}{\mu d_o \, \Delta T} \right]^{0.25} \tag{7.91}$$

where in each equation the temperature difference is that between the vapour temperature T and the surface temperature T_w, L is the length of vertical surface, d_o is the external tube diameter and the physical properties of the liquid are evaluated at the mean film temperature $(T_w + T)/2$. Experimental values for vertical surfaces are usually some 20% greater than those predicted by this theory and, therefore, the coefficient in Equation (7.90) is often quoted as 1.13.

Example 7.18

Water vapour at an absolute pressure of 20 kPa condenses on the outside of a horizontal thin metal tube 0.015 m in diameter. The tube contains water at a mean temperature of 25°C. What is the film heat transfer coefficient?

The temperature of saturated water vapour at 20 kPa is 60.1°C. The thermal resistance of the tube wall is small and it may be assumed that the wall temperature is 25°C and therefore the mean temperature for the evaluation of condensate properties is $(60.1 + 25)/2 \approx 42.5$°C. Thus

$$h_m = 0.725 \left[\frac{(991.2)^2 (0.635)^3 9.81 \times 2.358 \times 10^6}{6.23 \times 10^{-4} \times 0.015 \times (60.1 - 25)} \right]^{0.25} \text{W m}^{-2}\text{K}^{-1}$$

and

$$h_m = 8367 \, \text{W m}^{-2}\text{K}^{-1}$$

7.6. HEAT TRANSFER TO NON-NEWTONIAN FLUIDS

Very many non-Newtonian food fluids have a high apparent viscosity and therefore flow is usually laminar with fully developed turbulent flow being rare. Consequently the correlations for heat transfer to non-Newtonian fluids are based on those which have been developed for the laminar flow of viscous Newtonian fluids. Most attention has been given to power law and Bingham fluids because these are the most commonly encountered examples. The majority of correlations involve the Graetz number (Gz) which does not require a value of viscosity, which by definition is difficult to specify for non-Newtonian fluids. However the parameters in the relevant rheological models need to be measured, for example the flow behaviour index n, the consistency coefficient K and a yield stress for Bingham fluids. For power law fluids in laminar flow within a circular cross-section the Metzner and Gluck relationship is used. This takes the form

$$Nu = 1.75\delta^{0.33}(Gz)^{0.33}\left(\frac{K_b}{K_w}\right)^{0.14} \tag{7.92}$$

where the group δ is given by Equation (7.93)

$$\delta = \frac{3n+1}{4n} \tag{7.93}$$

and the Graetz number is defined by

$$Gz = \frac{mc_p}{kL} \tag{7.94}$$

in which L is the length of the heated pipe section and m is the mass flow rate. Although Gz contains the tube length, the Nusselt number is based upon the tube diameter; the fluid physical properties are evaluated at the mean bulk temperature, K_b is the consistency coefficient at the mean bulk temperature and K_w is the consistency coefficient at the wall temperature T_w. The Metzner and Gluck correlation predicts the mean heat transfer coefficient h_m and is intended to be used with an arithmetic temperature difference. Hence

$$q = h_m \Delta T \tag{7.95}$$

where

$$\Delta T = \left[T_w - \frac{(T_{b_i} + T_{b_o})}{2}\right] \tag{7.96}$$

and T_{b_i} and T_{b_o} are the inlet and outlet bulk temperatures, respectively. Equation (7.92) requires a knowledge of the variation of K with temperature which may not always be available and further implies that an iterative solution is required for problems where the outlet temperature from a heating section is to be calculated. However the term $(K_b/K_w)^{0.14}$ can be ignored without too great an error being introduced. In this form Equation (7.92) is known as the Pigford correlation. Note that this is valid only where there is no natural convection and therefore at $Gz > 50$.

Example 7.19

A fruit puree is heated from 25°C to 50°C in a 0.035 m diameter thin-walled pipe by steam at 120°C. The puree is shear thinning with a flow behaviour index of 0.50, a mean heat capacity of 3,200 J kg^{-1} K^{-1} and a thermal conductivity of 1.35 W m^{-1} K^{-1}. Use the Pigford correlation to estimate the length of the heating section required for a mass flow rate of 650 kg s^{-1}.

The Graetz number is a function of the unknown length of the heating section and

$$Gz = \frac{650 \times 3200}{3600 \times 1.35L}$$

or

$$Gz = \frac{428}{L}$$

The group δ becomes

$$\delta = \frac{(3 \times 0.50) + 1}{4 \times 0.50}, \qquad \delta = 1.25$$

Therefore, from the Pigford correlation, the mean film heat transfer coefficient is

$$h_{\mathrm{m}} = \frac{1.35}{0.035} 1.75(1.25)^{0.33} \left(\frac{428}{L}\right)^{0.33} \mathrm{W\,m^{-2}\,K^{-1}}$$

or

$$h_{\mathrm{m}} = \frac{536.6}{L^{0.33}} \mathrm{W\,m^{-2}\,K^{-1}}$$

An energy balance, equating the rate of heat gain by the fluid to the rate of heat transfer from the wall, gives

$$mc_{\mathrm{p}}(T_{\mathrm{b_i}} - T_{\mathrm{b_o}}) = h_{\mathrm{m}} \pi\, dL\, \Delta T$$

which on substitution of the data gives

$$\frac{650}{3600} 3200(50 - 25) = \frac{536.6}{L^{0.33}} \pi 0.035L \left[120 - \frac{(50 + 25)}{2} \right]$$

which can be solved to give

$$L^{0.67} = 2.97, \qquad L = 5.07\,\mathrm{m}$$

The calculation is valid because substituting $L = 5.07\,\mathrm{m}$ into the expression for Gz yields

$$Gz = \frac{428}{5.07}$$

and thus $Gz = 84.4$.

For Bingham fluids, Hirai suggested a correction to the basic form of the laminar flow correlation

$$Nu = 1.75(Gz)^{0.33} \left[\frac{3(1 - c)}{c^4 - 4c + 3} \right]^{0.33} \qquad (7.97)$$

where c is the ratio of the yield stress to the shear stress at the wall. Again this is valid only at relatively high Graetz numbers and for

$$(Gz) \left[\frac{3(1 - c)}{c^4 - 4c + 3} \right] > 100 \qquad (7.98)$$

It should be noted that in all these correlations the heat transfer coefficient is proportional to the Graetz number which in turn is inversely proportional to the length of the heating section. Consequently shorter lengths of heating pipe are more effective for heat transfer to non-Newtonian fluids and L should be minimised.

7.7. PRINCIPLES OF RADIATION

Radiation is the term given to the transfer of energy by electromagnetic waves; specifically, thermal radiation is the transfer of energy by that part of the electromagnetic spectrum with wavelengths between 10^{-7} and 10^{-4} m. The electromagnetic spectrum is illustrated in Figure 7.15. All wave forms are propagated at what is known as the velocity of light c, which has a value of approximately 3×10^8 m s^{-1}. The relationship between the velocity of propagation, wavelength λ and the frequency υ is then

$$c = \lambda \upsilon \tag{7.99}$$

where frequency is measured in Hz. Note that wavelengths are often quoted in nanometers (nm) or in Angstroms (Å) where $1\,\text{Å} = 10^{-10}$ m. The wave forms in Figure 7.15 differ in their wavelength and in their properties. For example, visible light is simply that part of the spectrum to which the human eye is sensitive; X-rays have the property of being able to pass through solid objects and expose photographic plates.

All bodies at a temperature above absolute zero emit thermal radiation but the quantities involved only become significant at temperatures well above ambient. Unlike conduction and convection, radiation can be propagated through a vacuum and the most important example for human life is that of solar radiation; that is, thermal radiation travelling through space (as an electromagnetic wave) from a body with a surface temperature of about 6,000 K. On reaching the earth this energy is then dissipated by conduction and convection. In many industrial examples of heat transfer conduction and convection occur simultaneously with radiation, despite the very dissimilar mechanisms involved. However thermal radiation is the dominant heat transfer mechanism in a number of food processing applications such as drying, especially where the product is in a granular form with a relatively low initial moisture content. Some examples are tunnel or conveyor driers, vacuum band or shelf driers used for the manufacture of 'puffed' or expanded cereal products and freeze driers. Radiation is also the principal mechanism in solar drying which is used extensively in the developing world for the drying of food in the open air. In addition mention may be made of baking, roasting and heat shrink packaging.

7.7.1. Absorption, Reflection and Transmission

All radiation falling upon a surface is either absorbed, reflected or transmitted. The fraction of incident radiation which is absorbed is called the absorptivity of that surface and is

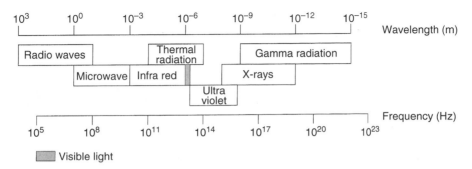

Figure 7.15. Electromagnetic spectrum.

denoted by α. Similarly, the reflectivity ρ and transmissivity τ of the surface are the fractions of incident radiation which are reflected from the surface or transmitted through the body, respectively. It therefore follows that:

$$\alpha + \rho + \tau = 1 \qquad (7.100)$$

Most solids, except those that are transparent to visible light (e.g., quartz, glass, and some plastics) do not transmit radiation. Hence,

$$\alpha + \rho = 1 \qquad (7.101)$$

For gases, reflectivities are zero and therefore

$$\alpha + \tau = 1 \qquad (7.102)$$

At this point it is essential to understand that the emission, absorption and reflection of radiation are surface phenomena. The rate at which thermal radiation is emitted is a function of the surface temperature and surface characteristics such as colour. Equally, the colour of a surface determines its absorptivity. The absorption of radiation takes place only at the surface and thereafter heat transfer through the body is by conduction. The nature of a surface, especially its roughness, has a considerable effect upon reflection. For a specular reflector the angle of incidence equals the angle of reflection. This will be the case where the roughness dimension is very much smaller than the wavelength (Figure 7.16a). However, if the roughness dimension is considerably greater than the wavelength of the incident radiation the reflection will be diffuse, that is it is reflected in all directions and with varying intensities (Figure 7.16b). For a specular surface, for example polished metals, the reflectivity approaches unity and the absorptivity will be close to zero whereas for a diffuse surface the absorptivity approaches unity.

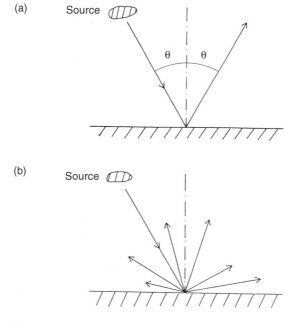

Figure 7.16. Surface characteristics: (a) specular surface, (b) diffuse surface.

7.7.2. Black Body Radiation

Emissive power E is defined as the total emitted radiant thermal energy leaving a surface per unit time and per unit area, over all wavelengths and in all directions. This should not be confused with radiosity which is the total radiant thermal energy leaving a surface, per unit time and per unit area, and which includes both emitted and reflected radiation. A black body is defined as one which absorbs all radiation falling upon it (regardless of directional or spectral differences) that is, $\alpha_B = 1$. The emissive power of a black body depends upon the fourth power of its absolute temperature, thus

$$E_B = \sigma T^4 \tag{7.103}$$

where σ is the Stefan–Boltzmann constant and has a value of $5.67 \times 10^{-8} \, \text{W m}^{-2} \, \text{K}^{-4}$.

Example 7.20

Determine the total emissive power of a black body at $1027°\text{C}$.

Emissive power depends upon the absolute temperature. Thus, from Equation (7.103),

$$E_B = 5.67 \times 10^{-8} \times (1300)^4 \, \text{W m}^{-2}, \qquad E_B = 161.9 \, \text{kW m}^{-2}.$$

The energy emitted by a black body is given out over a range of wavelengths and therefore, at any given temperature, there will be a wavelength λ_m at which energy emission is a maximum. This wavelength can be found from Wien's law which states that

> the wavelength for maximum monochromatic emissive power varies with the inverse of absolute temperature

or, in mathematical form,

$$\lambda_m T = 0.002897 \tag{7.104}$$

for wavelength in m and temperature in K. Thus, a plot of monochromatic emissive power $E_{B\lambda}$ against wavelength (Figure 7.17) shows a sharp peak at λ_m and the area under the curve, which represents the total emissive power, increases rapidly with temperature. Hence

$$\int_0^\infty E_{B\lambda} \, d\lambda = \sigma T^4 \tag{7.105}$$

$E_{B\lambda}$ can be obtained from Planck's equation

$$E_{B\lambda} = \frac{c_1 \lambda^{-5}}{\exp(c_2/\lambda T) - 1} \tag{7.106}$$

where the constants c_1 and c_2 have values of $3.742 \times 10^{-16} \, \text{W m}^2$ and $0.01439 \, \text{m K}$, respectively.

Example 7.21

At what wavelength is the monochromatic emissive power a maximum for a black body maintained at a temperature of $1,300 \, \text{K}$?

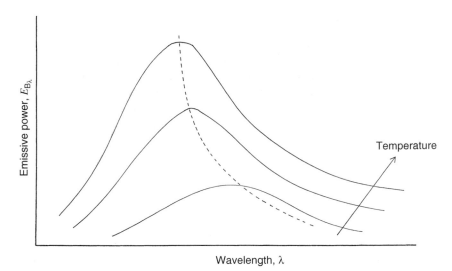

Figure 7.17. Wien's law.

Using Wien's law the wavelength λ_m is

$$\lambda_m = \frac{0.002897}{T}$$

and thus

$$\lambda_m = \frac{0.002897}{1300}\ \text{m}$$

and

$$\lambda_m = 2.23 \times 10^{-6}\ \text{m} \ \ \text{or} \ \ 2.23\ \mu\text{m}$$

7.7.3. Emissivity and Real Surfaces

The emissivity e of a surface is defined as the ratio of emissive power of the surface to that of a black body radiator at the same temperature, that is,

$$e = \frac{E}{E_B} \tag{7.107}$$

Values of emissivity must be known in order to calculated rates of radiative heat transfer between surfaces. Kirchoff's law states

the emissivity of any surface equals its absorbtivity when it is in thermal equilibrium with its surroundings

This can be shown to be the case by considering two bodies in an adiabatic enclosure as depicted in Figure 7.18. Let I be the incident radiation rate, A_1 and A_2 the respective areas of the two bodies and α_1 and α_2 the respective absorbtivities. Now at equilibrium the radiation falling on body 1 will equal the radiation emitted by body 1 and therefore

$$I A_1 \alpha_1 = E_1 A_1 \tag{7.108}$$

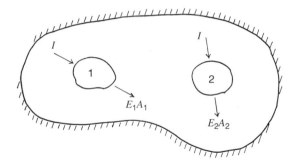

Figure 7.18. Kirchoff's law.

and for body 2

$$IA_2\alpha_1 = E_2A_2 \tag{7.109}$$

where E_1 and E_2 are the respective emissive powers. Thus, from Equations (7.108) and (7.109),

$$\frac{E_1}{\alpha_1} = \frac{E_2}{\alpha_2} = \frac{E}{\alpha} \tag{7.110}$$

In other words for all bodies the ratio of emissive power to absorptivity is the same and depends only on temperature. It must also be the case that

$$\frac{E}{E_B} = \frac{\alpha}{\alpha_B} \tag{7.111}$$

Now by definition α_B is equal to unity and therefore

$$e = \alpha \tag{7.112}$$

for all bodies.

A grey body is one which has a constant value of e over all wavelengths. However, most real surfaces do not have a constant emissivity and e should be quoted for a specified wavelength. In practice, emissivity usually increases with temperature and the concept of a grey body is useful for radiative heat transfer calculations. Table 7.5 lists values of emissivity for both food and non-food materials.

7.7.4. Radiative Heat Transfer

Imagine two infinite parallel plates 1 and 2. The rate at which radiant heat is *emitted by surface 1 and absorbed by surface 2* is $E_1A_1\alpha_2$ whilst heat is emitted by surface 2 and absorbed by surface 1 at a rate $E_2A_2\alpha_1$. The net rate of heat transfer between the two surfaces is then

$$Q_{1\rightarrow 2} = E_1A_1\alpha_2 - E_2A_2\alpha_1 \tag{7.113}$$

The emissive powers E_1 and E_2 are fractions of the emissive powers of a black body radiator and therefore

$$\left. \begin{aligned} E_1 &= e_1 E_{B_1} \\ E_2 &= e_2 E_{B_2} \end{aligned} \right\} \tag{7.114}$$

TABLE 7.5
Approximate Values of Emissivity for
Various Surfaces

	e
Dull black surface	1.0
Non metallic/non polished	0.90
Oxidised metals	0.60–0.85
Highly polished metals	0.03–0.06
Most foods	0.80–0.90
Dough	0.85
Lean beef	0.74
Beef fat	0.78
Water	0.95
Ice	0.97

Thus, on substitution, Equation (7.113) becomes

$$Q_{1 \to 2} = e_1 E_{B_1} A_1 \alpha_2 - e_2 E_{B_2} A_2 \alpha_1 \tag{7.115}$$

However from the Stefan–Boltzmann law

$$\left.\begin{aligned} E_{B1} = \sigma T_1^4 \\ E_{B2} = \sigma T_2^4 \end{aligned}\right\} \tag{7.116}$$

and consequently

$$Q_{1 \to 2} = \sigma \left(e_1 A_1 \alpha_2 T_1^4 - e_2 A_2 \alpha_1 T_2^4 \right) \tag{7.117}$$

Now the areas of surfaces 1 and 2 are equal (they are both infinite plates of area A) and the absorbtivities and emissivities are the same (e and α, respectively) for similar surfaces, hence

$$Q_{1 \to 2} = \sigma e A \alpha \left(T_1^4 - T_2^4 \right) \tag{7.118}$$

Thus the heat flux is given by Equation (7.119)

$$q_{1 \to 2} = \sigma e \alpha \left(T_1^4 - T_2^4 \right) \tag{7.119}$$

and for the case of black surfaces where $\alpha = e = 1$ this reduces to

$$q_{1 \to 2} = \sigma \left(T_1^4 - T_2^4 \right) \tag{7.120}$$

Example 7.22

Two infinite parallel plates are held at 850°C and 700°C, respectively. If each plate approximates to a black body, calculate the rate of heat exchange between them.

For a black body $e_B = \alpha_B = 1$ and therefore the net heat flux is given by Equation (7.120).

$$q = 5.67 \times 10^{-8}[(1123)^4 - (923)^4]\,\mathrm{W\,m^{-2}}, \qquad q = 49.03\,\mathrm{kW\,m^{-2}}$$

7.7.5. View Factors

For cases other than transfer between two infinite surfaces, the rate of energy exchange between two bodies depends upon the radiation leaving one surface and *reaching* the other and therefore is a function of geometry. This necessitates the use of view factors (sometimes called angle, shape or configuration factors). The view factor F_{mn} is defined as the fraction of radiant energy leaving surface m and reaching surface n. Thus the energy emitted by surface 1, reaching surface 2 and absorbed by surface 2 is $F_{12}E_1A_1\alpha_2$ whilst the energy emitted by surface 2, reaching surface 1 and absorbed by surface 1 is $F_{21}E_2A_2\alpha_1$. Consequently for heat exchange between two non-infinite surfaces

$$Q_{1\rightarrow 2} = F_{12}E_1A_1\alpha_2 - F_{21}E_2A_2\alpha_1 \tag{7.121}$$

Substituting for E_1 and E_2 from Equation (7.114), and then for E_{B_1} and E_{B_2} from Equation (7.116) gives

$$Q_{1\rightarrow 2} = F_{12}e_1A_1\alpha_2\sigma T_1^4 - F_{21}e_2A_2\alpha_1\sigma T_2^4 \tag{7.122}$$

View factors can be determined from first principles by considering solid geometry; however this is outside the scope of this text. By far the easiest way of obtaining view factors for relatively simple geometries is to use chart solutions, for example Figure 7.19 which has been prepared for identical parallel directly opposed flat plates.

The expression for the rate of radiant heat transfer [Equation (7.122)] can be simplified by using the reciprocity theorem. Consider two black body surfaces. The energy leaving surface 1 and arriving at surface 2 is $A_1F_{12}E_{B_1}$ and by the same token the energy leaving surface 2 and arriving at surface 1 is $A_2F_{21}E_{B_2}$. If the surfaces are at the same temperature the net exchange of heat will be zero and therefore

$$A_1F_{12}E_{B_1} - A_2F_{21}E_{B_2} = 0 \tag{7.123}$$

Equal surface temperatures implies that

$$E_{B_1} = E_{B_2} \tag{7.124}$$

and therefore

$$A_1F_{12} = A_2F_{21} \tag{7.125}$$

Thus in general for any two surfaces m and n the reciprocity theorem holds. That is

$$A_m F_{mn} = A_n F_{nm} \tag{7.126}$$

Applying this to Equation (7.122) produces

$$Q_{1\rightarrow 2} = A_1 F_{12}\sigma\left(e_1\alpha_2 T_1^4 - e_2\alpha_1 T_2^4\right) \tag{7.127}$$

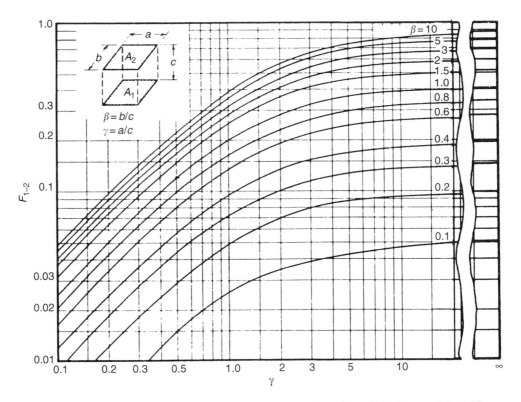

Figure 7.19. View factor for parallel directly opposed flat plates (from D.R. Pitts and L.E. Sissom, Heat Transfer, Schaum's outline series, McGraw-Hill, 1977, reproduced with the permission of The McGraw-Hill Companies).

Example 7.23

A continuous length of biscuit 0.20 m wide is baked in an oven which consists of a conveyor belt in a square section enclosure, the roof of which contains a radiative element 0.20 m wide. The biscuit is baked at 180°C and is conveyed directly beneath the element with a vertical separation of 0.40 m. The emissivities of element and biscuit are 0.95 and 0.80, respectively. If the rate of heat exchange is 900 W m^{-1}, what will be the temperature of the element?

Let subscript 1 refer to the element and subscript 2 to the biscuit. Consider a 1 m length of the element from which the radiant emission is 900 W. Using Figure 7.19, $a = 0.2$ m, $b = 1.0$ m and $c = 0.4$ m. Thus $\beta = 2.5$ and $\gamma = 0.5$ and from the chart $F_{12} = 0.17$. Now

$$e_1 = \alpha_1 = 0.95, \qquad e_2 = \alpha_2 = 0.80$$

Therefore, substituting into Equation (7.127)

$$900 = (0.20 \times 1.0)0.17 \times 5.67 \times 10^{-8} \times 0.80 \times 0.95\left(T_1^4 - 453^4\right)$$

from which $T_1 = 870$ K.

7.8. MICROWAVE HEATING OF FOODS

7.8.1. Microwaves

Microwaves are non-ionising electromagnetic waves and constitute that part of the electromagnetic spectrum with frequencies between 300 MHz and 300 GHz, corresponding to wavelengths in the range 0.001–1 m (Figure 7.15). Commercial microwave heating applications use frequencies of 915 or 2450 MHz. Microwave energy is not thermal energy like the thermal radiation described in section 7.7, but microwaves can generate heat on absorption by, and interaction with, certain dielectric materials. A dielectric material (e.g., air or glass) has a high resistance to the flow of an electrical current but does allow an electrostatic or magnetic field to pass through it. A perfect vacuum forms the perfect dielectric. Heating is the only observable effect when foodstuffs are exposed to microwave radiation. Any effect on micro-organisms is due to the generation of heat and not because they are susceptible to microwave radiation.

Microwaves are reflected by metals, tend to be transmitted through glass, ceramics and thermoplastics but are absorbed by water and by carbon. They are refracted on passing between different materials rather as visible light is refracted and because they are reflected by metals microwaves can be transmitted via hollow metal tubes. It is also possible to focus microwaves into a beam. Two mechanisms by which microwaves generate heat can be identified.

(*a*) *Ionic polarisation*   If an electric field is applied to ions in solution they will then orientate themselves in the field, undergo acceleration and increase their kinetic energy. Kinetic energy is then converted to heat when ions collide with one another. When the field is reversed, for example at 915×10^6 times per second (i.e., a frequency of 915 MHz), the number of collisions increases as ions rapidly change their direction. In foods the presence of sodium, potassium or calcium in salts will generate cations whilst chlorine will produce anions. The ionic polarisation effect increases with greater density or concentration of solution as the ions are more frequently in collision. In gases the spacing between molecules is too great for significant collisions.

(*b*) *Dipole rotation*   Similarly, polar molecules will line up in the direction of an applied electric field depending upon its polarity (Figure 7.20), much as a compass needle will align

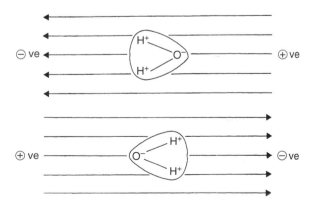

Figure 7.20. Dipole rotation in a water molecule.

itself in a magnetic field, and if a high frequency alternating field is applied then molecules will experience a rotating force (or torque) and will rotate at that frequency. This rotation generates heat. Water is the most important polar molecule encountered in foodstuffs; the hydrogen atoms are positively charged and the oxygen atoms negatively charged. This mechanism depends on the ability of molecules to move, so that for example ice is less susceptible to microwave heating because the water molecules are less able to move than when in the liquid state. Dipole rotation is far more significant than ionic polarisation, except for foods with a very high salt content.

7.8.2. Generation of Microwaves

Microwaves are generated by a cavity magnetron which itself emerged during the development of radar in the 1940s. The magnetron consists of an anode made from an evacuated copper tube which contains a number of copper vanes, usually 12, extending radially inwards. Alternate vanes are connected electrically. At the centre of the tube is a cavity, about 1 cm in diameter, and in the centre of the cavity, and coaxial with it, is a wire filament which acts as the cathode. A high negative voltage is applied to the cathode and generates a stream of electrons which are attracted to the anode and move radially outwards to it. A magnetic field is applied parallel with the axis of the tube which causes the electrons to follow a curved path to the anode. Suitable adjustment of the magnetic field produces an alternating charge on a vane which changes at very high frequency (between 300 MHz and 300 GHz). One vane is connected to an antenna which then radiates a high frequency microwave signal. The signal is transferred to the point of application by a wave guide which consists of either a metal tube or a coaxial cable.

7.8.3. Energy Conversion and Heating Rate

The conversion of microwave energy to heat and the consequent rate of heating depends upon the properties of the energy source and the properties of the dielectric. In addition the heat capacity of the food and other physical properties are important because once microwave energy has been absorbed heat is then transferred throughout the food mass by conduction or convection. The microwave properties are the field strength E, usually quoted in $V\,cm^{-1}$, and the microwave frequency υ in Hz. The food to be heated must be characterised in terms of the dielectric constant and the dielectric loss tangent. The relative dielectric constant (or relative permittivity) ε' is a measure of the ability of a material to store electrical energy and is defined as the ratio of capacitance (or permittivity) of the dielectric material to the capacitance (or permittivity) of a vacuum. Air approximates to a perfect vacuum and therefore the dielectric constant for air is approximately unity. It is a dimensionless quantity. The loss tangent $\tan \delta$ is a measure of how a material dissipates electrical energy as heat. These properties are often combined to give the relative dielectric loss ε'' (which again is dimensionless) and therefore

$$\varepsilon'' = \varepsilon' \tan \delta \qquad (7.128)$$

The power dissipation or rate of energy conversion per unit volume Q/V (measured in $W\,cm^{-3}$) is given by

$$\frac{Q}{V} = 5.56 \times 10^{-13} E^2 \upsilon \varepsilon' \tan \delta \qquad (7.129)$$

The absorption of microwave energy is thus increased by the frequency of the waveform and especially by the field strength. The greater the value of ε'' the greater will be the dissipation

TABLE 7.6

Dielectric Properties of Some Foods[a]

	Temperature (°C)	Frequency (MHz)	Relative dielectric constant	Relative dielectric loss
Beef (frozen)	−40	915	3.6	0.21
	−40	2450	3.5	0.13
	−20	915	4.8	0.53
	−20	2450	4.4	0.51–0.53
Beef (raw)	25	915	54.5	22.4
	25	2450	50.8–52.4	16–17.0
	40	2450	45.2	12.5
	80	915	45.7	32.9
	80	2450	44.1	19.5
Beef (cooked)	25	2450	31.0–35.4	10.0–11.6
Bread	25	2450	4.0	2.0
Butter (unsalted)	25	2450	3.0	0.1
Carrots (raw)	25	915	73.0	20.0
	25	2450	65.0	15.0
Carrots (cooked)	25	2450	71.0–72.0	15.0–17.8
Chicken (raw)	25	2000	53.0	18.0
Chicken (frozen)	−20	2450	4.0	0.5
Ice	0	2450	3.2	0.003
Potato (raw)	25	2450	62.0–64.0	14.0–17.0
Turkey (cooked)	15	2450	39.0	16.0
Water	25	2450	78.0	12.5

[a] After Buffler, Decareau and Peterson, and Rao and Rizvi.

of thermal energy and the more suitable is the food for microwave heating. Table 7.6 gives the values of relative dielectric constant and relative dielectric loss factor for a range of foods.

However the suitability of a food for microwave heating is crucially dependant upon the penetration characteristics. Figure 7.21 shows microwave power as a function of penetration depth. Curve (a) is obtained with poor attenuation, that is, poor absorption and volumetric heating and is typical of ice and frozen foods. The other extreme condition is curve (b) where the incident energy is absorbed at or near the surface giving poor penetration and only surface heating and is similar to the effect of conventional radiative heating. The ideal curve would be a linear relationship between power and depth, however the most usual characteristic is an exponential curve of the form of curve (c) in Figure 7.21. The loss of power with depth can be expressed in terms of the penetration depth which is defined as the distance from the surface at which the power has decreased to a fraction of e^{-1} of the incident power (about 36.8%).

The power P at the penetration depth z is given by

$$P = P_o e^{-2\alpha' z} \tag{7.130}$$

where P_o is the incident power and α' is the attenuation factor defined by

$$\alpha' = \frac{2\pi}{\lambda} \left[\frac{\varepsilon'}{2} \left(\sqrt{1 + \tan^2 \delta} - 1 \right) \right]^{0.5} \tag{7.131}$$

and λ is the wavelength. A very large value of z suggests poor absorption and a low value indicates that heating will take place only at the surface. Energy at a frequency of 915 MHz

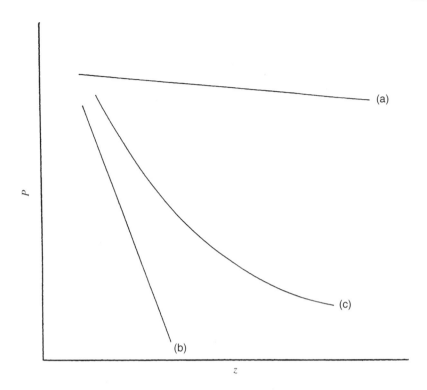

Figure 7.21. Absorption of microwave power as a function of penetration depth.

will penetrate more deeply than energy at 2,450 MHz and will be more appropriate for thicker food blocks.

7.8.4. Microwave Ovens and Industrial Plant

In the domestic microwave oven (Figure 7.22) the power supply is converted to a high DC voltage of the order of 4,000 V. The magnetron generates an oscillating high energy microwave field which is transmitted to the oven cavity via an antenna placed inside the wave guide. The wave guide is usually a rectangular metal tube, the internal surface of which reflects the waveform; there is no energy loss in the short wave guides used for domestic ovens. Thus the alternating electric field is produced inside the cavity of the oven and extends in three orthogonal directions. Food is often anisotropic, that is, the physical properties and composition are not the same in each direction, and therefore it is important to obtain a uniform distribution of microwave energy throughout the oven to ensure even heating. Consequently a rotating metal stirrer, from which microwaves are again reflected, is placed in the metal lined oven cavity and the food is rotated on a turntable. The length of the oven cavity should be a multiple of half the wavelength.

Industrial microwave plant is either a larger version of the domestic oven and is used for batch processing (e.g., tempering) or, for continuous processes, takes the form of a tunnel with food conveyed on a belt inside an enclosure. The wave guides used in industrial plant may be several metres long and the energy losses are estimated at about 5% per 30 m of wave

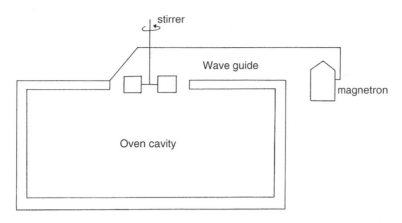

Figure 7.22. Microwave oven.

guide length. The use of the 915 MHz frequency has a number of advantages at the industrial scale. First, the output of single 2,450 MHz units is limited to about 20 kW whereas 915 MHz units can be rated up to 100 kW. Energy losses occur in the magnetron, which is either air or water cooled, and the efficiency of a 2,450 MHz unit is slightly lower at 70–80% than that for 915 MHz which can be up to 90%. Second, penetration depths at 915 MHz are greater by a factor of three which is more appropriate for large frozen meat blocks. In addition, both the capital and operating costs at 915 MHz are lower. However the greater absorption rate obtainable at 2,450 MHz is particularly important in dry frying and combined baking applications.

7.8.5. Advantages and Applications of Microwave Heating

The most important advantage of microwave heating is that microwaves are capable of penetrating very deeply into certain materials which results in very significant reductions in process heating times for drying, thawing, sterilisation and so on. Microwave heating is more uniform avoiding large temperature gradients and the overheating of food surfaces; when used for drying, microwave heating is far less likely to give case hardening. The geometry of the food to be heated is an important consideration; conduction and convection may often take far longer to heat a given mass of food. This could mean that a batch heating process which uses microwaves may be able to be integrated into a continuous line with all the advantages of continuous operation. There are overall energy savings which can be made by using microwaves and the total plant volume needed is likely to be far smaller. Microwave heating is also particularly relevant to food processing because of the ability to avoid overheating with temperature sensitive materials. A major disadvantage might be the capital cost of microwave equipment and therefore microwaves are more likely to be used with high value products.

A major use of microwaves in the food industry is the tempering and thawing of frozen foods, especially meat, fish, butter, and fruit. Tempering raises the temperature of frozen food to around $-4°C$ and this allows other operations such as cutting to be carried out much faster. Microwave tempering has a number of advantages over conventional methods: it takes minutes rather than days, reduces labour costs and requires a very much smaller space than the conventional refrigerated storage room. More even heating leads to lower drip loss from meat and there is an overall improvement in hygiene. However the very different dielectric properties of ice and water mean that great care must be taken during tempering to keep the temperature below the freezing point of water. The dielectric loss factor of ice is lower than that of water and ice is therefore much more transparent to microwaves and absorbs less energy. In both tempering and thawing a partial melting of the ice can lead to runaway heating because water heats more rapidly than ice. In thawing operations this may lead to cooking of part of the food whilst the remainder is still frozen. One way of preventing this is to pulse the application of microwave energy and allow the temperature to come to equilibrium in the interval.

In drying operations microwaves are used particularly to remove the last few percent of moisture; the selective heating of water then avoids overcooking of the surface. This has been found to be especially useful with pasta to avoid case hardening. Microwaves also find application in the drying of snack foods and in the freeze drying of meat and vegetables. Microwave cooking at an industrial scale is used for meat and poultry (especially bacon and chicken) because the shape and size of meat is better retained. For the baking of bread, biscuits and snack foods microwaves give a reduction in baking time without overheating the surface and changing surface colour. The microwave element is usually located at the end of tunnel ovens to remove the final few percent of moisture. When microwaves are used in conjunction with conventional heating it is possible to combine thorough heating and sterilisation with browning at the surface for products where an attractive appearance is important, for example, cakes, and sponges. The use of combined microwave heating also increases throughput considerably. Other applications of microwaves include sterilisation, especially in plastic pouches which are transparent to microwaves, the roasting of coffee beans and nuts and the blanching of vegetables.

NOMENCLATURE

A Area
c Velocity of light; ratio of the yield stress to the shear stress at the wall
c_p Heat capacity at constant pressure
c_1 Constant in Planck's equation
c_2 Constant in Planck's equation
C Coefficient
d Diameter
e Emissivity
E Emissive power; field strength
F Correction factor for multi-pass heat exchangers
F_{mn} View factor, fraction of radiation leaving surface m and arriving at surface n
g Acceleration due to gravity
Gr Grashof number
Gz Graetz number
h Film heat transfer coefficient
h_m Mean film heat transfer coefficient
h_{fg} Enthalpy of vaporisation
I Incident radiation rate
k Thermal conductivity
K Consistency coefficient
L Pipe length; linear dimension characteristic of heat transfer geometry

m	Mass flow rate; index
n	Number of tubes; flow behaviour index; index
Nu	Nusselt number
P	Pressure; power
P_o	Incident power
Pr	Prandtl number
q	Heat flux
q_{max}	Peak heat flux
$q_{1 \to 2}$	Net radiant heat flux between surfaces 1 and 2
Q	Rate of heat transfer
$Q_{1 \to 2}$	Net rate of radiant heat transfer between surfaces 1 and 2
r	Radius
r_a	Arithmetic mean radius
$r_{critical}$	Critical radius of insulation
r_{lm}	Logarithmic mean radius
r_p	Inner radius of insulation
R	Thermal resistance per unit area per unit length of pipe
R_f	Fouling factor
Re	Reynolds number
T	Temperature
$T(r)$	Temperature at radius r
U	Overall heat transfer coefficient
V	Volume
x	Length or thickness in the x direction
X	Dimensionless temperature difference
Y	Dimensionless temperature difference
z	Penetration depth

GREEK SYMBOLS

α	Absorptivity; thermal diffusivity
α'	Attenuation factor
β	Coefficient of linear expansion
ΔT	Temperature difference
ΔT_{lm}	Logarithmic mean temperature difference
δ	Group defined by Equation (7.93)
$\tan \delta$	Loss tangent
ε'	Relative dielectric constant
ε''	Relative dielectric loss
λ	Wavelength
λ_m	Wavelength at which energy emission is a maximum
μ	Viscosity
θ	Tube-side temperature
ρ	Density; reflectivity
σ	Surface tension; Stefan–Boltzmann constant
τ	Transmissivity
υ	Frequency
Ψ	Group defined by Equation (7.88)

SUBSCRIPTS

B	Black body
b	Bulk
c	'Cold' fluid
h	'Hot' fluid
i	Inner fluid; inlet
L	Liquid
m	Surface m
n	Surface n
o	Outer fluid; outlet
V	Vapour
w	Wall

PROBLEMS

7.1. A bakery oven is maintained at a temperature of 850°C. The wall of the oven is 0.10 m thick and has a thermal conductivity of 0.9 W m^{-1}K^{-1}. If no insulation is provided, what will be the rate of heat loss from the oven per m^2 of wall surface? Assume that the inner surface of the oven wall is at 850°C and that the outer surface of the wall is at the ambient air temperature of 18°C.

7.2. The wall of an oven consists of three layers of brick. The inside is built of 20 cm firebrick, thermal conductivity = 1.8 W m^{-1}K^{-1}, surrounded by 10 cm of insulating brick, k = 0.26 W m^{-1}K^{-1}, and an outside layer of building brick, thickness 15 cm, k = 0.70 W m^{-1}K^{-1}. The oven operates at 1150 K and it is anticipated that the outer side of the wall can be maintained at 320 K by the circulation of air. How much heat will be lost from the oven and what are the temperatures at the interfaces of the layers?

7.3. A cold store is constructed from 20 cm thick concrete slabs, 20 cm thick insulation and 3 cm thick wooden cladding (thermal conductivities 1.10, 0.020 and 0.20 W m^{-1}K^{-1}, respectively). Calculate the rate of heat penetration to the cold store if the external air temperature is 10°C and the internal air temperature is −25°C. Assume that the wall temperatures are at the respective air temperatures.

7.4. The dimensions of the cold store in problem 7.3 are 15 m × 25 m × 6 m. On a particular occasion the store is 50% full by volume when the refrigeration circuit fails for a period of 1 h. Neglecting the effect of the frozen food, and assuming the floor to be perfectly insulated, estimate the rise in temperature of the air. Take the physical properties of air at −25°C.

7.5. An industrial furnace is constructed from 0.20 m of firebrick (thermal conductivity 1.40 W m^{-1}K^{-1}) and is covered on the outer surface with insulation having a thermal conductivity of 0.09 W m^{-1}K^{-1}. If the inner surface temperature is 1150°C and the outer surface of the insulation is maintained at 30°C, determine the thickness of insulation necessary to reduce the heat loss to $1,000\,W\,m^{-2}$.

7.6. A wooden box of external dimensions 0.40 m × 0.50 m × 0.60 m deep is made of 0.01 m thick wood (thermal conductivity $0.15\,W\,m^{-1}\,K^{-1}$) and lined with 0.02 m of polystyrene (thermal conductivity $0.025\,W\,m^{-1}\,K^{-1}$). The box is filled with flake ice (density $600\,kg\,m^{-3}$) and a lid of the same construction is placed on top. If the box is transported by lorry for 6 h, calculate the mass of ice likely to be delivered. Assume the latent heat of fusion of ice to be $320\,kJ\,kg^{-1}$ and the air temperature to be 293 K.

7.7. A mild steel pipe of internal diameter 50 mm and wall thickness 3 mm is lagged with a 10 mm layer of cork. Calculate the heat loss from the pipe if the inner surface of the pipe is maintained at 420 K and the outer surface of the cork at 310 K. The thermal conductivities of mild steel and cork are $45\,W\,m^{-1}\,K^{-1}$ and $0.06\,W\,m^{-1}\,K^{-1}$, respectively.

7.8. A pipe carrying a hot fluid has an outside diameter of 40 mm. The pipe is insulated with two layers, each 10 mm thick, one insulating material having a thermal conductivity three times greater than the other. Two cases are to be considered: (a) with the better insulator touching the pipe and the poorer insulator outside, and (b) with the poorer insulator touching the pipe and the better insulator outside. If the overall temperature difference is the same for each configuration, determine the ratio of heat loss from the pipe in case (a) to that for case (b).

7.9. The interior of a thin spherical vessel, of radius 0.5 m, is maintained at a constant temperature of 200°C. The vessel is lagged with 75 mm of insulation (thermal conductivity $0.05\,W\,m^{-1}\,K^{-1}$) and the outside surface temperature is 40°C. What is the rate of heat loss from the vessel?

7.10. A horizontal storage tank, 1 m in diameter, has hemispherical ends and an overall length of 2 m. It contains a refrigerated liquid at a temperature of −15°C and the surrounding atmosphere is at 10°C. The tank is covered with lagging 15 cm thick having a thermal conductivity of $0.07\,W\,m^{-1}\,K^{-1}$. Neglecting the thermal resistance of the wall, and assuming an air film heat transfer coefficient of $6\,W\,m^{-2}\,K^{-1}$, calculate the rate of heat penetration to the tank.

7.11. A condenser tube has outside and inside diameters of 23 and 20 mm, respectively. The thermal conductivity of the tube wall is $200\,W\,m^{-1}\,K^{-1}$ and the surface heat transfer coefficients on the steam and water sides are 10,000 and $5,000\,W\,m^{-2}\,K^{-1}$, respectively. The tube is 7 m long and the mean overall temperature difference between the steam and water is 80°C. If the water flows at a velocity of $4\,m\,s^{-1}$ calculate: (a) the rate of heat transfer to the water and (b) the temperature rise of the water during its passage through the tube.

7.12. Water is heated in the annular space between two concentric tubes whose outside diameters are 105 and 51 mm, respectively. The wall thickness of the outer pipe is 3 mm. The water enters at 10°C and leaves at 80°C with a mean velocity 0.8 m s^{-1}. Evaluate the water-side film heat transfer coefficient.

7.13. Oil enters a single pass heat exchanger at 383 K and leaves at 338 K. The coolant enters at 290 K and leaves at 320 K. What is the logarithmic mean temperature difference for both co-current and counter-current operation?

7.14. A fruit puree (mean heat capacity, 3.71 kJ kg^{-1} K^{-1}) is cooled from 75°C to 25°C in a scraped surface heat exchanger, at a flow rate of 100 kg h^{-1}. The scraped surface is the inside of a tube 0.15 m in diameter and the overall heat transfer coefficient based upon this surface is 600 W m^{-2} K^{-1}. Cooling water enters at 10°C, flows counter-currently to the puree, and leaves at 21°C. Calculate the length of the heat exchanger.

7.15. Oil flows at the rate of 7,250 kg h^{-1} and is cooled from 110°C to 60°C in a shell and tube heat exchanger, with a heat transfer surface area of 4.5 m^2, which is operated counter-currently. Water enters the tubes at 20°C and leaves at 35°C. Determine the mass flow rate of water and the overall heat transfer coefficient. The mean heat capacity of the oil is 2.0 kJ kg^{-1} K^{-1}.

7.16. Water, flowing at a rate of 0.60 kg s^{-1}, is heated in a thin walled jacketed pipe, 25 mm diameter, from 20°C to 40°C. The jacket contains condensing steam at 95°C. The film heat transfer coefficients for water and steam are 3 and 7 kW m^{-2} K^{-1}, respectively. Calculate the overall heat transfer coefficient and the length of pipe required.

7.17. Water flows at 1.5 m s^{-1} through a 2.5 m length of a 25 mm diameter tube. If the tube wall is maintained at 320 K and the water enters and leaves at 291 and 294 K respectively, what is the value of the heat transfer coefficient?

7.18. Water vapour from an evaporator operating at 0.20 bar is condensed in a single pass tubular condenser containing 200 tubes, 2.0 m long, of external diameter 12.5 mm and internal diameter 11.5 mm. The cooling water enters at 20°C and leaves at 30°C. The surface heat transfer coefficients are 8 kW m^{-2} k^{-1} on the steam side and 4 kW m^{-2} K^{-1} on the water side and the thermal conductivity of the tube wall is 20 W m^{-1} K^{-1}. Estimate the condensation rate.

7.19. A food fluid flowing at a rate of 5 m^3 h^{-1} is to be sterilised at 140°C in a plate heat exchanger using steam at 144°C. The food stream enters at 110°C from a pre-heater. The overall heat transfer coefficient may be assumed to be 2.3 kW m^{-2} K^{-1}. Determine the number of plates necessary if the heat transfer surface area per plate is 0.15 m^2. The food has a mean heat capacity of 3.85 kJ kg^{-1} K^{-1} and a density of 900 kg m^{-3}.

7.20. A salt solution (boiling point, 108°C) is boiled at atmospheric pressure. Calculate the film heat transfer coefficient and the temperature of the surface required to maintain a heat flux of 80% of the peak value. Data for the solution at 108°C: enthalpy of vaporisation = 2.2 × 10^6 J kg^{-1}; density = 1,150 kg m^{-3}; viscosity = 1.2 × 10^{-3} Pa s; thermal conductivity = 0.50 W m^{-1} K^{-1}; heat capacity = 3.4 kJ kg^{-1} K^{-1}; vapour density = 0.77 kg m^{-3}; surface tension = 0.045 N m^{-1}.

7.21. Saturated steam at a pressure of 0.20 bar condenses on the outside of a thin walled horizontal tube, 12 mm in diameter, fabricated from a high thermal conductivity alloy. Water, at a mean temperature of 9.9°C, flows in the tube at a velocity such that the inside film heat transfer coefficient is 1,500 W m^{-2} K^{-1}. Determine the mean heat transfer coefficient for the condensate film and the condensation rate per unit length of tube.

7.22. A power law food fluid with a flow behaviour index of 0.80 enters a 0.05 m diameter, 4 m long, heated pipe at 20°C with a mass flow rate of 900 kg h^{-1}. If the pipe wall temperature is 130°C estimate the outlet temperature. For the fluid assume a mean heat capacity of 3,800 J kg^{-1} K^{-1} and a thermal conductivity of 1.0 W m^{-1} K^{-1}.

7.23. The total incident radiant energy falling on a body which partially reflects, partially absorbs and partially transmits radiant energy is 2,200 W m^{-2}. Of this 450 W m^{-2} is reflected and 900 W m^{-2} is absorbed. What is the transmissivity of the body?

7.24. An object (which can be assumed to be a black body radiator) has a total emissive power of 40 kW m^{-2}. What is the temperature of the object?

7.25. Determine the temperature of a black body which emits maximum radiation at a wavelength of 1.5 μm.

7.26. Two infinite parallel plates, each having an emissivity of 0.80, are held at 1000°C and 500°C, respectively. Calculate the net radiant heat exchange between them.

7.27. Two rectangular parallel plates, 2.5 m by 0.4 m, are spaced 0.5 m apart. One plate with an emissivity of 0.90 is held at 1000°C whilst the other plate has an emissivity of 0.70 and is maintained at 800°C. What is the net rate of radiant heat exchange between the plates?

FURTHER READING

C. R. Buffler, *Microwave Cooking and Processing*, Van Nostrand Reinold, New York (1993).

K. Cornwell, *The Flow of Heat*, Van Nostrand Reinhold, New York (1977).

R. V. Decareau and R. A. Peterson, *Microwave Processing and Engineering*, Ellis Horwood, Chichester (1986).

B. Hallstrom, C. Skjoldebrand, and C. Tragardh, *Heat Transfer and Food Products*, Elsevier, Amsterdam (1988).

D. R. Heldman and D. B. Lund, *Handbook of Food Engineering*, Marcel Dekker, New York (1992).

D. Q. Kern, *Process Heat Transfer*, McGraw Hill, New York (1950).

P. J. Fryer, D. L. Pyle, and C. D. Rielly (eds.), *Chemical Engineering for the Food Industry*, Chapman and Hall, London (1997).

A. T. Jackson and J. Lamb, *Calculations in Food and Chemical Engineering*, MacMillan, New York (1981).

Mass Transfer

8.1. INTRODUCTION

Mass transfer is concerned with the movement of material in fluid systems, that is both gases and liquids, under the influence of a concentration gradient. As we saw in chapter five, this is analogous to the movement of heat under the influence of a temperature gradient. For example, in drying operations water is removed in vapour form from either a liquid or a solid food into a warm gas stream (usually air). Thus the mass transfer of water occurs because there is a high concentration of water in the food and a lower concentration of water in the air. Most examples of mass transfer in food processes involve the transfer of a given component from one phase across an interface to a second phase. Some examples are listed in Table 8.1.

In humidification water is transferred from the liquid phase into the vapour or gaseous phase. However in the case of pure liquid water there is no concentration gradient in the liquid phase and hence no diffusion through this phase. Extraction involves the transfer of a liquid solute from a liquid phase into a second liquid which is a selective solvent. The aeration of a fermentation broth requires the transfer of oxygen through air bubbles to the surface of the bubble and thence into the liquid phase. Mass transfer describes both the transfer of (usually) water through a semi-permeable membrane and, crucially, the transfer of solute molecules from the bulk feed stream to the membrane surface in ultrafiltration or reverse osmosis. In the case of crystallisation a solute is transferred through the liquid phase, the magma or mother liquor, up to a solid/liquid interface where it is assimilated into the crystal lattice.

TABLE 8.1
Examples of Mass Transfer in Food Processes

	Transfer of	From	To
Drying	Water	Liquid/Solid	Gas
Humidification	Water	Liquid	Gas
Extraction	Liquid Solute	Liquid	Liquid
Leaching	Solid Solute	Solid	Liquid
Fermentation (Aeration)	Oxygen	Gas	Liquid
Membrane processes	Water	Liquid	Liquid
Crystallisation	Liquid Solute	Liquid	Solid

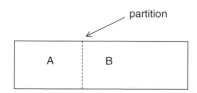

Figure 8.1. Diffusion due to a concentration gradient.

Often mass transfer is accompanied by heat transfer. In this chapter the principles of mass transfer are set out and are illustrated, for example by application to simple drying problems. The analysis will be limited to binary systems. At this stage it will be helpful to repeat and expand upon the illustration used in section 5.4.

Consider a fluid consisting of two components but with an unequal spatial distribution of the two components. It may be helpful to visualise a box with a removable partition and a different gas placed in each compartment (Figure 8.1). Immediately on removing the partition each gas will be concentrated at one end of the box. However after a time the concentrations of the two gases will become uniform throughout the entire box. Each component will flow so as to reduce the concentration gradient. In a stagnant fluid mass transfer occurs because of the random motion of molecules or *molecular diffusion*. Diffusion is slow and it will take some time for the concentration gradients to diminish. The obvious way to reduce this 'mixing time' is to agitate the gas and to introduce turbulence. In a turbulent fluid, mass transfer is due to both molecular diffusion and to transfer by convection currents or eddies. This is known as *convective mass transfer*. The introduction of convective mass transfer does not eliminate diffusion but it is usually far more significant. Diffusion still takes place whenever there is a concentration gradient.

8.2. MOLECULAR DIFFUSION

8.2.1. Fick's Law

Molecular diffusion diminishes the concentration gradients in a fluid system and is independent of any convection currents which may be present. In a mixture of A and B, the rate of transfer of A is determined by the diffusion of A and by the *presence and behaviour of B*. Thus the molar flux of A, J_A, is proportional to the concentration gradient and

$$J_A \propto \frac{dC_A}{dz} \tag{8.1}$$

Fick's first law can then be written as

$$J_A = -D_{AB}\frac{dC_A}{dz} \tag{8.2}$$

where the constant of proportionality D_{AB} is the diffusivity of A in B. This equation predicts the mass flux when A and B diffuse in opposite directions at the same rate. This condition is known as equimolar counter diffusion.

8.2.2. Diffusivity

Diffusivity is defined as the constant of proportionality in Fick's first law and has the dimensions of $L^2 T^{-1}$. Tables 8.2 and 8.3 list the diffusivities of some commonly encountered species in gaseous and dilute aqueous phases, respectively. Note that diffusivities in the liquid phase are some four orders of magnitude smaller than those in the gaseous phase.

TABLE 8.2
Diffusivities of Gases at Atmospheric Pressure: Binary Mixtures

	Temperature (K)	Diffusivity $(m^2\,s^{-1}) \times 10^5$
Air–hydrogen	273	6.11
	298	4.10
Air–oxygen	273	1.78
	298	2.06
Air–ammonia	298	2.36
Air–water	298	2.56
	333	3.05
Air–carbon dioxide	273	1.38
	298	1.64
Air–ethanol	298	1.19
Carbon dioxide–nitrogen	298	1.65
Carbon dioxide–oxygen	293	1.60
Carbon dioxide–water	307.5	2.02

TABLE 8.3
Diffusivities in Dilute Aqueous Solution

	Temperature (K)	Diffusivity $(m^2\,s^{-1}) \times 10^9$
Oxygen	293	1.80
	298	2.10
Carbon dioxide	293	1.50
	298	1.92
Nitrogen	293	1.64
	298	1.88
Ammonia	293	1.76
	298	1.64
Ethanol	293	1.00
	298	1.24
Glycerol	293	0.72
Glucose	293	0.60
	298	0.69
Sucrose	293	0.45
	298	0.56
Acetic acid	293	0.88
	298	1.21
Hydrogen	293	5.13

TABLE 8.4
Effective diffusivities in solid foods[a]

	Temperature (K)	Diffusivity $(m^2 s^{-1}) \times 10^{11}$
Water–starch gel (moisture content 30%)	298	2.3
Water–blanched potato	327	26.0
Water–fish muscle (moisture content 30%)	303	34.0
Water–minced beef (moisture content 60%)	333	10.0
Sucrose-0.79% agar gel	278	25.0
Sodium chloride–cheese	293	188.0
Sodium chloride–meat muscle	275	220.0
Sodium chloride–fish	293	230.0

[a] Data taken from Rao and Rizvi.

As was explained in section 5.5.3, it is possible to quantify an effective diffusivity which describes the transfer of gas or liquid within the solid phase. Table 8.4 lists the effective diffusivities for some food systems. It should be noted that these values can be up to a further two orders of magnitude smaller than liquid phase diffusivities.

8.2.3. Concentration

Concentrations can be expressed in a number of ways and the units differ considerably. This variation gives rise to the perceived difficulties of mass transfer. Conceptually mass transfer is no more difficult than heat transfer, indeed it might be thought easier because it deals with the movement of material rather than of the more abstract energy. The key to understanding mass transfer is to become thoroughly conversant with the units of concentration and the consequent units of mass transfer coefficients. The units of concentration which will be used initially are molar concentration and partial pressure.

(*a*) *Molar concentration* C_A, the molar concentration (or molar density) of A, is the molar mass of A per unit volume of the mixture and has units of $kmol\ m^{-3}$. This measure can be used both for liquids and gases.

(*b*) *Partial pressure* In a gas mixture the partial pressure p_A which is exerted by component A is a measure of the concentration of A. For a gas which can be assumed to behave ideally, molar concentration and partial pressure are related by Equation (3.13).

The concentration of component A can be expressed in a number of other ways, including:

(*c*) *Mass concentration* This is simply the density (or mass density) ρ_A and is given by the mass of component A per unit volume of the mixture. Density and molar concentration are related by

$$C_A = \frac{\rho_A}{M_A} \tag{8.3}$$

where M_A is the molecular weight of component A.

(*d*) *Mass fraction*   This was defined in section 4.2.2; the symbol x_A is used for liquids and y_A for gases. Note that mass fraction can be expressed in terms of the mass density of the mixture ρ

$$x_A = \frac{\rho_A}{\rho} \tag{8.4}$$

(*e*) *Mole fraction*   Similarly mole fraction (with symbols x_A for liquids and y_A for gases) is equal to the molar concentration of A divided by the total molar concentration

$$x_A = \frac{C_A}{C} \tag{8.5}$$

The relationship between pressure and molar concentration in the gas phase, which was set out at some length in chapter three, is the key to understanding a good deal of the material on mass transfer. It is important to grasp that this relationship is true for both *total* pressure and molar concentration and for *partial* pressures and molar concentrations of the given species.

8.3. CONVECTIVE MASS TRANSFER

In most applications of mass transfer to food process engineering problems, mass is transferred between distinct phases across a phase boundary or interface. The rate of mass transfer then depends upon four principal factors:

a   the driving force, or concentration difference,
b   the physical properties of each phase,
c   the interfacial area,
d   the hydrodynamics within each fluid phase, that is, the degree of turbulence.

In a given process little can be done to change either the overall concentration difference or the physical properties of the materials involved and the rate of mass transfer can be maximised only by increasing either or both of the interfacial area and the degree of turbulence in each fluid. It should be noted that only rarely can the area be determined independently in real situations.

8.3.1. Whitman's Theory

Whitman's two-film theory is the earliest and most generally applicable theory to account for inter-phase mass transfer. It is developed here in terms of the transfer of a gaseous solute from a gas phase to a liquid phase (e.g., the aeration of a fermenter) but is equally applicable in reverse or to the transfer of mass between two liquid phases. This example has been chosen to show the effect upon mass transfer coefficients of the various units available for gas and liquid concentrations. Consider Figure 8.2 which represents the interface between gas and liquid. Component A is transferred from a high concentration in the gas to a lower concentration in the liquid phase. In practice, the concentration gradient is approximately linear very close to the interface but reduces as the distance from the interface increases. This applies to both phases. Whitman's theory, however, assumes that two laminar films exist, one on each side of the interface, and that, first, all the resistance to mass transfer is contained within the two films and second, that the concentration gradients across each film are linear.

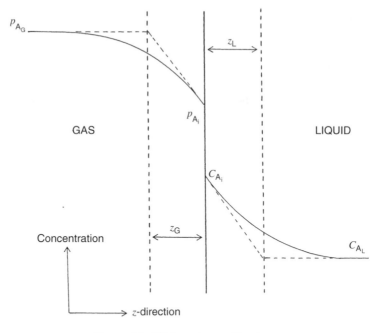

Figure 8.2. Whitman's two-film theory.

Thus, convection is responsible for mass transfer from the bulk phase up to the film boundary, where turbulence dies out. Mass transfer within the films is assumed to be due solely to molecular diffusion and thus Fick's law can be applied directly to each film. Whitman's theory makes three further assumptions: that there is no resistance to mass transfer at the interface, that the concentrations in each phase at the interface are related by a simple equilibrium condition such as Henry's law, and finally that the time taken to establish concentration gradients is much less than the time over which mass transfer takes place, in other words that mass transfer is effectively a steady-state process.

Writing Fick's law, in terms of a partial pressure difference, for the transfer of a solute A in the z direction across the gas film gives:

$$J_A = -D_G \frac{dp_A}{dz} \tag{8.6}$$

where D_G is the diffusivity of A in the gas phase. Integrating over the linear concentration gradient gives

$$J_A = \frac{D_G}{z_G}(p_{A_G} - p_{A_i}) \tag{8.7}$$

where z_G is the thickness of the gas film and $(p_{A_G} - p_{A_i})$ is the concentration difference across the film. The partial pressure of A in the bulk gas is denoted by p_{A_G} and the partial pressure of A in the gas phase at the interface by p_{A_i}. Similarly for the liquid film

$$J_A = \frac{D_L}{z_L}(C_{A_i} - C_{A_L}) \tag{8.8}$$

where C_{A_L} is the concentration of A in the bulk liquid, C_{A_i} is the concentration of A in the liquid *at the interface*, D_L is the diffusivity of A in the liquid phase and z_L is the liquid film thickness.

8.3.2. Film Mass Transfer Coefficients

The laminar films proposed in Whitman's theory are hypothetical and consequently the thicknesses z_G and z_L cannot be measured or predicted independently. In order to overcome this difficulty in predicting the molar flux of A, the term in D_G/z_G Equation (8.7) is replaced by a gas film mass transfer coefficient, k_G. Mass transfer coefficients are equivalent to heat transfer coefficients and depend upon physical properties, hydrodynamics and geometry. At the same time the symbol J for molar flux is replaced by N. Thus

$$N_A = k_G(p_{A_G} - p_{A_i}) \tag{8.9}$$

Molar flux depends therefore upon the product of a mass transfer coefficient and a concentration difference. In the same way the molar flux of A across the liquid film is given by

$$N_A = k_L(C_{A_i} - C_{A_L}) \tag{8.10}$$

where k_L is the liquid film mass transfer coefficient. It is important to understand that Equations (8.9) and (8.10) *define* k_G and k_L, respectively. The use of the symbol N for molar flux indicates that mass transfer takes place due to convection; the symbol J being reserved for mass transfer solely by diffusion.

The units of k_G and k_L are derived as follows: For $[N_A] = \text{kmol m}^{-2}\,\text{s}^{-1}$, $[p_A] = \text{Pa}$ and $[C_A] = \text{kmol m}^{-3}$, then

$$[k_G] = \frac{(\text{kmol m}^{-2}\,\text{s}^{-1})}{(\text{N m}^{-2})} = \text{kmol N}^{-1}\,\text{s}^{-1}$$

and

$$[k_L] = \frac{(\text{kmol m}^{-2}\,\text{s}^{-1})}{(\text{kmol m}^{-3})} = \text{m s}^{-1}$$

It would be equally possible to express the concentration difference across the gas film as a molar density which would give the gas film coefficient the units of m s^{-1}. Clearly both of these possibilities are SI units. The problem of alternative units for mass transfer coefficients is covered later in section 8.3.7.

Example 8.1

In an experiment to measure the gas film mass transfer coefficient, a model food material saturated with water was formed into an horizontal insulated slab with an exposed surface measuring 10 cm × 10 cm. The slab was placed in a warm air stream of absolute humidity 0.005 kg/kg dry air and the temperature of the slab was measured as 21°C. Over a 20 min period the loss in mass from the sample was 3.6 g. Assuming constant rate drying, calculate the film mass transfer coefficient.

In this problem only one film resistance, that of the gas film, must be taken into account. Water passes freely to the surface of the food sample and therefore the air immediately in contact with the surface is saturated with water vapour such that the partial pressure p_{wo} at this point is equal to the pure component vapour pressure of water. The partial pressure of water in the bulk air stream p_w must be calculated from the humidity. Adapting Equation (8.9),

$$N_w = k_G(p_{wo} - p_w)$$

Now the flux of water vapour N_w is the evaporation rate divided by the cross-sectional area of the slab $(0.01\ m^2)$. Hence

$$N_w = \frac{3.6 \times 10^{-3}}{18 \times 20 \times 60 \times 0.01}\ kmol\,m^{-2}\,s^{-1}$$

and

$$N_w = 1.67 \times 10^{-5}\ kmol\,m^{-2}\,s^{-1}$$

From steam tables the pure component vapour pressure of water p_{wo} at $21°C$ is 2,486 Pa. From Dalton's law the partial pressure of water in the bulk air stream is the product of the mole fraction of water vapour y_w and atmospheric pressure, that is, $p_w = y_w P$. Converting the humidity (a mass ratio) to a mole fraction, and taking the molecular weights of water and air to be 18 and 29, respectively, gives

$$y_w = \frac{(0.005/18)/(1/29)}{1 + (0.005/18)/(1/29)} = 0.008$$

Thus the partial pressure of water vapour is

$$p_w = 0.008 \times 1.013 \times 10^5\ Pa, \qquad p_w = 809.5\ Pa$$

Therefore the gas film mass transfer coefficient is

$$k_G = \frac{1.67 \times 10^{-5}}{2486 - 809.5}\ kmol\,N^{-1}\,s^{-1}$$

and

$$k_G = 9.96 \times 10^{-9}\ kmol\,N^{-1}\,s^{-1}$$

8.3.3. Overall Mass Transfer Coefficients

The molar concentration C_{A_i} and the partial pressure p_{A_i} cannot be determined independently because of the uncertainty of the position of the interface in most real mass transfer problems; liquid droplets and gas bubbles do not remain stationary in food processes. Even in cases where the interface remains in a predictable location, the measurement of interfacial concentration is a virtual impossibility. Therefore an *overall* gas mass transfer coefficient is introduced based upon a concentration difference which *can* be determined. The overall coefficient K_G is defined by

$$N_A = K_G(p_{A_G} - p_A^*) \tag{8.11}$$

where p_A^* is the partial pressure of A in the gas phase which is in equilibrium with the bulk liquid concentration C_{A_L}. The equilibrium relationship is then

$$p_A^* = HC_{A_L} \tag{8.12}$$

In this way, both p_{A_G} and p_A^* are known quantities, K_G can be determined and thus the molar flux can be predicted. Similarly, for the liquid film, C_{A_i} is replaced by C_A^* such that:

$$N_A = K_L(C_A^* - C_{A_L}) \tag{8.13}$$

where K_L is the overall liquid mass transfer coefficient and C_A^* is the molar concentration of A in the liquid phase which is in equilibrium with the bulk gas partial pressure p_{A_G}. Again, Henry's Law defines the equilibrium concentration and

$$p_{A_G} = HC_A^* \tag{8.14}$$

It should be noted that the interface concentrations are also related by Henry's constant

$$p_{A_i} = HC_{A_i} \tag{8.15}$$

The units of K_G and K_L will of course be the same as the units of k_G and k_L, respectively.

8.3.4. Addition of Film Mass Transfer Coefficients

In a manner analogous to the addition of film heat transfer coefficients, film mass transfer coefficients can be added together to determine the overall mass transfer coefficient.
From the definition of K_G

$$\frac{1}{K_G} = \frac{p_{A_G} - p_A^*}{N_A} \tag{8.16}$$

and

$$\frac{1}{K_G} = \frac{p_{A_G} + (p_{A_i} - p_{A_i}) - p_A^*}{N_A} \tag{8.17}$$

or

$$\frac{1}{K_G} = \frac{(p_{A_G} - p_{A_i}) + (p_{A_i} - p_A^*)}{N_A} \tag{8.18}$$

substituting from Equation (8.9) simplifies this to

$$\frac{1}{K_G} = \frac{1}{k_G} + \left(\frac{p_{A_i} - p_A^*}{N_A}\right) \tag{8.19}$$

and further substitution from Henry's law gives

$$\frac{1}{K_G} = \frac{1}{k_G} + H\frac{(C_{A_i} - C_{A_L})}{N_A} \tag{8.20}$$

Replacing the final term of Equation (8.20) with the reciprocal of the liquid film mass transfer coefficient yields

$$\frac{1}{K_G} = \frac{1}{k_G} + \frac{H}{k_L} \tag{8.21}$$

The significance of Equation (8.21) is that overall resistance to the transfer of A from the gas phase to the liquid phase is equal to the sum of the resistances within the individual phases. The term $1/k_G$ represents the *resistance* to mass transfer in the gas film (and in the gas phase since the film is assumed to contain all the resistance) and that H/K_L represents the resistance to mass transfer in the liquid phase. The overall resistance to the transfer of A from gas to liquid phases is given by $1/K_G$.

Similarly, an expression can be derived for the overall liquid mass transfer coefficient K_L in terms of the film coefficients and Henry's constant

$$\frac{1}{K_L} = \frac{1}{H k_G} + \frac{1}{k_L} \tag{8.22}$$

Combining the expressions for $1/K_G$ and $1/K_L$ and eliminating the film coefficients yields the following:

$$\frac{1}{K_G} = \frac{H}{K_L} \tag{8.23}$$

Example 8.2

The growth of a particular micro-organism in a fermenter requires a critical dissolved oxygen concentration of $2.5 \times 10^{-5}\,\text{kmol m}^{-3}$ and an oxygen utilisation rate of $4.0 \times 10^{-5}\,\text{kmol s}^{-1}$ per m^3 of fermenter volume. Air is sparged into the fermenter at a pressure of 4 bar. Hydrodynamic studies suggest that the bubble hold up should be sufficient to give a mass transfer surface area of $120\,\text{m}^2$ per m^3 of fermenter volume and that an overall liquid mass transfer coefficient of $9.0 \times 10^{-4}\,\text{m s}^{-1}$ can be achieved. Demonstrate by calculation whether the mass transfer rate will be sufficient to maintain growth. Take Henry's constant for oxygen and water to be $7.0 \times 10^7\,\text{J kmol}^{-1}$.

The molar flux of oxygen is given by Equation (8.13), that is

$$N_{O_2} = K_L \left(C_{O_2}^* - C_{O_{2L}} \right)$$

Now the required flux of oxygen is equal to the rate of consumption divided by the surface area over which transfer takes place that is, $4 \times 10^{-5}/120$ or $3.33 \times 10^{-7}\,\text{kmol m}^{-2}\,\text{s}^{-1}$. The concentration of oxygen in water which is in equilibrium with the oxygen in an air bubble can be calculated from Henry's law, assuming that the partial pressure of oxygen is 21% of the total pressure. Thus

$$C_{O_2}^* = \frac{0.21 \times 1.4 \times 10^5}{7 \times 10^7}\,\text{kmol m}^{-3}$$

and

$$C_{O_2}^* = 4.2 \times 10^{-4}\,\text{kmol m}^{-3}$$

The maximum achievable flux of oxygen depends upon the mass transfer characteristics; it is the product of the overall coefficient and the difference between the equilibrium concentration and the actual dissolved oxygen concentration. Thus

$$N_{O_2} = 9 \times 10^{-4}\,(4.2 \times 10^{-4} - 2.5 \times 10^{-5})\,\text{kmol m}^{-2}\,\text{s}^{-1}$$

or

$$N_{O_2} = 3.56 \times 10^{-7} \, \text{kmol m}^{-2} \, \text{s}^{-1}$$

This is greater than the required flux of $3.33 \times 10^{-7} \, \text{kmol m}^{-2} \, \text{s}^{-1}$ and consequently the actual rate of mass transfer is sufficient to maintain growth of the micro-organisms.

8.3.5. Resistances to Mass Transfer in Food Processing

In food processing the liquid phase usually presents a greater resistance to mass transfer than the gas phase and therefore becomes rate controlling. Four categories of liquid phase resistance can be identified:

   i   close to or at a gas–liquid interface, because of the relatively low solubility of most gases (e.g., oxygen) in the (usually aqueous) liquid phase;

   ii   close to or at a solid–liquid interface between solids suspended in a liquid (e.g., starch granules in water). In this case dissolution of the solid into the liquid produces a high concentration liquid film, of high viscosity, around the particle;

   iii   within bulk liquids. This is particularly the case if mixing within the bulk is poor due to high viscosity or to non-Newtonian behaviour; and

   iv   within liquids which are contained within the pore spaces of porous solids, for example, within an immobilised enzyme carrier or a particle from which a solute is being leached. Here the added complication of diffusion through the pores and poor liquid mixing contribute to the high resistance.

8.3.6. Effect of Solubility on Mass Transfer Coefficients

The driving forces for mass transfer are represented diagrammatically in Figures 8.3 and 8.4, where the solid line represents the equilibrium relationship.

(a) *Very soluble gases*   For a very soluble gas (Figure 8.3), the value of Henry's constant is small. In other words only a relatively small concentration of solute in the gas is required to produce a large concentration of that solute in the liquid phase and the equilibrium curve lies close to the C_A axis. Thus the overall driving force approximates to that across the gas film and

$$\frac{N_A}{K_G} \simeq \frac{N_A}{k_G} \tag{8.24}$$

or

$$K_G \cong k_G \tag{8.25}$$

and the overall coefficient approximates to the gas film coefficient.

(b) *Almost insoluble gases*   For an almost insoluble gas (Figure 8.4), Henry's constant is large (a high gas concentration is necessary to produce only a modest liquid concentration) and the equilibrium curve lies close to the p_A axis. Thus the overall driving force will be approximately equal to the liquid film driving force. Therefore

$$\frac{N_A}{K_L} \simeq \frac{N_A}{k_L} \tag{8.26}$$

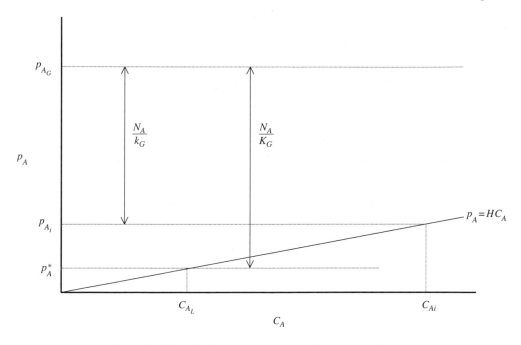

Figure 8.3. Driving forces for mass transfer: very soluble gas.

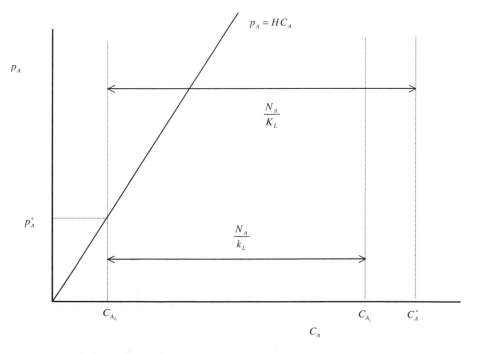

Figure 8.4. Driving forces for mass transfer: almost insoluble gas.

and

$$K_{\rm L} \cong k_{\rm L} \qquad\qquad (8.27)$$

and the overall coefficient approximates to the liquid film coefficient.

8.3.7. Alternative Units for Mass Transfer Coefficients

Because of the many possible units for concentration, a variety of units for mass transfer coefficients exist. There are far too many variations to cover in a chapter of this length, however the most important and commonly encountered alternatives for *gas* coefficients are set out below. This is especially important in correlating coefficients with physical properties and process variables in terms of dimensionless groups.

If the concentration difference in the gas phase is expressed in terms of molar concentration, or molar density ΔC (kmol m^{-3}), the gas film coefficient is given a different symbol $k_{\rm g}$ and is defined by

$$N_{\rm A} = k_{\rm g}\,\Delta C \qquad\qquad (8.28)$$

By inspection of Equation (8.28) the units of $k_{\rm g}$ must now be m s^{-1}; clearly if the units of concentration change the units of the mass transfer coefficient must change also if molar flux is to remain in kmol m^{-2} s^{-1}. Equally, in any given problem, changing the units of concentration will change the numerical value of the mass transfer coefficient as well. Thus $k_{\rm G} \neq k_{\rm g}$.

The relationship between $k_{\rm G}$ and $k_{\rm g}$ can now be obtained by equating expressions for $N_{\rm A}$ and therefore

$$k_{\rm G}\,\Delta p = k_{\rm g}\Delta C \qquad\qquad (8.29)$$

where Δp represents the driving force in terms of a difference in partial pressure. Now, from the ideal gas law,

$$\frac{\Delta p}{RT} = \Delta C \qquad\qquad (8.30)$$

and therefore

$$k_{\rm G} = \frac{k_{\rm g}}{RT} \qquad\qquad (8.31)$$

As well as partial pressure and molar density, concentration can be expressed as a mole fraction and driving forces as differences in mole fraction Δy such that

$$N = k_y\,\Delta y \qquad\qquad (8.32)$$

As before, the units and magnitude of k_y cannot be equal to $k_{\rm G}$. Again, equating expressions for molar flux gives

$$k_y\,\Delta y = k_{\rm G}\,\Delta p \qquad\qquad (8.33)$$

Mole fraction and partial pressure are related by Dalton's law and

$$\Delta y = \frac{\Delta p}{P} \qquad\qquad (8.34)$$

which on substitution into Equation (8.33) gives

$$k_G = \frac{k_y}{P}$$

(8.35)

Equations (8.32) and (8.35) are satisfied by units of $kmol\,m^{-2}\,s^{-1}$ for k_y.

Obviously partial pressures cannot be used to express liquid concentrations and hence the most commonly encountered alternative to molar density for liquid phase concentrations is mole fraction. Thus,

$$N_A = k_x\,\Delta x$$

(8.36)

where k_x has units of $kmol\,m^{-2}\,s^{-1}$.

In summary, for gas film mass transfer coefficients,

$$k_G = \frac{k_g}{RT} = \frac{k_y}{P}$$

(8.37)

Thus, substituting for $k_g = D_G/z_G$ and for $k_L = D_L/z_L$, the expression for addition of film mass transfer coefficients [Equation (8.21)] becomes

$$\frac{1}{K_G} = \frac{z_G\,RT}{D_G} + \frac{H z_L}{D_L}$$

(8.38)

Example 8.3

Air is bubbled into a fermentation broth held at $37°C$ so as to maintain the minimum dissolved oxygen concentration of $5.5 \times 10^{-5}\,kmol\,m^{-3}$ necessary for organism viability. The gas and liquid film mass transfer coefficients are estimated to be 0.01 and $1.6 \times 10^{-4}\,m\,s^{-1}$, respectively. Determine the overall mass transfer coefficient and hence the necessary molar flux of oxygen. Assume that Henry's constant for oxygen and water is $6.6 \times 10^7\,J\,kmol^{-1}$.

No overall mass transfer coefficient is quoted in the data given and therefore K_G must be obtained by adding together the mass transfer resistances. However it is first necessary to convert k_g to k_G. From Equation (8.31)

$$k_G = \frac{0.01}{8314 \times 310}\,kmol\,N^{-1}\,s^{-1}$$

and

$$k_G = 3.88 \times 10^{-9}\,kmol\,N^{-1}\,s^{-1}$$

The overall gas mass transfer coefficient [from Equation (8.21)] is now

$$K_G = \frac{1}{3.88 \times 10^{-9}} + \frac{6.6 \times 10^7}{1.6 \times 10^{-4}}\,kmol\,N^{-1}\,s^{-1}$$

and

$$K_G = 2.42 \times 10^{-12}\,kmol\,N^{-1}s^{-1}$$

TABLE 8.5
Alternative Units for Henry's Constant

	Gas concentration	Liquid concentration	Units of Henry's constant
$p = HC$	Partial pressure (Pa)	Molar density (kmol m^{-3})	J kmol^{-1}
$p = Hx$	Partial pressure (Pa)	Mole fraction (−)	Pa (mole fraction)$^{-1}$
$y = Hx$	Mole fraction (−)	Mole fraction (−)	Dimensionless
$C_G = HC_L$	Molar density (kmol m^{-3})	Molar density (kmol m^{-3})	Dimensionless

The driving force in this problem is the difference between the partial pressure of oxygen in the bubble, equal to 21% of atmospheric pressure, and the pressure in equilibrium with the dissolved oxygen concentration. The former can be found by using Dalton's law:

$$p_{O_2} = 0.21 \times 1.1013 \times 10^5 \, \text{Pa}$$

and

$$p_{O_2} = 21273 \, \text{Pa}$$

The equilibrium concentration is

$$p_{O_2}^* = 6.6 \times 10^6 \times 5.5 \times 10^{-5} \, \text{Pa}$$

and

$$p_{O_2}^* = 3630 \, \text{Pa}$$

Substituting p_{O_2} and $p_{O_2}^*$ into Equation (8.11) gives the oxygen flux as

$$N_{O_2} = 2.42 \times 10^{-12}(21273 - 3630) \, \text{kmol m}^{-2} \, \text{s}^{-1}$$

and therefore

$$N_{O_2} = 4.27 \times 10^{-8} \, \text{kmol m}^{-2} \, \text{s}^{-1}$$

8.3.8. Units of Henry's Constant

Henry's law relates equilibrium concentrations in the gas and liquid phase, respectively, and therefore because there are a number of possible measures of both gas and liquid concentration there is considerable variety in the possible units for Henry's constant. This should not concern unduly the reader of this book but for completeness these are summarised in Table 8.5.

8.4. BINARY DIFFUSION

8.4.1. General Diffusion Equation

The general diffusion equation relates the mass fluxes of the components in a binary system to concentration gradients and diffusivities. It is the starting point for the analysis of mass transfer in any particular geometry, such as mass transfer from a drying droplet or from a gas bubble.

Consider the box of Figure 8.1 divided into unequal sized segments by the removable partition. Liquid A is placed into section I and an equal molar mass of B (of different molar density) is placed into section II so as to give equal depths of liquid when the two compartments are of different size. After removing the partition, and when diffusion has ceased, the concentrations of A and B will be uniform throughout the entire box. Defining the direction I → II as positive, the flux of A relative to the partition is positive and the flux of B relative to the partition is negative. A diffuses to the right (I → II) and there is also a *net* movement to the right (I → II). At steady state, the net flux is the sum of the fluxes of A and B

$$N = N_A + N_B \qquad (8.39)$$

using the symbol N for molar flux, the symbol J being retained only for the molar flux due to molecular diffusion described by Fick's law. The movement of A is made up of two parts:

i   That resulting from the bulk motion which is equal to the product of the mole fraction of A and the net molar flux, that is, $x_A N$
ii That resulting from molecular diffusion, that is, J_A

Adding these together gives

$$N_A = x_A N + J_A \qquad (8.40)$$

and substituting for both N and x_A gives

$$N_A = \frac{C_A}{C}(N_A + N_B) + J_A \qquad (8.41)$$

whilst further substituting for J_A from Fick's law yields

$$N_A = \frac{C_A}{C}(N_A + N_B) - D_{AB}\frac{dC_A}{dz} \qquad (8.42)$$

This is one form of the general diffusion equation for a binary system.

In summary, the molar flux of A relative to fixed co-ordinates is made up of two parts: the 'bulk motion contribution', equal to $(C_A/C)(N_A + N_B)$, which arises from A being carried in the bulk flow of the fluid and second, the 'concentration gradient contribution' $-D_{AB}dC_A/dz$ which is a result of the concentration gradients in A and B.

8.4.2. Other Forms of the General Diffusion Equation

The general diffusion equation can be put into a number of different forms based on the various ways of expressing concentration. First, a simple algebraic rearrangement of Equation (8.42) gives

$$N_A\left(1 - \frac{C_A}{C}\right) - \frac{C_A}{C}N_B = -D_{AB}\frac{dC_A}{dz} \qquad (8.43)$$

$$N_A\frac{C_B}{C} - N_B\frac{C_A}{C} = -D_{AB}\frac{dC_A}{dz} \qquad (8.44)$$

and finally

$$N_A C_B - N_B C_A = -D_{AB}C\frac{dC_A}{dz} \qquad (8.45)$$

Second, substituting partial pressures for molar concentrations in Equation (8.45) gives

$$N_A p_B - N_B p_A = -D_{AB} \frac{P}{RT} \frac{dp_A}{dz} \tag{8.46}$$

Third, rewriting in terms of mole fractions yields (for a gaseous system)

$$N_A y_B - N_B y_A = -C D_{AB} \frac{dy_A}{dz} \tag{8.47}$$

Subsequent derivations for particular circumstances may use any of these forms as a starting point.

8.4.3. Diffusion through a Stagnant Gas Film

Diffusivity can be measured very simply using an Arnold cell. Figure 8.5 represents a narrow vertical tube partially filled with a pure volatile liquid A and maintained at a constant pressure and temperature. Gas B, which is both chemically inert with respect to A and insoluble in A, flows across the open end of the tube. Compound A vaporises and diffuses up through the gas in the tube. Now consider a small element of the tube of dimension δz in the direction of diffusion. A material balance at steady state gives

$$N_A \delta x \, \delta y - \left(N_A + \frac{dN_A}{dz} \delta z \right) \delta x \, \delta y = 0 \tag{8.48}$$

where $(dN_A/dz)\delta z$ is the change in molar flux across the element and $\delta x \, \delta y$ represents the cross-sectional area of the tube. Dividing through by the cross-sectional area gives

$$\frac{dN_A}{dz} = 0 \tag{8.49}$$

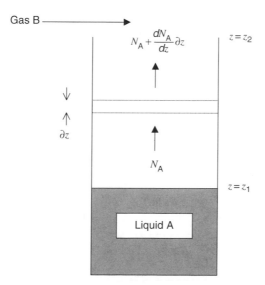

Figure 8.5. Diffusion through a stagnant gas film.

and thus N_A is constant throughout the tube. Similarly N_B must be constant. However at the plane $z = z_1$ (the liquid surface) the flux of B must be zero because B cannot be absorbed by the liquid A (being insoluble and unreactive) and consequently, as the flux of B is constant throughout the column, $N_B = 0$ throughout.

Now rearranging the general diffusion equation [Equation (8.47)]

$$N_A = y_A(N_A + N_B) - CD_{AB}\frac{dy_A}{dz} \tag{8.50}$$

and putting $N_B = 0$ gives

$$N_A = y_A N_A - CD_{AB}\frac{dy_A}{dz} \tag{8.51}$$

or

$$N_A = \frac{-CD_{AB}}{(1 - y_A)}\frac{dy_A}{dz} \tag{8.52}$$

The relationship between flux, concentration and distance along the diffusion path is now obtained by integrating between the boundary conditions, $y_A = y_{A_1}$ at $z = z_1$ and $y_A = y_{A_2}$ at $z = z_2$. Thus the differential equation becomes

$$N_A \int_{z_1}^{z_2} dz = -CD_{AB} \int_{y_{A_1}}^{y_{A_2}} \frac{dy_A}{(1 - y_A)} \tag{8.53}$$

for which the solution is

$$N_A = \frac{CD_{AB}}{(z_2 - z_1)} \ln\left(\frac{1 - y_{A_2}}{1 - y_{A_1}}\right) \tag{8.54}$$

Equation (8.54) allows the molar flux through the tube to be determined as a function of the concentrations at either end of the diffusion path. This relationship is often put into the form

$$N_A = \frac{CD_{AB}(y_{A_1} - y_{A_2})}{(z_2 - z_1)y_{B_{1m}}} \tag{8.55}$$

where $y_{B_{1m}}$ is the logarithmic mean concentration difference in B defined by

$$y_{B_{1m}} = \frac{y_{B_2} - y_{B_1}}{\ln(y_{B_2}/y_{B_1})} \tag{8.56}$$

or, noting that for a binary system the sum of the concentrations of A and B at any point must be unity,

$$y_{B_{1m}} = \frac{(1 - y_{A_2}) - (1 - y_{A_1})}{\ln[(1 - y_{A_2})/(1 - y_{A_1})]} \tag{8.57}$$

The flux of A can be determined by weighing the contents of the cell over a sufficiently long period, the concentration of A at the liquid surface is given by the pure component vapour pressure and the flow of B over the open end of the tube ensures that y_A is zero at that point. Consequently the diffusivity of A in B can be calculated.

Equation (8.55) can also be written in terms of partial pressures, thus

$$N_A = \frac{D_{AB}P(p_{A_1} - p_{A_2})}{RT(z_2 - z_1)p_{B_{1m}}} \tag{8.58}$$

where $p_{B_{1m}}$ is the logarithmic mean partial pressure of B.

Example 8.4

In an experiment to measure the diffusivity of ethanol vapour in air at 20°C, liquid ethanol was placed in an open-ended cylinder, 10 mm in diameter, with its axis vertical and air blown gently across the open end. The interface was 25 mm below the open end of the cylinder and the rate of evaporation was measured by weighing the cylinder and its contents. After five minutes the measured loss in mass was 1.2 mg. Calculate the diffusivity if the vapour pressure of ethanol at 20°C is 5.65 kPa.

The molar flux of ethanol can be determined from the mass evaporation rate, the relative molecular mass of the liquid and the cross-sectional area of the tube. Therefore, writing the subscript A for ethanol and B for air,

$$N_A = \frac{1.2 \times 10^{-6} \times 4}{46 \times 300 \times \pi (0.01)^2} \text{ kmol m}^{-2} \text{ s}^{-1}$$

$$N_A = 1.107 \times 10^{-6} \text{ kmol m}^{-2} \text{ s}^{-1}$$

The partial pressure of ethanol at $z = z_1$ (the liquid surface) is given by the pure component vapour pressure of ethanol at 20°C, thus $p_{A_1} = 5.65$ kPa and, in a binary system, with a total pressure of 101.3 kPa, $p_{B_1} = (101.3 - 5.65) = 95.65$ kPa. At the top of the tube the concentration of ethanol is negligible because of the stream of air blown across the open end and therefore $p_{A_2} = 0$ and $p_{B_2} = 101.3$ kPa. Thus the logarithmic mean partial pressure of air is

$$p_{B_{lm}} = \frac{101.3 - 95.65}{\ln(101.3/95.65)} \text{ kPa}, \qquad p_{B_{lm}} = 98.45 \text{ kPa}$$

Now from Equation (8.58), the diffusivity of A in B is

$$D_{AB} = \frac{1.107 \times 10^{-6} \times 8314 \times 293 \times 0.025 \times 98.45 \times 10^3}{101.3 \times 10^3 \times (5.65 - 0) \times 10^3} \text{ m}^2 \text{ s}^{-1}$$

and

$$D_{AB} = 1.16 \times 10^{-5} \text{ m}^2 \text{ s}^{-1}$$

8.4.4. Particles, Droplets and Bubbles

Mass transfer to and from spheres, or bodies which approximate to spheres, has important implications in food processing. Many examples may be quoted: the evaporation of water from atomised droplets in a spray drier, solvent extraction from dispersed oil droplets, the dissolution of spheroidal particles or granules and the transfer of oxygen from air bubbles to a fermentation broth. Consequently important insights into convective mass transfer can be gained by a close study of diffusion from a sphere.

Figure 8.6 represents a sphere of volatile material A which is placed in a stagnant environment of fluid B where there is no convective mass transfer. Diffusion will take place radially outwards and thus the general diffusion equation can be written for this case as

$$N_A C_B - N_B C_A = -D_{AB} C \frac{dC_A}{dr} \tag{8.59}$$

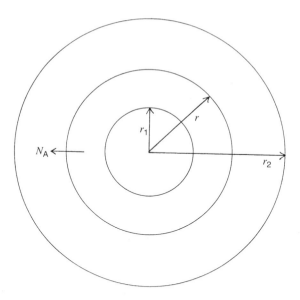

Figure 8.6. Mass transfer from a sphere.

In a stagnant environment the flux of B will be zero and therefore

$$N_A = -\frac{D_{AB}C}{C - C_A}\frac{dC_A}{dr} \tag{8.60}$$

The molar flux of A from the sphere is equal to the rate of evaporation ω divided by the surface area of the sphere and is therefore a function of radius

$$N_A = \frac{\omega}{4\pi r^2} \tag{8.61}$$

Substituting Equation (8.61) into Equation (8.60) and integrating gives

$$\frac{\omega}{4\pi}\int_{r_1}^{r_2}\frac{dr}{r^2} = -D_{AB}\int_{C_1}^{C_2}\frac{C}{C - C_A}dC_A \tag{8.62}$$

where the concentrations of A are C_1 and C_2 at the specific radii r_1 and r_2, respectively. Thus

$$\frac{\omega}{4\pi}\left(\frac{1}{r_1} - \frac{1}{r_2}\right) = D_{AB}C\ln\left(\frac{C - C_2}{C - C_1}\right) \tag{8.63}$$

If now A diffuses from the sphere into an infinite medium then r_2 must equal infinity and

$$\omega = 4\pi\,r_1\,D_{AB}C\ln\left(\frac{C - C_2}{C - C_1}\right) \tag{8.64}$$

This expression can be simplified by making the reasonable assumption that both C_1 and C_2 are significantly smaller than C which is the total molar concentration in the system, in other words if the concentration of A at any point is very much smaller than that of B. It follows that Equation (8.64) can be reduced[*] to

$$\omega = 4\pi \, r_1 \, D_{AB} \, (C_1 - C_2) \tag{8.65}$$

What follows next is an extremely important result for the prediction of mass transfer coefficients involving spherical bubbles, droplets, and particles. If the radius of the sphere is now r_1 then the molar flux at the surface becomes, from Equations (8.61) and (8.65),

$$N_A = \frac{D_{AB}}{r_1} (C_1 - C_2) \tag{8.66}$$

Flux of course depends upon the product of concentration difference and mass transfer coefficient and therefore from Equation (8.66) the mass transfer coefficient k_g is given by

$$k_g = \frac{D_{AB}}{r_1} \tag{8.67}$$

This is better put in the form

$$\frac{k_g d}{D_{AB}} = \frac{D_{AB} d}{D_{AB} r_1} \tag{8.68}$$

The left-hand side of Equation (8.68) is the Sherwood number which is defined by

$$Sh = \frac{k_g L}{D} \tag{8.69}$$

and where in the case of a sphere the characteristic length L is the diameter d. The right-hand side of Equation (8.68) is equal to the diameter divided by the radius and this result can be summarised as

$$Sh = 2 \tag{8.70}$$

The Sherwood number is the mass transfer equivalent of the Nusselt number in heat transfer and is a measure of the ratio of convective mass transfer, represented by the mass transfer coefficient, to molecular diffusion represented by the diffusivity. Equation (8.70) indicates that the limiting value of Sh for mass transfer from a sphere is equal to 2. This is the minimum position with no convection; any positive relative velocity between the sphere and its surroundings gives rise to convection currents which increase Sh, increase the value of the film mass transfer coefficient and improve mass transfer.

Example 8.5

During pilot plant tests, water is atomised within a spray drying chamber to form 200 μm droplets. The temperature of the air in the drier is 88°C and the absolute humidity is such that the temperature of the water droplets is 30°C. Conditions are such that the Sherwood number for a droplet is 10. Determine the film mass transfer coefficient and hence estimate the evaporation rate from a single droplet and the corresponding drying time assuming that the evaporation rate remains constant and is not a function of droplet diameter. Assume the diffusivity of water in air to be $3.0 \times 10^{-5} \, \mathrm{m^2 \, s^{-1}}$.

---

[*]Using $\ln(1 - x) = -x - \frac{x^2}{2} + \frac{x^3}{3} - \cdots$

As in Example 8.1, only one film resistance is involved here. The droplet is pure water and therefore the only resistance to mass transfer is the air film around the droplet across which the water vapour concentration falls from the saturated value to that of the surrounding air. As before, the humidity of the air is converted to a mole fraction y_w

$$y_w = \frac{(0.002/18)/(1/29)}{1 + (0.002/18)/(1/29)} = 0.00321$$

giving the partial pressure of water in the bulk air stream (using Dalton) as

$$p_w = 0.00321 \times 1.013 \times 10^5 \, \text{Pa}, \qquad p_w = 325.4 \, \text{Pa}$$

From steam tables the partial pressure at saturation p_{wo} is 4242 Pa.

The film mass transfer coefficient can be obtained from the Sherwood number, thus

$$k_g = \frac{Sh\,D}{d}, \qquad k_g = \frac{10 \times 3 \times 10^{-5}}{200 \times 10^{-6}} \, \text{m s}^{-1}$$

or

$$k_g = 1.5 \, \text{m s}^{-1}$$

Converting this to k_G using Equation (8.31) gives

$$k_G = \frac{1.5}{8314 \times 361} \, \text{kmol N}^{-1} \, \text{s}^{-1}$$

or

$$k_G = 5.0 \times 10^{-7} \, \text{kmol N}^{-1} \, \text{s}^{-1}$$

and therefore the molar flux of water becomes

$$N_w = 5.0 \times 10^{-7} (4242 - 325.4) \, \text{kmol m}^{-2} \, \text{s}^{-1}$$

or

$$N_w = 1.958 \times 10^{-3} \, \text{kmol m}^{-2} \text{s}^{-1}$$

The molar evaporation rate ω is therefore the product of the molar flux and the surface area of the droplet πd^2 and

$$\omega = 1.958 \times 10^{-3} \pi \times (200 \times 10^{-6})^2 \, \text{kmol s}^{-1}$$

or

$$\omega = 2.46 \times 10^{-10} \, \text{kmol s}^{-1}$$

Assuming that the rate of evaporation is not a function of droplet size (which is not strictly correct) then the time for complete evaporation t is equal to the molar mass of the droplet divided by the evaporation rate. Taking the density of water to be $1,000 \, \text{kg m}^{-3}$ then

$$t = \frac{1000}{18} \frac{\pi}{6} \frac{(200 \times 10^{-6})^3}{2.46 \times 10^{-10}} \, \text{s}, \qquad t = 0.95 \, \text{s}$$

8.5. CORRELATIONS FOR MASS TRANSFER COEFFICIENTS

Film mass transfer coefficients can be correlated with physical properties and process variables in a manner which is analogous to that for heat transfer. The relevant dimensionless groups are the Sherwood number, the Reynolds number and the Schmidt number which is defined by

$$Sc = \frac{\mu}{\rho D} \tag{8.71}$$

The Schmidt number is the ratio of kinematic viscosity to diffusivity and is therefore a measure of the effectiveness of momentum transfer to mass transfer by diffusion. In chapter seven we saw that heat transfer correlations for given geometries can be developed on the basis of dimensional analysis and experimentation. In the same way mass transfer correlations usually take the form

$$Sh = a Re^b Sc^c \tag{8.72}$$

and, for similar geometries, the exponents on Re and Sc for mass transfer are often numerically similar to the exponents on Re and Pr, respectively, in heat transfer. For example, for mass transfer between a fluid and the walls of a circular cross-section tube, such as a tube in a climbing film evaporator, an expression similar in form to the Dittus–Boelter equation can be used:

$$Sh = 0.023 Re^{0.8} Sc^{0.4} \tag{8.73}$$

As we have seen, for a sphere in stagnant surroundings $Sh = 2$ and the mass transfer coefficient is proportional to diffusivity as predicted by the two-film theory. This diffusional component to mass transfer is always present but is dominated by convection in turbulent systems. For a particle in a turbulent fluid the correlation becomes

$$Sh = 2 + 0.13 Re^{0.75} Sc^{0.33} \tag{8.74}$$

where the second term represents convective mass transfer and the term $Sh = 2$ is negligible in comparison and is often therefore omitted. Typical correlations for particles, droplets and bubbles are summarised in Table 8.6.

Examples which may be classed as spheres with rigid interfaces are solid particles, cells and small gas bubbles (<2.5 mm in diameter) in aqueous solution. Spheres with mobile interfaces might include oscillating liquid droplets and larger gas bubbles. When a particle or droplet is in free suspension, that is, in natural or free convection, the drag and buoyancy

TABLE 8.6

Correlations for Mass Transfer to Spheres

	Rigid interfaces	Mobile interfaces
Stagnant environment	$Sh = 2$	$Sh = 2$
Free suspension	$Sh = 2 + 0.31(Gr\,Sc)^{0.33}$	$Sh = 2 + 0.42(Gr^{0.33}\,Sc^{0.5})$
Turbulent flow	$Sh = 2 + 0.13 Re^{0.75} Sc^{0.33}$	$Sh = 2 + 1.31(Re\,Sc)^{0.5}$

forces on the sphere are in equilibrium and the Grashof number Gr for mass transfer is used rather than Re, where

$$Gr = \frac{L^3 \rho g \, \Delta\rho}{\mu^2} \qquad (8.75)$$

and $\Delta\rho$ is the density difference between fluid and particle.

The Ranz–Marshall equation is often used to describe mass transfer from freely falling droplets produced by atomisation in pressure or twin-fluid nozzles, for example in spray drying. This relationship takes the form

$$Sh = 2 + 0.60 \, Re^{0.50} \, Sc^{0.33} \qquad (8.76)$$

Example 8.6

Use the Ranz–Marshall correlation to estimate the film gas mass transfer coefficient for the evaporation of water from atomised milk droplets, 300 µm in diameter, falling in air at 77°C with a relative velocity between the air and the droplets of $0.5 \, \mathrm{m \, s^{-1}}$.

The Reynolds number for a droplet is given by

$$Re = \frac{\rho u d}{\mu}$$

where the length d is the droplet diameter and the density and viscosity are those of air; at 77°C (350 K), $1.009 \, \mathrm{kg \, m^{-3}}$ and $2.075 \times 10^{-5} \, \mathrm{Pa \, s}$ respectively. Thus

$$Re = \frac{1.009 \times 0.5 \times 300 \times 10^{-6}}{2.075 \times 10^{-5}}, \qquad Re = 7.29$$

The diffusivity of water vapour in air is needed in order to find the Schmidt number; from Table 8.2 the diffusivity at 333 K is $3.05 \times 10^{-5} \, \mathrm{m^2 \, s^{-1}}$. Diffusivities in the gas phase are proportional to absolute temperature to the power 1.5 and therefore at 350 K

$$D = \left(\frac{350}{333}\right)^{1.5} 3.05 \times 10^{-5} \, \mathrm{m^2 \, s^{-1}}$$

and

$$D = 3.29 \times 10^{-5} \, \mathrm{m^2 \, s^{-1}}$$

Thus the Schmidt number is

$$Sc = \frac{2.075 \times 10^{-5}}{1.009 \times 3.29 \times 10^{-5}} \quad \text{or} \quad Sc = 0.625$$

The Ranz–Marshall correlation [Equation (8.76)] gives the Sherwood number as

$$Sh = 2 + 0.60 \left[(7.29)^{0.50} (0.625)^{0.33} \right], \qquad Sh = 3.39$$

Hence the gas film coefficient is

$$k_g = \frac{D}{d} Sc$$

from which

$$k_g = \left[\frac{3.29 \times 10^{-5}}{300 \times 10^{-6}} \right] 3.39 \, \mathrm{m \, s^{-1}}, \qquad k_g = 0.372 \, \mathrm{m \, s^{-1}}$$

NOMENCLATURE

a	Index
b	Index
c	Coefficient
C	Molar concentration
C_{A_i}	Concentration of A (liquid phase) at the interface
C_{A_L}	Concentration of A in the bulk liquid
C_A^*	Concentration of A in equilibrium with the bulk gas partial pressure
d	Diameter
D	Diffusivity or diffusion coefficient
D_{AB}	Diffusivity of A in B
D_G	Diffusivity in the gas phase
D_L	Diffusivity in the liquid phase
g	Acceleration due to gravity
Gr	Grashof number
H	Henry's constant
J_A	Molar flux of A
k_g	Film gas mass transfer coefficient
k_G	Film gas mass transfer coefficient
K_G	Overall gas mass transfer coefficient
k_L	Film liquid mass transfer coefficient
K_L	Overall liquid mass transfer coefficient
k_x	Film liquid mass transfer coefficient
k_y	Film gas mass transfer coefficient
L	Characteristic length
M_A	Molecular weight of component A
N	Molar flux
N_w	Molar flux of water vapour
p_A	Partial pressure of A
p_{A_G}	Partial pressure of A in bulk gas
p_A^*	Partial pressure in equilibrium with bulk liquid concentration
p_{A_i}	Partial pressure of A at interface
p_w	Partial pressure of water vapour
p_{wo}	Partial pressure of water vapour at saturation
$p_{B_{lm}}$	Logarithmic mean partial pressure difference in B
P	Pressure
Pr	Prandtl number
r	Radius
R	Universal gas constant
Re	Reynolds number
Sc	Schmidt number
Sh	Sherwood number
T	Absolute temperature
z	Dimension in the z direction
z_G	Thickness of gas film
z_L	Thickness of liquid film

x Mass or mole fraction in liquid phase
y Mass or mole fraction in gas phase
$y_{B_{lm}}$ Logarithmic mean concentration difference in B

GREEK SYMBOLS

ΔC Concentration difference
Δp Partial pressure difference
Δx Mole fraction difference (liquid phase)
Δy Mole fraction difference (gas phase)
$\Delta \rho$ Density difference
μ Viscosity
ρ Mass density
ρ_A Mass density of A
ω Molar rate of evaporation

SUBSCRIPTS

A   Component A
B   Component B
w   Water

PROBLEMS

8.1. Convert the gas film mass transfer coefficient in Example 8.1 to units of $m\,s^{-1}$ if the air temperature is $47°C$.

8.2. The mass transfer characteristics of a drier were determined in an experiment in which mashed potato, saturated with water, was formed into a horizontal insulated slab with its upper surface exposed to a stream of warm air. The wet and dry bulb temperatures of the air at steady-state were $25°C$ and $60°C$, respectively. Assuming constant rate drying, and a film heat transfer coefficient of $120\,W\,m^{-2}\,K^{-1}$, calculate the heat flux between the air and the slab, the molar flux of water from the slab and the gas film mass transfer coefficient (in molar units). Take atmospheric pressure to be $101.3\,kPa$.

8.3. A $2\,m^3$ fermenter is sparged with air at 1.8 bar such that oxygen is transferred into the liquid phase at a rate of $1.6 \times 10^{-5}\,kmol\,s^{-1}$. The fermentation broth is held at $37°C$ and agitation is sufficient to give a bubble surface area of $100\,m^2$ per m^3 of fermenter volume and film mass transfer coefficients of $k_g = 0.009\,m\,s^{-1}$ and $k_L = 1.6 \times 10^{-4}\,m\,s^{-1}$. Determine by calculation whether the resultant dissolved oxygen concentration meets the target of $7 \times 10^{-5}\,kmol\,m^{-3}$ to maintain micro-organism viability. Assume Henry's constant for oxygen/solution to be $6.8 \times 10^7\,J\,kmol^{-1}$.

8.4. Calculate the mass evaporation rate from a 50% by mass aqueous solution of sucrose (molecular weight = 342) maintained at a temperature of $60°C$ in a 5 cm diameter beaker if the liquid surface is 3 cm below the top of the vessel. The surrounding air has a relative humidity of 20% and atmospheric pressure is 101.3 kPa.

FURTHER READING

E. L. Cussler, *Diffusion: Mass Transfer in Fluid Systems*, Cambridge University Press, Cambridge (1984).

J. M. Kay, and R. M. Nedderman, *Fluid Mechanics and Transfer Processes*, Cambridge University Press, Cambridge (1985).

M. A. Rao, and S. S. H. Rizvi, *Engineering Properties of Foods*, Marcel Dekker, New York (1995).

R. E. Treybal, *Mass Transfer Operations*, McGraw Hill, New York (1981).

9

Psychrometry

9.1. INTRODUCTION

Psychrometry is concerned with the behaviour of humid air and the prediction of its properties. More strictly it covers the behaviour of any vapour (not just water vapour) when mixed with a gas (not just air). However, because the air/water system is of huge importance, not least to food processing, and is the most commonly encountered gas/vapour mixture, the terminology employed often appears to be specific to air and water. Prediction of the properties of moist air, for example humidity, maximum possible humidity, temperature, density and so on, is especially important in drying operations. The psychrometric chart is a simple graphical method of presenting this information, and the principles behind the chart are explained in this chapter. The outcome of a drying process can often be followed more easily by monitoring the condition of the air with the aid of a psychrometric chart than by directly measuring the moisture content of the substance being dried. Of equal interest is the establishment of the correct atmospheric conditions for food storage.

9.2. DEFINITIONS OF SOME BASIC QUANTITIES

9.2.1. Absolute Humidity

Absolute humidity H is the mass of water vapour associated with unit mass of dry air. This is often referred to as simply *humidity* or sometimes as *moisture content* although the latter is more appropriate for the water content of solids. Humidity is a mass ratio and as such is a dimensionless quantity. However the units are often quoted as kg of water vapour per kg of dry air, that is, $kg\,kg^{-1}$, to emphasise the difference between humidity and a mass fraction. Assuming ideal behaviour, the mass of vapour per unit volume of air is given by the ideal gas law and

$$\text{mass of vapour} = \frac{p_w M_w}{RT} \tag{9.1}$$

where p_w is the partial pressure of water vapour and M_w is the molecular weight of water. If the total system pressure is P, then the partial pressure of the air will be $P - p_w$ and the mass of dry air is then

$$\text{mass of dry gas} = \frac{(P - p_w)M_A}{RT} \tag{9.2}$$

where M_A is the mean molecular weight of the air. Thus humidity becomes

$$H = \frac{p_w M_w}{(P - p_w) M_A} \qquad (9.3)$$

Example 9.1

A sample of humid air has an absolute humidity of 0.01 kg kg^{-1} dry air. What is the partial pressure of the water vapour if atmospheric pressure is 101.3 kPa?

Assuming the molecular weights of water and air to be 18 and 29, respectively, Equation (9.3), based on the ideal gas law gives

$$0.01 = \frac{p_w 18}{(101.3 - p_w) 29}$$

from which

$$p_w = 1.606 \, \text{kPa}$$

9.2.2. Saturated Humidity

At any given temperature there is a limit to the quantity of water vapour which air can hold before the vapour condenses back into liquid water. Saturated humidity H_o is the absolute humidity of air when it is saturated with water vapour at a given temperature; saturated humidity increases with temperature. If p_{wo} is the partial pressure of water vapour at saturation then, from Equation (9.3) saturated humidity is given by

$$H_o = \frac{p_{wo} M_w}{(P - p_{wo}) M_A} \qquad (9.4)$$

Example 9.2

Calculate the humidity of air saturated with water vapour at 30°C and standard atmospheric pressure.

At 30°C the pure component vapour pressure of water (i.e., p_{wo}) is 4.242 kPa. Therefore, from Equation (9.4),

$$H_o = \frac{4.242 \times 18}{(101.3 - 4.242) 29}, \qquad H_o = 0.0271 \, \text{kg kg}^{-1}$$

9.2.3. Percentage Saturation

This is a measure of the humidity of air compared to the saturated humidity at the same temperature. It is defined by

$$\text{percentage saturation} = \frac{H}{H_o} \times 100 \qquad (9.5)$$

and is also called *percentage absolute humidity* or *percentage humidity*. It should not be confused with relative humidity.

9.2.4. Relative Humidity

There are two distinct way of expressing humidity relative to saturation conditions. Relative humidity is variously called *percentage relative humidity* or simply *%RH* and should not be confused with the previous definition. Again it is a measure of the humidity of air relative to saturation conditions, but it is defined in terms of partial pressures:

$$\text{relative humidity} = \frac{p_\text{w}}{p_\text{wo}} \times 100 \tag{9.6}$$

Percentage saturation and percentage relative humidity are *not* the same and the terms are *not* interchangeable.

9.2.5. Relationship between Percentage Saturation and Relative Humidity

The relationship between percentage saturation and relative humidity can be obtained simply, as follows: substitution of the definitions of H and H_o into Equation (9.5) gives

$$\text{percentage saturation} = \left(\frac{P - p_\text{wo}}{P - p_\text{w}}\right) \frac{p_\text{w}}{p_\text{wo}} \times 100 \tag{9.7}$$

It can be seen from Equation (9.7) that percentage saturation is equal to %RH multiplied by the quantity $(P - p_\text{wo})/(P - p_\text{w})$. In practice the values do not differ greatly but are equal only when this quantity is equal to unity. This occurs at two conditions:

   i   when the gas is almost, or completely, saturated and $p_\text{w} \approx p_\text{wo}$.
  ii   when the gas is almost, or completely, dry and $p_\text{w} \approx p_\text{wo} \approx 0$.

Example 9.3

The partial pressure of water vapour in a storage area, maintained at a temperature of 25°C and a pressure is 101.3 kPa, is 2 kPa. Calculate the relative humidity and the percentage saturation.

At 25°C the pure component vapour pressure of water (i.e. p_wo) is 3.166 kPa. Therefore from the definition of relative humidity

$$\%\text{RH} = \frac{2}{3.166} \times 100 \ \text{ or } \ \%\text{RH} = 63.2\%$$

Now the definition of percentage saturation, Equation (9.7), gives

$$\text{percentage saturation} = \left(\frac{101.3 - 3.166}{101.3 - 2}\right) \frac{2}{3.166} \times 100$$

and therefore

$$\text{percentage saturation} = 62.4\%$$

This example highlights the significance of the difference in relative humidity and percentage saturation for a given mixture of water vapour and air.

9.2.6. Humid Heat

This is simply the heat capacity of moist air. On some psychrometric charts it is possible to read values of humid heat from an auxiliary line. Alternatively, the heat capacity of moist air ($kJ\,kg^{-1}\,K^{-1}$) can be estimated from

$$c_p = 1.005 + 1.88H \qquad (9.8)$$

9.2.7. Humid Volume

Humid volume is the volume occupied by unit mass of dry gas and its associated vapour. This is sometimes called *specific volume* and is simply the reciprocal of density.

9.2.8. Dew point

The dew point is the temperature at which the gas is saturated with vapour. Consider an unsaturated mixture of air and water vapour which is cooled. As the temperature falls the relative quantities of air and water vapour remain unchanged, in other words if the total pressure is constant the partial pressure of water will remain constant. However when the temperature is such that the pure component vapour pressure of the liquid equals the existing partial pressure of the vapour then the gas will be saturated. This temperature is known as the dew point. Further cooling, below the dew point, will result in condensation.

9.3. WET BULB AND DRY BULB TEMPERATURES

9.3.1. Definitions

One of the simplest ways to determine absolute humidity is to measure the wet and dry bulb temperatures of the gas and read off absolute humidity from a psychrometric chart. Section 9.4 covers the details of how to read the chart but it is first necessary to define the terms wet bulb and dry bulb. Figure 9.1 represents a stream of unsaturated air passing over the surface of a mass of water. Due to the difference in water vapour concentration, water will evaporate and increase the humidity of the air. As evaporation takes place the temperature of the water will fall (i.e., 'evaporation causes cooling') and there will be a temperature difference between the air and the water. This temperature difference results in heat transfer to the water which in turn causes further evaporation. At equilibrium, the heat which is transferred from the air is just sufficient to vaporise the water and the temperature of both air and water become constant. The equilibrium temperature attained by the water is known as the wet bulb temperature, T_w, and that attained by the air is called the dry bulb temperature, T. Unless the air is saturated (when the two temperatures will be equal), the wet bulb is always lower than the dry bulb temperature.

9.3.2. The Wet Bulb Equation

Humidification is an example of a process involving simultaneous heat and mass transfer. The basis of psychrometry is the relationship between temperature and humidity which can be obtained by combining the relevant heat and mass transfer equations. The result is the wet

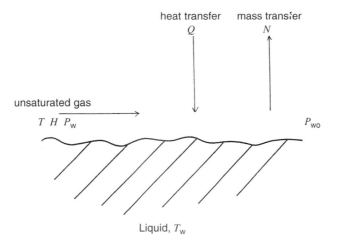

Figure 9.1. Humidification of an unsaturated gas.

bulb equation from which humidity can be calculated as a function of the wet and dry bulb temperatures.

The heat flux resulting from the temperature difference between air and water is described by

$$q = h(T - T_w) \tag{9.9}$$

where h is the relevant film heat transfer coefficient and the mass flux due to the diffusion of water vapour from just above the liquid surface, where the partial pressure of water vapour is p_{wo}, to the air stream is

$$N = \frac{k_g M_w}{RT}(P_{wo} - p_w) \tag{9.10}$$

The molecular weight of water is included here to convert from a molar to a mass flux. Next it is necessary to replace partial pressure as a measure of concentration with humidity. From Equation (9.3) the partial pressure of water vapour is

$$p_w = H(P - p_w)\frac{M_A}{M_w} \tag{9.11}$$

Substituting for p_w and, in the same way, for p_{wo} from Equation (9.4) into the expression for mass flux gives

$$N = \frac{k_g M_A}{RT}[H_o(P - p_{wo}) - H(P - p_w)] \tag{9.12}$$

To simplify this expression the mean partial pressure of the humid gas p_g can be substituted for both $(P - p_w)$ and $(P - p_{wo})$ to give

$$N = \frac{k_g M_A p_g}{RT}[H_o - H] \tag{9.13}$$

in which the driving force term is now a difference in humidity. If the ideal gas law is used to replace the mean partial pressure with the mean gas density ρ_g, that is,

$$\rho_g = \frac{p_g M_A}{RT} \tag{9.14}$$

then the mass transfer equation can be reduced to the form

$$N = k_g \rho_g (H_w - H) \tag{9.15}$$

In Equation (9.15) H_o has been replaced with H_w to represent the saturated humidity at the wet bulb temperature. The heat and mass fluxes are related by the latent heat of vaporisation of water such that

$$q = N h_{fg} \tag{9.16}$$

and therefore the relationship between temperature and humidity becomes

$$h(T - T_w) = k_g \rho_g (H_w - H) h_{fg} \tag{9.17}$$

Rearranging this to be explicit in H gives the wet bulb equation:

$$H - H_w = -\frac{h(T - T_w)}{k_g \rho_g h_{fg}} \tag{9.18}$$

The measurement of T and T_w and a knowledge of the other quantities in Equation (9.18) allow the humidity to be determined. Wet and dry bulb temperatures can be measured very easily with two similar mercury-in-glass thermometers, one of which has a piece of saturated muslin wrapped around the bulb, thus simulating the experiment illustrated in Figure 9.1. The sling hygrometer makes use of two such thermometers mounted on a rattle which is rotated rapidly in the air. Alternatively, and particularly for use in process streams, thermocouples may be substituted for the glass thermometers.

9.3.3. Adiabatic Saturation Temperature

If the gas is passed over the liquid surface at such a rate that equilibrium can be established, the gas becomes saturated and both phases attain the same temperature. However in an adiabatic (i.e., thermally insulated) system the sensible heat of the gas will fall by an amount equal to the latent heat of the liquid being evaporated. The equilibrium temperature of the liquid is now known as the adiabatic saturation temperature T_s. Whereas in deriving the wet bulb equation the *rates* of heat and mass transfer were equated, in an adiabatic system it is appropriate to write an energy balance. Hence, per unit mass of material,

$$c_p(T - T_s) = h_{fg}(H_s - H) \tag{9.19}$$

where c_p is the heat capacity (humid heat) of the mixture, which is almost constant for small changes in H, and H_s is the saturated humidity at T_s. Rearranging Equation 9.19 produces

$$(H - H_s) = -\frac{c_p}{h_{fg}}(T - T_s) \tag{9.20}$$

which is an approximately linear relationship between humidity and temperature for all mixtures having the same adiabatic saturation temperature. A plot of humidity against temperature is called an adiabatic cooling line and, for different temperatures, a series of approximately linear adiabatic cooling lines can be drawn, each with a gradient equal to $-c_p/h_{fg}$.

9.3.4. Relationship between Wet Bulb Temperature and Adiabatic Saturation Temperature

It will have been noticed that the wet bulb and adiabatic saturation equations have a similar form. Indeed the wet bulb and adiabatic saturation temperatures are equal when the gradients in Equations (9.18) and (9.20) are equal, that is, when

$$c_p = \frac{h}{k_g \rho_g} \tag{9.21}$$

The quantity $h/k_g \rho_g c_p$, the ratio of the two gradients, is called the psychrometric ratio and has a value of approximately unity for the air/water system but has significantly higher values for non-air/water mixtures. For example, for mixtures of organic liquids and air the values are usually in the range 1.3–2.5. Thus for air/water mixtures only $T = T_s$ and 'cooling at a constant wet bulb temperature' and 'adiabatic cooling' are synonymous.

9.4. THE PSYCHROMETRIC CHART

9.4.1. Principles

A psychrometric chart can be used to follow changes in the condition of moist air during processing operations such as drying or humidification which involve mixtures of air and vaporised water. The chart is a way of representing graphically the adiabatic cooling (or wet bulb) line. Essentially therefore it is a plot of absolute humidity (vertical axis) against dry bulb temperature (horizontal axis) with a series of curves representing particular values of percentage saturation superimposed. On some charts relative humidity is used in place of percentage saturation. The wet bulb temperature scale is shown along the saturated humidity curve. The chart is represented diagrammatically in Figure 9.2 and a detailed chart is reproduced in Figure 9.3. It is possible to add a great deal of other information but this can serve to make the chart difficult to read. The example in Figure 9.3 has been chosen because of its clarity; it includes a series of steeply sloping lines which are marked 'specific volume m^3/kg' and an enthalpy scale. The latter is used by aligning the point of interest with two scale marks showing equal values, one on either side of the chart. Note that interpolation may be necessary with both the percentage saturation and specific volume scales. A knowledge of any two quantities (e.g., wet bulb temperature, percentage saturation, etc.) is required to specify the condition of humid air and from a point on the chart representing this condition it is possible to read off any other property.

Example 9.4

A sample of air has a dry bulb temperature of 35°C and a wet bulb temperature of 30°C. Using a psychrometric chart, determine the absolute humidity, percentage saturation, dew point and specific volume of the air.

The point representing the sample is located at the intersection of a vertical line from $T = 35°C$ on the dry bulb temperature axis with a line representing $T_w = 30°C$. From this

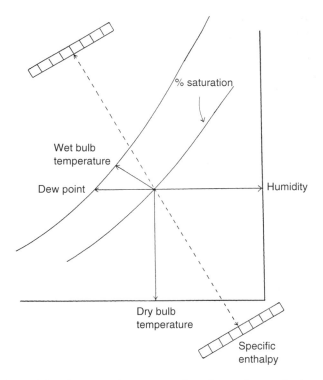

Figure 9.2. Diagrammatic representation of a psychrometric chart.

point the following values can be read:

$$H = 0.025\,\text{kg kg}^{-1}$$

$$\text{percentage saturation} = 68\%$$

$$\text{dew point} = 28.5^{\circ}\text{C}$$

$$\text{specific volume} = 0.908\,\text{m}^{3}\,\text{kg}^{-1}$$

Example 9.5

Use a psychrometric chart to find the wet bulb temperature and specific enthalpy of air with an absolute humidity of 0.015 kg per kg and a dry bulb temperature of 39°C.

A point at the intersection of $T = 39^{\circ}$C and $H = 0.015$ has a specific enthalpy of 78 kJ kg^{-1} and wet bulb temperature (by interpolation) of 25.6°C.

Only four processes can be represented on a psychrometric chart (if the mixing of two air streams is excluded). These processes are represented in Figure 9.4.

i *Heating (A → B)*

The addition of sensible heat to moist air increases its temperature but the absolute humidity remains constant. However, the percentage saturation will decrease during heating because the air has a greater capacity to hold moisture as the temperature is increased.

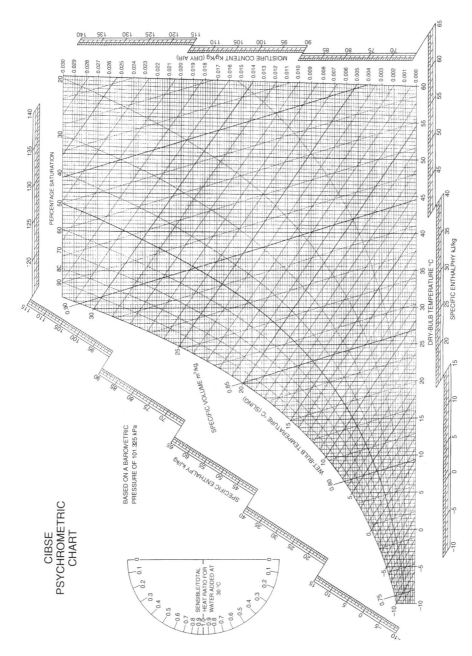

Figure 9.3. Psychrometric chart (reproduced by permission of the Chartered Institution of Building Services Engineers, London).

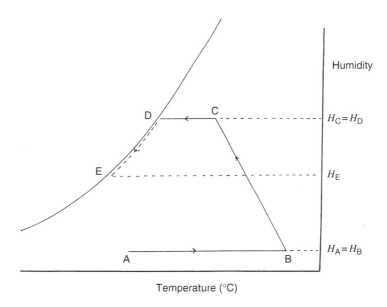

Figure 9.4. Psychrometric chart: heating, humidification, cooling and condensation processes.

ii *Humidification (B → C)*

The humidification of air can be followed along an adiabatic cooling line (or line of constant wet bulb temperature). As air picks up moisture the dry bulb temperature decreases and the percentage saturation increases. Humidification can be continued until the air is saturated at which point (on the saturation curve) the wet and dry bulb temperatures will be equal. However in the example of Figure 9.4 humidification ceases at a point C before saturation is reached. The mass of water vapour taken up by unit mass of dry air is then $H_C - H_B$.

iii *Cooling (C → D)*

If heat is removed from moist air the dry bulb temperature will fall, at a constant absolute humidity, until the saturation curve is reached. The temperature at this point is the dew point, the temperature at which droplets of water begin to condense from the air (point D in Figure 9.4).

iv *Condensation (D → E)*

If further heat is removed from a stream of saturated air both cooling and continued condensation will result. As water is condensed the air remains saturated and therefore this process can be traced along the saturation curve. The mass of water vapour condensed from unit mass of dry air is then $H_D - H_E$.

Example 9.6

Air, originally at 40°C and 10% humidity, is cooled adiabatically by contacting it with water, which is at the wet bulb temperature of the gas. What is the lowest temperature to which the air may be cooled and how much water is vaporised per kg of dry air in reaching this temperature?

The lowest temperature to which the air may be cooled is the wet bulb temperature of the air, that is, 19°C. At this point the air is saturated and the mass of water vaporised per kg of dry air is equal to the difference in humidity which equals (0.0138–0.0049) kg kg^{-1} or 0.0089 kg water per kg of dry air.

9.4.2. Mixing of Humid Air Streams

If two humid air streams (1 and 2, respectively) are mixed the humidity of the mixture (stream 3) can be determined from a material balance. Hence

$$m_1 H_1 + m_2 H_2 = m_3 H_3 \tag{9.22}$$

where m is the mass flow rate of *dry* air and

$$m_1 + m_2 = m_3 \tag{9.23}$$

Such problems can be solved graphically using the *inverse lever rule*. Referring to Figure 9.5, the point which represents the mixture of two streams must lie on a straight line joining the points representing the two unmixed streams. Further, the mass flow rate of an unmixed stream is inversely proportional to the distance between the points representing itself and the mixture, respectively. Therefore

$$\frac{m_1}{m_2} = \frac{a}{b} \tag{9.24}$$

Whilst this rule can be proved algebraically, it is necessary only to apply common sense to verify the truth of it. If m_1 is large compared to m_2 then point 3 (the mixture) will approach point 1 and the ratio a/b will increase correspondingly. The reverse is true also; as m_2 increases, point 3 approaches point 2 and the ratio becomes zero.

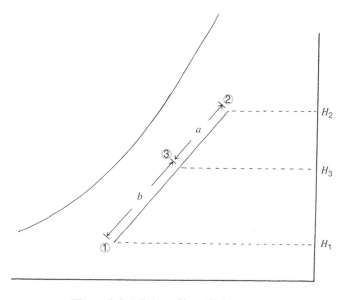

Figure 9.5. Mixing of humid air streams.

Example 9.7

Air with a dry bulb temperature of 14°C, a wet bulb temperature of 9°C and a *dry air* mass flow rate of $0.02 \, \text{kg s}^{-1}$, is mixed with an air stream with corresponding values of 42°C, 32°C and $0.04 \, \text{kg s}^{-1}$, respectively. Determine, graphically and by calculation, the humidity of the mixed stream.

From the psychrometric chart the humidities of streams 1 and 2 can be found as follows: $H_1 = 0.005 \, \text{kg per kg}$ and $H_2 = 0.0262 \, \text{kg per kg}$. The mixed stream has a dry air flow rate of $0.06 \, \text{kg s}^{-1}$ and therefore the component material balance becomes

$$(0.02 \times 0.005) + (0.04 \times 0.0262) = 0.06 H_3$$

from which

$$H_3 = 0.0191 \, \text{kg kg}^{-1}.$$

By the lever rule

$$\frac{m_1}{m_2} = \frac{0.02}{0.04} = \frac{a}{b}$$

On a full size chart the length of the line $a+b$ which joins streams 1 and 2 is 202 mm. Therefore $a = 67.3 \, \text{mm}$ and $b = 134.7 \, \text{mm}$ and the point representing stream 3 can be located on the line at a point 67.3 mm from stream 2. This again gives $H_3 = 0.0191 \, \text{kg kg}^{-1}$.

9.5. APPLICATION OF PSYCHROMETRY TO DRYING

Commonly foodstuffs are dried in simple driers in which the food is placed on a series of shelves and a hot air stream is passed over the surface. It is possible to arrange the flow of air so that it is reheated between each pass over a shelf. Whilst far more complex and efficient drier arrangements are possible (see chapter twelve), the simple example which follows shows how a drying process can be plotted on a psychrometric chart.

Example 9.8

Air, originally at 14°C and 70% saturation, is heated to 48°C and then passed consecutively over 2 shelves in a tray drier. In passing over each shelf the air regains its original percentage humidity but is reheated again to 48°C by heaters between the shelves. Assuming that the material on each shelf reaches the wet bulb temperature and that heat losses can be neglected, determine the temperature of the material on each shelf and the rate of removal of water (kg s^{-1}) if $3 \text{m}^3 \, \text{min}^{-1}$ of moist air enters the drier.

The change in humidity and temperature of the air is shown diagrammatically in Figure 9.6. The inlet air (point A) has an absolute humidity of $0.007 \, \text{kg per kg}$ and a specific volume of $0.822 \, \text{m}^3 \, \text{kg}^{-1}$. After heating to 48°C (point B) the air is humidified at a constant wet bulb temperature of 23°C to 70% saturation (point C). It is then reheated to 48°C (point D) and is humidified at a wet bulb temperature of 28.3°C until it again reaches 70% saturation and an absolute humidity of $0.0227 \, \text{kg kg}^{-1}$ (point E). The air is then discharged from the drier. Consequently, the temperature on the first tray is 23°C and on the second tray 28.3°C.

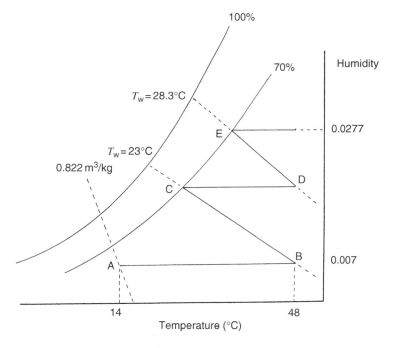

Figure 9.6. Solution to example 9.8.

As the air passes through the drier, the humidity increases from 0.007 to 0.0227 kg kg^{-1} and therefore each kg of *dry air* removes $(0.0227 - 0.007) = 0.0157$ kg of moisture. Now, the mass flow rate of moist air entering the drier is $3/0.822 \times 60$ kg s^{-1}. However the mass fraction of dry air in this stream ($H = 0.007$ kg kg^{-1}) is $1/(1 + 0.007)$ and the rate of water removal by the air, or the drying rate, is then:

$$\text{rate of water removal} = \left[\frac{1}{1.007} \times \frac{3}{0.822} \right] \times (0.0227 - 0.007) \text{ kg s}^{-1}$$

$$= 9.48 \times 10^{-4} \text{ kg s}^{-1}$$

Note that the quantity in square brackets is the mass flow rate of *dry air*.

NOMENCLATURE

a	Length
b	Length
c_p	Heat capacity
h	Film heat transfer coefficient
h_{fg}	Latent heat of vaporisation of water
H	Humidity
H_o	Saturated humidity
H_s	Saturated humidity at adiabatic saturation temperature
H_w	Saturated humidity at wet bulb temperature
k_g	Film mass transfer coefficient

m Mass flow rate of dry air
M_A Mean molecular weight of air
M_w Molecular weight of water
N Mass flux of water vapour
p_g Mean partial pressure of the humid gas
p_w Partial pressure of water vapour
p_{wo} Partial pressure of water vapour at saturation
P Total pressure
q Heat flux
R Universal gas constant
T Dry bulb temperature
T_s Adiabatic saturation temperature
T_w Wet bulb temperature

GREEK SYMBOLS

ρ_g mean density of humid gas

PROBLEMS

9.1. The partial pressure of water vapour in a humid air stream at atmospheric pressure is 1000 Pa. What is the absolute humidity?

9.2. If the absolute humidity of air is $0.020 \, \text{kg kg}^{-1}$ dry air at a pressure of 1 bar, what is the partial pressure of water vapour.

9.3. Air at a pressure of 101.3 kPa is maintained at 80%RH and 25°C. Express the humidity as percentage saturation.

9.4. Air has a percentage humidity of 60% and a dry bulb temperature of 38°C. What is the dew point?

9.5. The wet bulb temperature of air was found to be 24°C when the dry bulb temperature was 33°C. Determine the absolute humidity, percentage humidity and specific volume.

9.6. Air at 20°C and 30% humidity is humidified at a constant wet bulb temperature until the air is saturated. What is the saturation temperature? If the air is then reheated to 40°C, what are the absolute and percentage humidities?

9.7. An air chilling unit takes in air with wet and dry bulb temperatures of 32°C and 35°C, respectively. Water vapour is condensed before the air is reheated to the desired condition of 21°C and 50% humidity. What is the lowest temperature to which the air must be cooled and how much water is removed from 100 kg of dry air?

9.8. A stream of air (dry air mass flow rate $0.50 \, \text{kg s}^{-1}$) at 20°C and 20% saturation is to be humidified to an absolute humidity of 0.022 with a temperature of 45°C. Humidification is to be achieved by mixing with a second air stream at a mass ratio of 1:3. What must be the temperature of the second stream?

9.9. Hot air is passed, in series, through a two-stage drying process. Air at 10% saturation and 55°C enters the first stage and leaves saturated after being cooled adiabatically. The air is then reheated to its original temperature before passing to a second stage. Drying is again adiabatic and the air leaves at 50% saturation before passing to a cooler/condenser and leaving fully saturated at 14°C. Calculate the quantity of water evaporated per kg of dry air in each stage and the quantity of water removed in the condenser per kg of dry air.

10

Thermal Processing of Foods

10.1. UNSTEADY-STATE HEAT TRANSFER

10.1.1. Introduction

Chapter seven dealt only with steady-state heat transfer where conditions are constant with time. This is sufficient for an understanding of heat exchange in a continuous process, for example a heat exchanger where the flow rates of the two fluids and their inlet temperatures remain constant. However an analysis of unsteady-state heat transfer is required for a proper understanding of both freezing and sterilisation which are batch processes and where heating or cooling is stopped when a pre-determined temperature is reached. At steady-state the temperature profile in a body through which heat is being transferred remains constant with time and the rate of heat transfer is also constant. Consider as an example the slab shown in Figure 7.1. The temperature at each surface is constant with time and therefore the temperature gradient (equal in this case to the temperature difference divided by the slab thickness) remains constant. However if one surface of the slab is suddenly increased in temperature the temperature profile will change immediately and adjust over a period until equilibrium is re-established. Figure 10.1 shows the temperature profile in a food block at successive times t_1, t_2, and t_3 after the surface temperature is changed from T_0 to T_1 at time $t = 0$. Thus the temperature at any given point changes with time and it becomes important to be able to predict this change and to determine how long is required for thermal equilibrium to be regained.

10.1.2. The Biot Number

An important quantity in the analysis of unsteady-state behaviour is the dimensionless group called the Biot number. The Biot number is defined as internal resistance to heat transfer by conduction (equal to L/k, where L is a characteristic length) divided by the surface resistance to heat transfer (equal to $1/h$, where h is the film heat transfer coefficient). Thus

$$Bi = \frac{hL}{k} \tag{10.1}$$

The characteristic length is usually, but not always, defined as the volume of the object divided by its surface area. Therefore for a sphere L is $d/6$ and for a cylinder $d/4$. For a long square section rod the characteristic length is $L/4$, where L is the thickness of the rod. However there are a number of possible variations in the definition of the characteristic length and this can lead to considerable confusion. For a cylinder or sphere L may also be equal to the diameter and it is this definition which is used for the chart solutions presented in section 10.2.4.

235

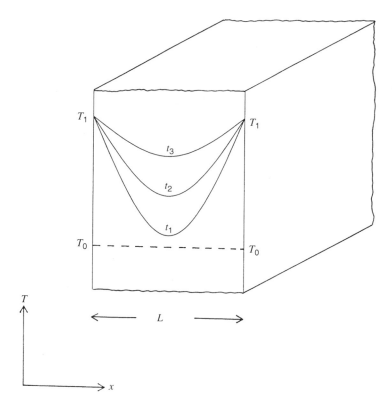

Figure 10.1. Unsteady-state heat transfer: heating an infinite slab.

The significance of the Biot number can be understood by referring to an object, initially at a temperature T_0, which is suddenly surrounded by a fluid at a temperature T_1. As Bi approaches zero, and it may be assumed that values of Bi less than 0.1 approximate to zero, all the resistance to heat transfer lies in the boundary layer within the fluid. The temperature T of the object is uniform at all times and the temperature drop across the film is $T_1 - T_0$. This is shown in Figure 10.2a. Under these circumstances the so-called 'lumped analysis', which is set out below, is valid. However at high values of Bi, greater than about 40, the convective resistance is small, conduction is rapid, and the surface temperature of the object will be equal to the fluid temperature (Figure 10.2b). The analysis of this case is rather more complex.

10.1.3. Lumped Analysis

If a mass of material has a low internal resistance to heat transfer and is subjected to a sudden change in external temperature then the temperature of the material at any given time can be found relatively easily. An example of what is known as *lumped analysis* might be a solid block with a very high thermal conductivity or a batch of liquid which is sufficiently well mixed such that the convective resistance within the liquid is low and at any given time t the liquid is at a uniform temperature T.

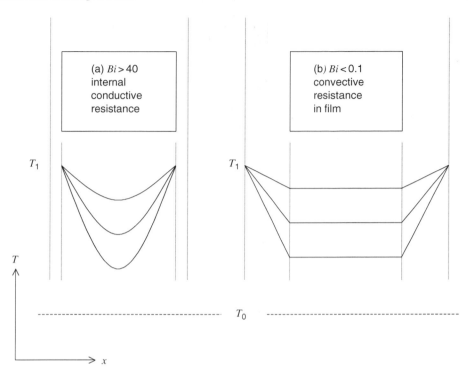

Figure 10.2. Effect of Biot number on temperature profiles.

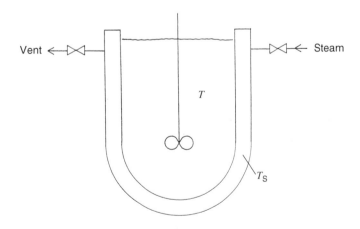

Figure 10.3. Lumped analysis: steam jacketed vessel.

Consider a batch of liquid in an agitated jacketed vessel heated by steam (Figure 10.3). The rate of heat transfer from the steam at any instant is dQ/dt and

$$\frac{dQ}{dt} = UA(T_S - T) \tag{10.2}$$

where T_S is the steam temperature and the overall heat transfer coefficient U is assumed to be constant. The heat added to the liquid results in a change of temperature at a rate dT/dt and therefore the enthalpy balance expression is

$$\frac{dQ}{dt} = mc_p \frac{dT}{dt} \tag{10.3}$$

where m is the mass of liquid and c_p is the heat capacity. Thus, combining Equations (10.2) and (10.3),

$$UA(T_S - T) = mc_p \frac{dT}{dt} \tag{10.4}$$

Separating the variables, and integrating between $T = T_0$ at $t = 0$ and $T = T$ at $t = t$, gives

$$\int_0^t dt = \frac{UA}{mc_p} \int_{T_0}^T \frac{dT}{(T_S - T)} \tag{10.5}$$

which results in

$$t = \frac{mc_p}{UA} \ln \left[\frac{T_S - T_0}{T_S - T} \right] \tag{10.6}$$

Alternatively, this can be rearranged to give the temperature at any time t and

$$\frac{T - T_S}{T_0 - T_S} = \exp \left[\frac{-UAt}{mc_p} \right] \tag{10.7}$$

If the liquid temperature is recorded as a function of time then the overall heat transfer coefficient can be obtained from the gradient of a plot of the logarithm of

$$\left[\frac{T_S - T_0}{T_S - T} \right]$$

against time.

Example 10.1

Water is to be heated in a jacketed stirred vessel using steam at a temperature of $120°C$. The vessel has a hemispherical bottom, is 0.75 m in diameter and is filled to a depth of 1.125 m. If the overall heat transfer coefficient is 550 W m^{-1} K^{-1}, estimate the time required to heat the water from $10°C$ to $80°C$.

The volume of the vessel is the volume of a cylinder 0.75 m in diameter and 0.75 m deep plus the volume of a hemisphere 0.75 m in diameter. Thus the total volume of water is $\frac{\pi}{4}(0.75)^3 + \frac{\pi}{12}(0.75)^3 = 0.442$ m^3. Taking the average density of water over the temperature range to be 985.8 kg m^{-3}, the mass of water is 435.7 kg. If the heat transfer surface area is assumed to be that in contact with the water, the area is $\pi(0.75)^2 + \frac{\pi}{2}(0.75)^2 = 2.65$ m^2. The mean heat capacity of water over the range 10–$80°C$ is approximately 4190 J kg^{-1} K^{-1} and therefore the batch heating time, given by Equation (10.6), is

$$t = \frac{435.7 \times 4190}{550 \times 2.65} \ln \left[\frac{120 - 10}{120 - 80} \right] \text{s}, \qquad t = 1267 \text{ s}$$

or

$$t = 21.1 \text{ min}$$

10.2. UNSTEADY-STATE CONDUCTION

10.2.1. Fourier's First Law of Conduction

Fourier's first law of conduction can be stated as

$$\frac{\partial T}{\partial t} = \alpha \left(\frac{\partial^2 T}{\partial x^2} + \frac{\partial^2 T}{\partial y^2} + \frac{\partial^2 T}{\partial z^2} \right) \tag{10.8}$$

This relates the temperature variation in three orthogonal dimensions x, y, and z to time and is derived in full in Appendix F. The use of partial derivatives acknowledges that heat is transferred in each direction and that a temperature gradient exists in each direction. The physical properties of the material through which heat is transferred are contained in the thermal diffusivity α defined by

$$\alpha = \frac{k}{\rho c_p} \tag{10.9}$$

where k is the thermal conductivity, ρ is the density and c_p is the heat capacity. Thermal diffusivities are of the order of 10^{-7} m^2 s^{-1} for most foods.

The solution of Equation (10.8) is complex and in the first instance it will be simplified by considering conduction in the x direction only. Thus there is no temperature variation in the y and z directions, both $\partial T / \partial y = 0$ and $\partial T / \partial z = 0$, and therefore

$$\frac{\partial T}{\partial t} = \alpha \left(\frac{\partial^2 T}{\partial x^2} \right) \tag{10.10}$$

This equation must now be solved for a number of geometries which are of interest in food processing including flat plates or slabs, brick shapes, spheres and cylinders, all of which approximate to the shapes of various food products.

10.2.2. Conduction in a Flat Plate

Figure 10.1 now represents a slab with a finite thickness of L in the x direction and which is infinite in the y and z directions. The slab is initially at a uniform temperature T_0 and the two large parallel faces of the slab are suddenly brought to a temperature T_1. Considering only conduction in the x direction (i.e. there is no surface resistance and conduction in the y and z directions is ignored), the temperature profile develops as in Figure 10.1. In order to find the temperature at any point in the slab at any given time it is first necessary to define a dimensionless temperature Y by

$$Y = \frac{T_1 - T}{T_1 - T_0} \quad \text{or} \quad Y = \frac{T - T_1}{T_0 - T_1} \tag{10.11}$$

The boundary conditions for this problem now become

   i $T = T_0$, and therefore $Y = 1$, at $t = 0$, $x = x$

   ii $T = T_1$, and therefore $Y = 0$, at $t = t$, $x = 0$

   iii $T = T_1$, and therefore $Y = 0$, at $t = t$, $x = 2L$

The solution of the differential equation [Equation (10.10)] takes the form

$$Y = [\exp(-a^2\alpha t)][A \cos ax + B \sin ax] \tag{10.12}$$

where A and B are constants and a is a parameter. Substitution of the boundary conditions into Equation (10.12) gives a series solution

$$Y = \sum_{n=0}^{n=\infty} \frac{4}{(2n+1)\pi} \exp\left[\frac{-(2n+1)^2\pi^2\alpha t}{L^2}\right] \sin\left[\frac{(2n+1)\pi x}{L}\right] \tag{10.13}$$

which on expansion gives

$$Y = \frac{4}{\pi}\left[\exp\left(\frac{-\pi^2\alpha t}{L^2}\right)\sin\left(\frac{\pi x}{L}\right) + \frac{1}{3}\exp\left(\frac{-3^2\pi^2\alpha t}{L^2}\right)\sin\left(\frac{3\pi x}{L}\right)\right.$$
$$\left. + \frac{1}{5}\exp\left(\frac{-5^2\pi^2\alpha t}{L^2}\right)\sin\left(\frac{5\pi x}{L}\right) + \cdots\right] \tag{10.14}$$

In Equation (10.14) the series has been expanded to three terms but usually the first term only is necessary, unless the initial and final temperatures are quite close. The use of this solution presents a number of problems. First it is necessary to use an average thermal diffusivity. However food is usually non-isotropic, that is properties such as density, thermal conductivity, and heat capacity vary in each direction. A good example of a non-isotropic food is meat which consists of a non-uniform distribution of lean meat, fat, and bone all of which have different physical properties. Second the prediction of thermal properties depends upon proximate analysis which itself is not very accurate. Third, such physical properties often change during processing. For example the heat capacity of beef falls during freezing from about 2.5 kJ kg^{-1} K^{-1} in the unfrozen state to about 1.5 kJ kg^{-1} K^{-1} in the frozen state. Using the first term only of Equation (10.14), the time required to reach a given temperature can be obtained directly. However use of the second and subsequent terms necessitates a trial and error solution.

Example 10.2

A large block of frozen fish, 100 mm thick, is placed in a cold store such that its surface temperature is brought instantaneously to $-25°C$. If the initial temperature of the block is $-5°C$ calculate the time for the centre to reach $-20°C$. Take the density, thermal conductivity and heat capacity of the fish block to be 1,000 kg m^{-3}, 1.1 W m^{-1} K^{-1}, and 1,600 J kg^{-1} K^{-1}, respectively.

If more than one term of Equation (10.14) is used then the equation cannot be solved directly. As a first estimate of the time, the first term only will be used. The thermal diffusivity of the block is found from Equation (10.9):

$$\alpha = \frac{1.1}{1000 \times 1800} \text{ m}^2 \text{ s}^{-1}, \qquad \alpha = 6.11 \times 10^{-7} \text{ m}^2 \text{ s}^{-1}$$

The thickness of the block is L and therefore at the centre $x/L = 0.5$. Now, noting that $L = 0.01$ m, the first term of the series solution gives

$$\frac{T + 25}{-5 + 25} = \frac{4}{\pi} \left[\exp \left(\frac{-\pi^2 6.11 \times 10^{-7} t}{(0.01)^2} \right) \sin \left(\frac{\pi}{2} \right) \right]$$

Putting $T = 20°$C this can be solved to give t equal to 2,699 s. Now the two-term solution is

$$Y = \frac{4}{\pi} \left[\exp \left(\frac{-\pi^2 \alpha t}{L^2} \right) \sin \left(\frac{\pi x}{L} \right) + \frac{1}{3} \exp \left(\frac{-3^2 \pi^2 \alpha t}{L^2} \right) \sin \left(\frac{3\pi x}{L} \right) \right]$$

and, using $t = 2,700$ as a guide, this can be solved to find the centre temperature at, say, times of 2,600 s and 2,800 s. Thus for $t = 2,600$ s, $T = -19.69°$C and for $t = 2,800$ s, $T = -20.29°$C. Interpolating between these values now gives a time of $t = 2,703$ s for a centre temperature of $-20°$C. That is, there is very little advantage in using more than one term of Equation (10.13).

Example 10.3

If the thickness of the fish block in Example 10.2 is reduced to 60 mm, then what would be the equivalent time for the centre to reach $-20°$C?

For the same centre temperature and the same value of x/L the required time is proportional to L^2. Hence the new time t is

$$t = 2700 \left(\frac{60}{100} \right)^2 \text{ s}, \qquad t = 972 \text{ s}$$

10.2.3. The Fourier Number

A further important dimensionless group must now be introduced; the Fourier number represents the ratio of the rate of conduction of heat across a characteristic length L to the rate of storage of heat in a volume of L^3. It is defined by

$$Fo = \frac{\alpha t}{L^2} \tag{10.15}$$

The characteristic length L is the shortest distance from the surface to the centre of the object and is equal to one half of the thickness of a plate or slab and equal to the radius of a sphere or cylinder. Increasing values of Fo indicate a deeper penetration of heat into the object within a given time.

10.2.4. Gurney–Lurie Charts

The use of series solutions can be very tedious, even with the ready availability of computers. Chart solutions, prepared originally in a pre-electronic calculator age, are still useful for obtaining solutions to complex problems rapidly. The Gurney–Lurie chart is a plot of dimensionless temperature Y against Fourier number Fo as a function of two further variables

m and n. Of these, m is the reciprocal of Biot number, where for a cylinder or sphere the characteristic length is defined by the diameter, and n is a dimensionless measure of the position within the object defined by $n = x/L$. Thus $n = 0$ represents the centre point of the object and a value of $n = 1$ represents the surface of the object. Charts are available for a range of geometries. Figure 10.4 is that for a flat plate or infinite slab, finite in the x direction with a thickness of $2L$ and infinite in both the y and z directions. Figures 10.5 and 10.6 are the

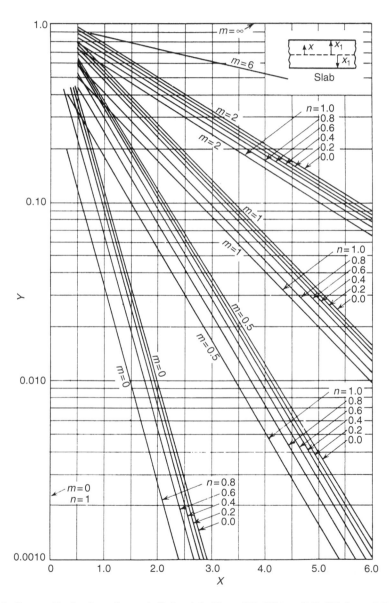

Figure 10.4. Gurney-Lurie chart for an infinite slab (from J.R. Welty, C.E. Wicks and R.E. Wilson, Fundamentals of Momentum, Heat and Mass Transfer, John Wiley, 1969, © John Wiley & Sons, Inc., reprinted by permission of John Wiley & Sons, Inc.).

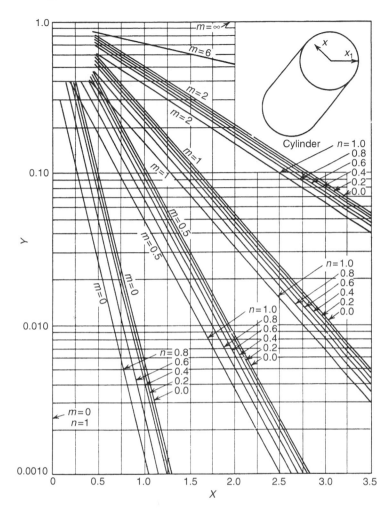

Figure 10.5. Gurney-Lurie chart for an infinite cylinder (from J.R. Welty, C.E. Wicks and R.E. Wilson, Fundamentals of Momentum, Heat and Mass Transfer, John Wiley, 1969, © John Wiley & Sons, Inc., reprinted by permission of John Wiley & Sons, Inc.).

Gurney–Lurie charts for cylinders of radius L and spheres of radius L, respectively. It should be stressed that these charts represent conduction in the x direction only but that they include the effect of convection by including a heat transfer coefficient in the variable m. However the Gurney–Lurie chart suffers from the disadvantage that it is sometimes difficult to read accurately and requires values to be interpolated. A more obvious drawback is that solutions are presented for specific regular shapes only and these may or may not approximate well to the shape of the desired food commodity.

Example 10.4

A meat product cut into slices 10 mm thick, and initially at a temperature of 80°C, is to be cooled by blowing cold air at 2°C across the surface. Estimate both the time for the centre of

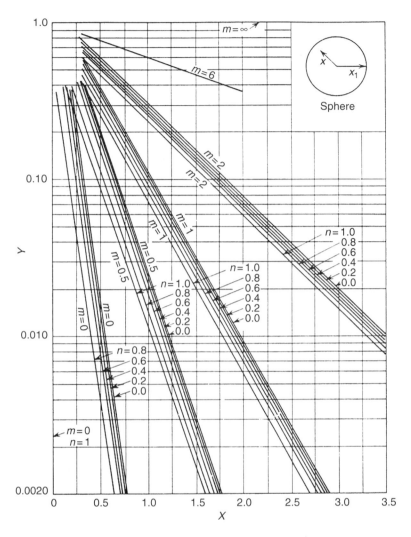

Figure 10.6. Gurney-Lurie chart for a sphere (from J.R. Welty, C.E. Wicks and R.E. Wilson, Fundamentals of Momentum, Heat and Mass Transfer, John Wiley, 1969, © John Wiley & Sons, Inc., reprinted by permission of John Wiley & Sons, Inc.).

a slice to reach $20°C$ and the temperature at a point 3 mm from the centre after three minutes. The meat has a thermal conductivity of $0.375 \text{ W m}^{-1} \text{ K}^{-1}$, a density of $1{,}250 \text{ kg m}^{-3}$ and a mean heat capacity of $2{,}130 \text{ J kg}^{-1} \text{ K}^{-1}$. Assume the film heat transfer coefficient between the meat and the air to be $150 \text{ W m}^{-2} \text{ K}^{-1}$.

The thermal diffusivity of the meat is given by

$$\alpha = \frac{0.375}{1250 \times 2130} \text{ m}^2 \text{ s}^{-1}$$

and therefore

$$\alpha = 1.41 \times 10^{-7} \text{ m}^2 \text{ s}^{-1}$$

The dimensionless temperature is then

$$Y = \frac{20 - 2}{80 - 2}, \qquad Y = 0.231$$

Now, for an infinite slab, $n = 0$ at the centre and the value of m is

$$m = \frac{0.375}{150(0.01/2)} \quad \text{or} \quad m = 0.50$$

Thus the Fourier number can be obtained from the chart (Figure 10.4) and $Fo = 1.4$. From the definition of the Fourier number the required time is

$$t = \frac{1.4(0.005)^2}{1.41 \times 10^{-7}} \text{ s}, \qquad t = 248 \text{ s}$$

After 3 min, the value of the Fourier number is

$$Fo = \frac{1.41 \times 10^{-7} \times 180}{(0.005)^2} \quad \text{or} \quad Fo = 1.015$$

A position 3 mm from the centre equates to $n = 0.6$ and from the chart $Y = 0.29$. Thus

$$0.29 = \frac{T - 2}{80 - 2}$$

and the temperature at this point is then

$$T = 24.6°C.$$

Of the Gurney–Lurie charts presented in Figures 10.4–10.6, only that for a sphere represents a finite body. Problems involving other finite bodies can be solved by assuming that the object of interest lies at the orthogonal intersection of two infinite shapes. For example, the traditional metal can is in effect a finite cylinder and lies at the intersection of an infinite cylinder and an infinite slab as shown in Figure 10.7. In this circumstance the dimensionless temperature for the can is obtained from

$$Y = Y_{\text{cylinder}} Y_{\text{slab}} \tag{10.16}$$

Note that the direction in which heat is conducted, the x direction, for the cylinder, is at right angles to the x direction for the slab. Further, the characteristic length L for the cylinder is the can radius r, and for the slab it is half the thickness that is, one half of the can height h. This concept can be extended to other shapes; for a regular cuboid or brick the dimensionless temperature is given by the orthogonal intersection of three infinite slabs in the x, y, and z directions, respectively, and therefore

$$Y = Y_x Y_y Y_z \tag{10.17}$$

Example 10.5

A can (of diameter 7.5 cm and height 11.5 cm) contains pea puree originally at 50°C and is heated in steam at 130°C. The density, thermal conductivity and mean heat capacity of the

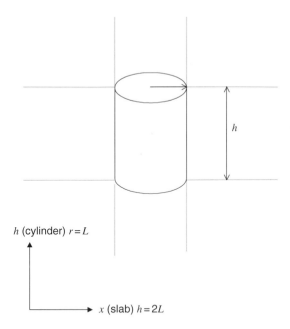

h (cylinder) $r=L$

x (slab) $h=2L$

Figure 10.7. Intersection of infinite cylinder and infinite slab.

puree are $1,100\,\mathrm{kg\,m^{-3}}$, $0.836\,\mathrm{W\,m^{-1}\,K^{-1}}$, and $3,800\,\mathrm{J\,kg^{-1}\,K^{-1}}$, respectively. Estimate the centre temperature of the can after 30 min, assuming that heat transfer is principally due to conduction.

The approach to this problem is to find the product of Y_{cylinder}, based upon the radius of the can, and Y_{slab}, based upon the half height of the can. Conduction is dominant, the Biot number is large and therefore $m = 0$. At the can centre $n = 0$. The thermal diffusivity of the puree is

$$\alpha = \frac{0.836}{1100 \times 3800}\,\mathrm{m^2\,s^{-1}}$$

and therefore

$$\alpha = 2.0 \times 10^{-7}\,\mathrm{m^2\,s^{-1}}$$

Now, for an infinite cylinder, the Fourier number is

$$Fo = \frac{2.0 \times 10^{-7} \times 1800}{(0.075/2)^2}$$

or $Fo = 0.256$ and from the Gurney–Lurie chart (Figure 10.5) $Y_{\mathrm{cylinder}} = 0.40$. For an infinite slab

$$Fo = \frac{2.0 \times 10^{-7} \times 1800}{(0.115/2)^2}$$

or $Fo = 0.109$ and therefore $Y_{\mathrm{slab}} = 1.0$. Consequently, the dimensionless temperature for the can is

$$Y_{\mathrm{can}} = 0.40 \times 1.0 \quad \mathrm{or} \quad Y_{\mathrm{can}} = 0.40$$

and the can centre temperature is then given by

$$0.40 = \frac{T - 130}{50 - 130}$$

from which $T = 98°C$.

10.2.5. Heisler Charts

In order to overcome the problem of inaccuracy in reading a Gurney-Lurie chart, a further series of charts have been prepared which plot the dimensionless temperature as a function of the Fourier number for a fixed position. These are known as Heisler charts and, because the temperature at the centre of a body is usually most important in food engineering problems, the most useful of these are drawn for $n = 0$. An example of a Heisler chart for the temperature at the centre of an infinite slab is given in Figure 10.8.

10.3. FOOD PRESERVATION TECHNIQUES USING HEAT

10.3.1. Introduction to Thermal Processing

The preservation of foods to ensure safety and to give acceptable shelf stability can be achieved by using a number of techniques:

i   High temperature
ii   Low temperature
iii Reduction of water activity
iv   Use of chemical preservatives
v   Control of storage gas composition
vi   Ionising radiation
vii High pressure

Of these, the first three might be considered to be the most widely used preservation methods. The use of low temperatures in freezing and chilling is covered in chapter eleven and the reduction of water activity by means of drying in chapter twelve. This chapter is concerned with the use of high temperatures in food preservation and principally with pasteurisation and sterilisation. These operations, which together may be termed thermal processing, are employed to reduce the number of micro-organisms (either vegetative cells, bacteria or spores) present in food by subjecting it to heat for sufficient time to ensure food safety or to reduce spoilage and increase shelf life.

Thermal processing is sometimes defined in relation to the heat treatment of *packaged* foods to bring about microbiological safety. Much of the analysis of thermal processing has been developed for foods placed in metal containers, usually the cylindrical metal can made from thin tin-plated steel or aluminium, and heated by steam. This process is often referred to as 'canning'. However the principles of thermal processing apply equally to a variety of packages including cans, bottles, and flexible and semi-rigid plastic pouches. Different packaging requires different handling methods; for example, glass bottles are heated by hot water to avoid breakage due to thermal shock. However we are concerned here with the principles of thermal processing and the necessary temperature–time history rather than with

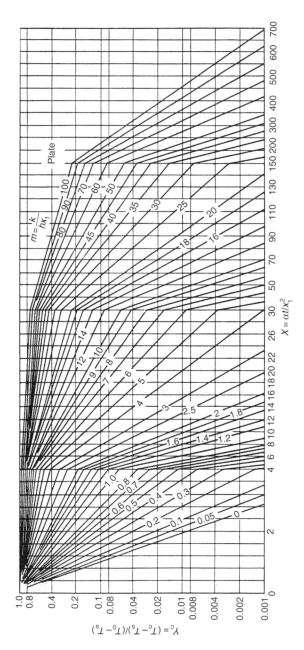

Figure 10.8. Heisler chart for an infinite slab (from J.R. Welty, C.E. Wicks and R.E. Wilson, Fundamentals of Momentum, Heat and Mass Transfer, John Wiley, 1969, © John Wiley & Sons, Inc., reprinted by permission of John Wiley & Sons, Inc.).

details of either the can structure or the details of retorts, although the major features of retorts will be described in a later section. Despite the foregoing, a wider definition of thermal processing would include the heat treatment of bulk liquids. The term aseptic processing refers to the sterilisation of the product and the packaging separately followed by a sterile packaging operation in a carefully controlled environment. Typically this makes use of the plate heat exchanger, for example, in the pasteurisation of milk.

Although pasteurisation and sterilisation are the principle areas of concern in this chapter, brief mention must be made of two other operations. Blanching is designed to denature enzymes in solid fruit and vegetable pieces, usually before freezing or canning or drying, and to prevent enzyme activity which might continue even at low-temperatures and low-moisture contents. In the case of canning, blanching reduces enzyme activity which may continue during the relatively long times required to heat and cool a product during sterilisation thus leading to a deterioration of quality. In particular, blanching is used in order to preserve colour. Blanching is normally carried out at atmospheric pressure using either steam or hot water and at temperatures between 90°C and 95°C held for about 1 or 2 min. Second, it should not be overlooked that cooking itself is a preservation technique using high temperatures: heating many foods so that the centre reaches a temperature of around 70°C for 2 to 3 min is sufficient to reduce food poisoning bacteria and to kill many parasites. However this regime will not destroy spore-forming bacteria and carefully designed hygienic handling subsequent to the cooking process is essential.

10.3.2. Pasteurisation

Pasteurisation is a form of thermal processing, for both bulk liquids and foods in containers, which uses moderate temperatures to extend shelf life by days, or at most a few weeks, but causes the minimum changes to the sensory properties or nutrient value of the food. Primarily it is designed to remove pathogenic bacteria and vegetative organisms but not heat resistant spores which are not destroyed at the temperatures employed. Pasteurisation is targeted at particular micro-organisms in particular foods but it is a milder treatment than commercial sterilisation and therefore does not give a safe shelf-stable product without subsequent storage at refrigerated temperatures. Second, pasteurisation is designed to reduce the enzymatic activity in the product.

The temperature/time combination used in pasteurisation depends to a degree upon the pH of the product concerned. For acidic foods, defined as those with a pH below 4.5, the main purpose is the inactivation of enzymes (e.g., in fruit juices) or the destruction of spoilage organisms such as wild yeasts in beer or residual yeasts from the fermentation. In low-acid foods, at pH greater than 4.5, the purpose is the destruction of particular pathogens such as *Brucella abortus* and *Mycobacterium tuberculosis* in milk or *Salmonella seftenburg* in egg. Table 10.1 gives some examples of the temperature/time treatments used for a variety of products. In the U.K., pasteurisation temperatures and storage temperatures are specified by the Food Safety Act (1990).

Pasteurisation can be achieved in a number of ways. For relatively small quantities of liquid a batch operated jacketed stirred vessel may be sufficient. Many liquids, especially juices and alcoholic beverages, are packaged and then pasteurised using steam or hot water retorts. However for bulk liquids at high throughputs (especially milk, but also fruit juices and beer) high-temperature short-time (HTST) processing may be used in order to reduce the loss of nutrients. Typically this procedure makes use of a plate heat exchanger (see Example 7.16) with three sections for heating, regeneration, and cooling, respectively, a holding tube to maintain the

TABLE 10.1
Typical Temperature/Time Treatments in
Pasteurisation

Food material	Temperature (°C)	Time (s)
Milk	72	15
Ice cream mix	80	20
Tomato juice	118	60
Honey	71	300
Fruit juice	88	15
Soft drinks	95	10

pasteurisation temperature for the required time and a flow diversion valve to ensure that liquid which has not reached the desired temperature is not discharged as finished product. It is important to ensure that rapid cooling occurs in order to prevent quality deterioration and to ensure a uniform residence time. The rate of heat transfer in HTST processing is far quicker than in a can which relies on either conduction, free convection or forced convection induced by rotation of the can. In a plate heat exchanger turbulence is induced in the liquid to be heat treated which increases the heat transfer coefficient and reduces overall processing times. Thus milk is held at a temperature of 71.5°C for 15 s; alternative treatments include 88°C for 1 s or 94°C for 0.1 s.

10.3.3. Commercial Sterilisation

In contrast to pasteurisation, commercial sterilisation is intended to give long shelf life (in excess of six months) to foods by destroying both microbial and enzyme activity. Sterilisation processes may be divided for convenience into two categories:

i   'In-container' sterilisation, which is used for foods which are placed in containers, traditionally the metal can but also bottles and increasingly plastic pouches of different shapes, before heat processing.

ii Continuous flow systems for ultra heat treatment (UHT) processes which precede packaging and generally use temperatures above 135°C held for a few seconds (in contrast to temperatures around 121°C maintained for several minutes). UHT processes have the advantage that less chemical damage is inflicted, for example, protein and vitamin denaturation, because these reactions are less temperature sensitive than is the destruction of microbial spores.

The former, the traditional canning operation, is an example of a non-steady-state heat transfer process in which a container is heated, usually by steam inside a pressure vessel, held at given temperature for a fixed period and then cooled. It is important to understand that the whole heating/hold/cooling cycle contributes to sterilisation. The heat treatment employed is severe and may be responsible for a considerable loss of quality and therefore it is important to keep processing times to the minimum required for any given objective. Further, in defining the concept of commercial sterility it is important to realise that after a sterilisation process viable micro-organisms are still present in the food but (i) they are at a level which is insignificant statistically and (ii) they exist under growth conditions which are not favourable and therefore they will not cause spoilage or affect quality during the given shelf life. In other words commercial sterility implies that there is a very low probability of the survival of organisms which are dangerous to human health.

10.4. KINETICS OF MICROBIAL DEATH

The destruction of micro-organisms in foods using heat is a well-known phenomenon which is exploited in the preservation techniques outlined above. However the temperature response of vegetative cells and spores is far from uniform. Spores tend to be more heat resistant than vegetative cells which in turn range widely in their heat resistance. Even individual bacteria within a population of a given species show a normal distribution of heat resistance. Thus it is possible to allow heat resistant (or thermoduric) organisms to survive by using a heating regime which is sufficient to destroy bacteria of low to intermediate heat resistance but which fails to kill thermoduric bacteria. These may then thrive within a processing unit, for example, a blancher, and increase the microbial load on a subsequent sterilisation operation.

The heat resistance of micro-organisms is also affected by a number of other factors such as:

i   the age of cells; younger cells are less heat resistant,

ii   the medium in which growth has occurred; a more nutritious medium increases heat resistance,

iii   moisture content; dry foods tend to require more severe heat treatment during sterilisation,

iv   the presence of sodium chloride, proteins and fats all increase heat resistance,

v   pH.

10.4.1. Decimal Reduction Time and Thermal Resistance Constant

The decline in the number of micro-organisms when subjected to heat is asymptotic with time and therefore it is not possible to eliminate all micro-organisms. There is a logarithmic relationship between the number of survivors of a given micro-organism n and time t at any given temperature (Figure 10.9). This is known as a survivor curve. The gradient of the survivor curve increases markedly with temperature.

The decimal reduction time D is defined as the time for a tenfold reduction in the number of survivors of a given micro-organism, in other words the time for one log cycle reduction in the microbial population. Higher values of D imply, at a given temperature, greater resistance of micro-organisms to thermal death. Because D depends upon temperature, the temperature in °C is appended as a subscript. Thus $D_{121.1}$ is the time required at 121.1°C to reduce a microbial population by 90%. A temperature of 121.1°C (or 250°F) is used as a common reference point and therefore, because of its importance, this is sometimes referred to as D_0.

The logarithmic decline in the number of organisms n is represented by

$$\frac{dn}{dt} = -kn \tag{10.18}$$

where k is a rate constant. Therefore the equation of the line in Figure 10.9 is represented by

$$t = D(\log n_1 - \log n_2) \tag{10.19}$$

or

$$t = D \log \left(\frac{n_1}{n_2} \right) \tag{10.20}$$

where n_1 and n_2 are the initial and final number of micro-organisms, respectively. Of course, the value of D is independent of the initial population of micro-organisms.

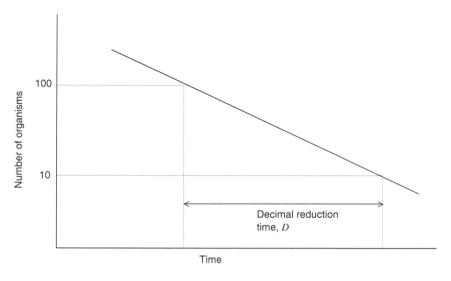

Figure 10.9. Survivor curve.

Example 10.6

A sample of fixed volume was held at a constant temperature and the number of micro-organisms in the sample measured as a function of time. Calculate the decimal reduction time.

Time (min)	Number of micro-organisms
1.0	2.00×10^5
2.0	4.31×10^4
4.5	6.32×10^3
6.0	2.00×10^3
7.5	6.32×10^2

A plot of the natural logarithm of the number of organisms against time gives a straight line of gradient 3 min. Hence $D = 3$ min.

The concept of decimal reduction time allows the probability of the survival of spores to be predicted. For example, if a process is sufficiently effective to produce 10 decimal reductions in the microbial population then, if a canned food which is to be sterilised contained initially 10^{10} spores per can, the final population would be one spore per can. Alternatively, for an initial population of 10^5 spores per can, the final population would be 10^{-5} spores per can. This latter figure is interpreted to mean that one can in 10^5 is likely to contain a spore. Such a process is referred to as a $10D$ process. Consequently there is a need to define commercial sterility, that is, to decide what is acceptable in any given process. This is based upon two

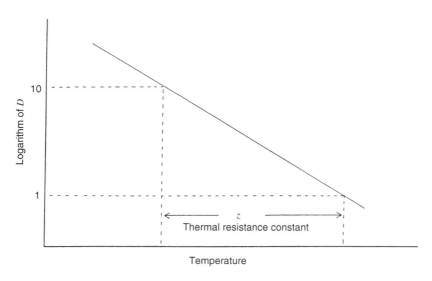

Figure 10.10. Thermal resistance curve.

criteria, as follows:

(*a*) *Reduction of* Clostridium botulinum   A minimum process must reduce the concentration of the spores of *Clostridium botulinum*, a pathogen responsible for the production of potent toxins under the anaerobic conditions experienced in a sealed container, by $12D$ in low acid foods that is, those with pH $>$ 4.5. This is called a $12D$ process and is based upon a probability of survival of *Clostridium botulinum* spores of one in 10^{12}. The decimal reduction time for *Clostridium botulinum* at $121°C$ is 0.25 min and a $12D$ process normally involves heating, at $121°C$, for 3 min. Acidic foods, in which the growth of bacteria are inhibited, do not require such drastic heat treatment and, for example, canned fruits need only to be pasteurised to inactivate the enzymes present.

(*b*) *Elimination of spoilage organisms*   Spoilage organisms may have a higher heat resistance than *Clostridium botulinum* and thus their inactivation becomes the basis for specifying the minimum heat treatment. In milk *Bacillus stearothermophilus* is a very heat resistant spore responsible for spoilage for which $D_{121.1}$ is approximately 4.0 min.

A plot of the logarithm of decimal reduction time against temperature is generally linear. This is known as a thermal resistance curve (Figure 10.10) from which a thermal resistance constant, or more commonly a z value, can be defined. The z value is the temperature change for a ten-fold change in decimal reduction time D and larger z values indicate greater heat resistance to higher temperatures. Thus, for an organism for which $z = 13$ K, an increase in temperature of 13 K will produce a decrease in the decimal reduction time of 90%. For *clostridium botulinum* the value of z is 10 K. The characterisation of the kinetics of microbial death in terms of decimal reduction time and thermal resistance constant is the first step in specifying a sterilisation process.

10.4.2. Process Lethality

Having established a method of describing microbial death rates it is necessary to find a way of characterising a sterilisation process so that its effectiveness for any given application

TABLE 10.2
Typical Values of F_0 for Particular
Foods

	F_0 (min)
Vegetables	3–6
Soups	4–5
Beans in tomato sauce	4–6
Milk puddings	4–10
Evaporated milk	5
Herring in tomato sauce	6–8
Meat in gravy	12–15
Pet food	15–18

can be judged. Because a range of temperature/time combinations can be used to achieve the same reduction in population of a given micro-organism, different sterilisation processes can be compared using a quantity known as total process lethality, F, which represents the total temperature/time combination to which a food is subjected. Less commonly this is called thermal death time.

F is the time required (usually expressed in minutes) to achieve a given reduction in a population at specified temperature. For example, a process lethality of $F = 2.5$ implies heating for two and a half minutes at the reference temperature and for a specified z value. The reference temperature is usually appended as a subscript and the z value as a superscript giving, for example, $F_{121.1}^{10}$. These particular conditions are used as a reference value of F which is designated as F_0. Table 10.2 gives typical values of F_0 for particular foods.

These data suggest that, in order to render safe it for a specified shelf life, evaporated milk should be subjected to a process which is equivalent to heating at 121.1°C for 5 min. Process lethalities include the effects of elevated temperature on micro-organisms during both heating and cooling cycles in addition to the effect whilst holding a steady temperature. Temperatures well below the maximum will make a significant contribution to the total lethality if the time is sufficiently long and thus the rates of heating and cooling are very important.

It follows that lethality and decimal reduction time are related by

$$F = D \log \left(\frac{n_1}{n_2} \right) \qquad (10.21)$$

and thus for the particular reference temperature of 121.1°C

$$F_0 = D_0 \log \left(\frac{n_1}{n_2} \right) \qquad (10.22)$$

Example 10.7

A total of 10^{12} cans, with an initial contamination of one spore per can, are to be sterilised. The decimal reduction time for the organism of interest at the processing temperature is 0.25 min. Determine the probability that any one can will retain a spore after a process equivalent to $F = 3$ min.

The initial concentration n_1 of micro-organisms is one spore per can. From Equation (10.21)

$$3 = 0.25 \log \left(\frac{10^{-12}}{n_2} \right)$$

which gives the final concentration n_2 as 10^{-12} spores per can, in other words one spore in 10^{12} cans.

10.4.3. Spoilage Probability

The relationship between process lethality, decimal reduction time and microbial population can be used to estimate the number of spoiled containers in a given batch. If now n_1 and n_2 are the initial and final number of micro-organisms *per container*, respectively, and the number of containers in the batch is r, then the total microbial load is rn and

$$F = D \log \left(\frac{rn_1}{rn_2} \right) \tag{10.23}$$

If the objective is to reduce the population to a single organism per batch then $rn_2 = 1$ and

$$F = D \log(rn_1) \tag{10.24}$$

from which the number of containers in the batch is given by

$$r = \frac{10^{F/D}}{n_1} \tag{10.25}$$

Example 10.8

Estimate the spoilage probability at the end of a process in which an initial microbial population of 2×10^5 organisms per container is subjected to heating at 118°C for 45 min. The decimal reduction time for the organism is $D_{118} = 3.5$ min.

From Equation (10.25) the number of containers r is

$$r = \frac{10^{45/3.5}}{2 \times 10^5}, \qquad r = 3.6 \times 10^7$$

10.5. THE GENERAL METHOD

The general method allows comparisons to be made between different processes and relies upon setting an arbitrary lethality at a given reference temperature. The base temperature for the comparison of microbial destruction rates (lethal rates L) is usually 121.1°C; the rate at this temperature is defined as unity that is, $L = 1$. The lethal rate is given by

$$L = 10^{-(121.1-T)/z} \tag{10.26}$$

and therefore at a temperature of 111.1°C the rate will be 0.10 and at 131.1°C it will be 10.0.

The experimental basis of the general method is to plot a curve of the temperature at the slowest heating point, or thermal centre, of the container against time (Figure 10.11). This is often called a heat penetration curve and its exact shape, for a packaged food, will depend upon the heating characteristics of the retort in which that container is placed. L is now calculated as function of temperature [from Equation (10.26)] and plotted against time (Figure 10.12). Now, because lethalities are additive, the area under this curve is a measure of the total destruction of micro-organisms over the whole process and, for a reference temperature of 121.1°C, is

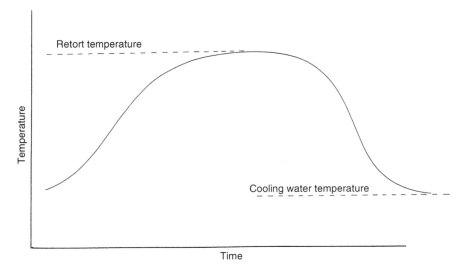

Figure 10.11. Heat penetration curve.

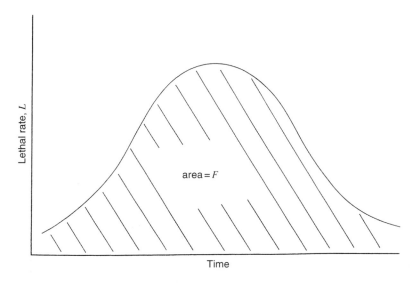

Figure 10.12. Lethal rate as a function of temperature.

called the F_0 value. Thus by definition

$$F_0 = \int_{t=0}^{t=t} L \, dt \qquad (10.27)$$

In most real cases the exact relationship between lethal rate and time is not known, the area under the curve cannot be determined analytically and it is necessary to use graphical integration techniques.

It is important to understand that F_0 is not an absolute measure of lethality and depends on the choice of base temperature, the definition of lethal rate as unity at 121.1°C and a z value of 10 K chosen because of the importance of *clostridium botulinum*. Thus the general method is relative only and F_0 simply allows comparisons to be made between different processes; the same F_0 value gives the same degree of sterilisation even if two processes produce different heat penetration curves and different temperature/time histories. F_0 has the dimensions of time with typical values between 1 and 10 min. For *clostridium botulinum* spores D_0 is about 0.25 min and therefore the standard $12D$ process equates to heating at 121.1°C for 3 min. However the destruction of spoilage organisms may require much larger F_0 values, up to tens of minutes. Tables which list lethal rates as a function of temperature are often used to simplify the calculation of F and can be constructed from Equation (10.26). An example, for $z = 10$ K and a reference temperature of 121.1°C is given in Table 10.3.

Note that there is no significant lethal effect until a temperature of 100°C is reached and even here L is below 0.01. The following example illustrates the calculation of F from the temperature history of a sample.

TABLE 10.3
Lethal Rates at a Reference Temperature of
121.1°C and $z = 10$ K

100	0.00776	118	0.48978
101	0.00977	119	0.61659
102	0.01230	120	0.77625
103	0.01549	121	0.97724
104	0.01950	122	1.23027
105	0.02455	123	1.54882
106	0.03090	124	1.94984
107	0.03890	125	2.45471
108	0.04898	126	3.09030
109	0.06166	127	3.89045
110	0.07762	128	4.89779
111	0.09772	129	6.16595
112	0.12303	130	7.76247
113	0.15488	131	9.77237
114	0.19498	132	12.30269
115	0.24547	133	15.48817
116	0.30903	134	19.49845
117	0.38905	135	24.54709

Example 10.9

The following data were generated from the thermal processing of canned peas.

Time (min)	Temperature (°C)
0	15
2	25
4	75
6	98
8	105
10	110
12	113
14	114
16	118
18	118
20	116
22	114
24	111
26	108
28	100

Determine the value of F_0 for the process.

A simple form of graphical integration can be obtained by assuming that the temperature–time relationship is linear between the recorded temperature intervals. Therefore a lethal rate can be calculated based on the arithmetic average temperature in each time interval or band. This generates the following table.

Time (min)	Temperature (°C)	Mean time (min)	Average temperature (°C)	Lethal rate
0	15			
2	25	1	20	7.76×10^{-11}
4	75	3	50	7.76×10^{-8}
6	98	5	86.5	3.47×10^{-4}
8	105	7	101.5	0.0110
10	110	9	107.5	0.0437
12	113	11	111.5	0.110
14	114	13	113.5	0.174
16	118	15	116	0.309
18	118	17	118	0.490
20	116	19	117	0.389
22	114	21	115	0.245
24	111	23	112.5	0.138
26	108	25	109.5	0.0692
28	100	27	104	0.0195
				$\sum = 2.0$

The process lethality in each band is then the product of this average lethal rate and the band width. If, as is desirable, the band width is constant then the total process lethality is simply the product of the sum of the lethal rates and the band width. In this example the sum of the lethal rates is 2.0, the band width is 2 minutes, and therefore F_0 is equal to $2 \times 2.0 = 4.0$ minutes.

The general method is very widely used and has the great advantage of relative simplicity but it should be realised that it presents only an approximation of the kinetics of microbial destruction and the method may very well underestimate the actual total lethality; in other words it can lead to considerable over-processing with the result that, although sterility is achieved, the quality of the food is impaired unnecessarily.

10.6. THE MATHEMATICAL METHOD

10.6.1. Introduction

The so-called mathematical method is a way of describing the heat penetration curve for in-container sterilisation and therefore of predicting the total process time required for a given total process lethality. It gives absolute process times, using the decimal reduction time for a given organism, whereas the general method is relative and simply compares one process with another. The mathematical method is particularly useful because it can be used to estimate the effect on the sterilisation process of changing the size of the container. The method assumes a logarithmic approach of the temperature at the thermal centre of the can to the retort temperature and is based upon a plot of the logarithm of 'temperature deficit', the difference between the retort or steam temperature and the can temperature, against time. An example of a heat penetration curve in this form is shown in Figure 10.13. The temperature deficit decreases with time as the can is heated but it is conventional to invert the logarithmic scale for temperature deficit simply to give the appearance of a positive gradient.

The plot is linear apart from the initial period of heating which is known as the 'come up time' (CUT). During this period the steam enters the retort and the retort pressure rises, becoming constant after a few minutes. Consequently the time scale on Figure 10.13 is adjusted to give an artificial zero at which point the retort is assumed to be at the desired temperature. Again by convention 42% of the come up time is assumed to be at a high enough temperature to contribute to the sterilising process and is therefore counted as heating time. Thus $t = 0$ is set at 0.58 of CUT.

The linear portion of Figure 10.13 is then described by

$$B = f_{\mathrm{H}} \log \left(\frac{j_{\mathrm{H}} I}{g} \right) \tag{10.28}$$

where B is the time to complete the process measured from the adjusted zero. The term $j_{\mathrm{H}} I$ represents the temperature deficit at the start of the linear portion and is expressed in terms of the actual initial temperature deficit I and a thermal lag factor j_{H} which is defined by

$$j_{\mathrm{H}} = \frac{pid}{I} \tag{10.29}$$

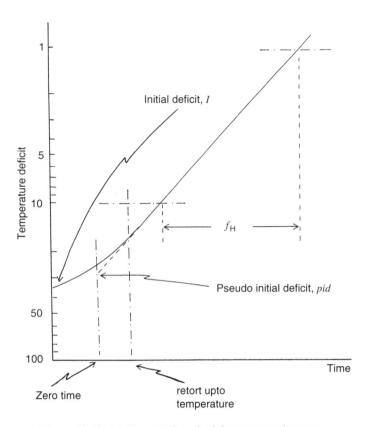

Figure 10.13. Mathematical method: heat penetration curve.

where pid is the pseudo initial temperature deficit. This quantity is defined as the point at which the extrapolated linear plot intersects the *adjusted* time scale at $t = 0$. g is the temperature deficit at the end of the process, after time B. Thus the logarithmic term in Equation (10.28) contains the beginning and end points of the heat penetration curve and the effective gradient is f_H, the heat penetration factor, which is defined as the time required for one log cycle of temperature deficit. Thus as f_H decreases the rate of heat penetration increases. It is usual, and convenient, to express both B and f_H in minutes.

10.6.2. The Procedure to Find Total Process Time

The procedure to find the total process time is usually attributed to Stumbo and is set out here. First the heat penetration curve is drawn and the values of f_H and j_H obtained. Second the required total process lethality F is calculated from Equation (10.21) as before. Third, the required temperature deficit at the end of the heating period g is determined. This is a function of the heat penetration factor f_H, the thermal resistance constant z and two other quantities j_C and U. j_C is the cooling lag factor and can be found from the cooling curve, that is a plot of temperature deficit against time during the cooling cycle. It is defined by

$$j_C = \frac{\text{cooling water temperature} - \text{pseudo initial can temperature}}{\text{cooling water temperature} - \text{initial can temperature}} \quad \text{(10.30)}$$

The lag factor for the cooling cycle is important here because a significant proportion of sterilisation takes place during the cooling period. In conduction especially there is a lag before the cooling water lowers the can temperature and therefore there is significant heating of the contents after the steam is turned off. However, in the absence of detailed information about the cooling cycle, it may be assumed that $j_H = j_C$. The final temperature deficit is also a function of the time for sterilisation, at the retort temperature T_R, which is equivalent to F_0. This is defined by

$$U = F 10^{(121.1 - T_R)/z} \quad \text{(10.31)}$$

Stumbo's empirical procedure now requires values of g to be tabulated as a function of f_H/U and j_C and a limited set of data is shown in Table 10.4. It is likely that required values will need to be interpolated from this data.

Finally the values of f_H, j_H, I and g are substituted in Equation (10.28) to give the process time B.

Example 10.10

Cans of a conductive food material yielded the following heat penetration data:

Time (min)	Temperature (°C)
0	75.0
5	75.5
10	76.0
15	82.3
20	91.1
30	99.0
40	106.0
50	109.6
60	111.6
70	112.9
80	113.7

The retort temperature reached 115°C after 5 min and remained constant thereafter during the heating cycle. The cooling cycle yielded a cooling lag factor j_C equal to 1.4. The target organism has a thermal resistance constant of $z = 10$ K and a total process lethality of 12 min is required. From a suitable plot of the heat penetration data, determine the heat penetration factor and the thermal lag factor and hence calculate the required process time.

TABLE 10.4
Values of g (K) as a Function of f_H/U and j_C for $z = 10$ K

f_H/U	j_C								
	0.4	0.6	0.8	1.0	1.2	1.4	1.6	1.8	2.0
0.20	2.27×10^{-5}	2.46×10^{-5}	2.64×10^{-5}	2.83×10^{-5}	3.02×10^{-5}	3.20×10^{-5}	3.39×10^{-5}	3.58×10^{-5}	3.76×10^{-5}
0.30	1.12×10^{-3}	1.19×10^{-3}	1.26×10^{-3}	1.33×10^{-3}	1.41×10^{-3}	1.48×10^{-3}	1.55×10^{-3}	1.63×10^{-3}	1.70×10^{-3}
0.40	7.39×10^{-3}	7.94×10^{-3}	8.44×10^{-3}	9.00×10^{-3}	9.50×10^{-3}	0.010	0.0106	0.0111	0.0116
0.50	0.0228	0.0246	0.0263	0.0281	0.0299	0.0317	0.0334	0.0352	0.0369
0.60	0.0483	0.0524	0.0567	0.0606	0.0644	0.0683	0.0728	0.0767	0.0806
0.70	0.0833	0.0906	0.0978	0.105	0.112	0.119	0.127	0.134	0.142
0.80	0.126	0.137	0.148	0.159	0.171	0.182	0.194	0.205	0.217
0.90	0.174	0.190	0.206	0.222	0.238	0.254	0.271	0.287	0.303
1.00	0.227	0.248	0.269	0.291	0.312	0.333	0.354	0.376	0.397
2.00	0.850	0.922	1.00	1.07	1.15	1.23	1.30	1.38	1.45
3.00	1.46	1.58	1.69	1.81	1.93	2.04	2.16	2.28	2.39
4.00	2.01	2.15	2.30	2.45	2.60	2.74	2.89	3.04	3.19
5.00	2.47	2.64	2.82	3.00	3.17	3.35	3.53	3.71	3.88
6.00	2.86	3.07	3.27	3.47	3.67	3.88	4.08	4.28	4.48
7.00	3.21	3.43	3.66	3.89	4.12	4.34	4.57	4.80	5.03
8.00	3.49	3.75	4.00	4.26	4.51	4.76	5.01	5.26	5.52
9.00	3.76	4.03	4.31	4.58	4.86	5.13	5.41	5.68	5.96
10.00	3.98	4.28	4.58	4.88	5.18	5.48	5.77	6.07	6.37

The first step is to calculate the temperature deficit, that is, the difference between the can centre and retort temperatures, and tabulate this as a function of time. This yields

Time (min)	Temperature deficit (K)
0	40.0
5	39.5
10	39.0
15	32.7
20	23.9
30	16.0
40	9.0
50	5.4
60	3.4
70	2.1
80	1.3

Next the heat penetration curve is plotted with an adjusted zero on the time scale at 58% of 5 min, that is, at 2.9 min. From this curve the initial temperature deficit is 40 K and the pseudo initial temperature deficit is 56 K. Hence the thermal lag factor is

$$j_H = \frac{56}{40} \quad \text{or} \quad j_H = 1.4$$

The gradient of the heat penetration curve gives the heat penetration factor as $f_H = 46.5$ min. Next U, the time for sterilisation at the retort temperature which is equivalent to F_0, is found from Equation (10.31). Thus

$$U = 12 \times 10^{(121.1-115)/10} \text{ min} \quad \text{or} \quad U = 48.9 \text{ min}$$

Consequently $f_H/U = 0.951$ and the final temperature deficit g can be found from Table 10.4. Note that interpolation is necessary and hence

$$g = 0.254 + \left[\frac{0.051}{0.10}(0.333 - 0.254) \right] \text{ K} \quad \text{or} \quad g = 0.294 \text{ K}$$

These quantities can now be substituted into Equation (10.28) to find the process time. Therefore

$$B = 46.5 \log \left(\frac{1.4 \times 40}{0.294} \right) \text{ min}$$

and $B = 106.0$ min. The total process time is now equal to B plus 2.9 min of come-up time that is, a total of 108.9 min.

10.6.3. Heat Transfer in Thermal Processing

Thermal processing is an example of unsteady-state heat transfer. Heating on the outside of the can is rapid with high heat transfer coefficients due to condensing steam and the thermal resistance at this point is not limiting. Thermal process calculations are based on the slowest heating point in the container (sometimes called the cold spot). In cans where conduction is the dominant mode of heat transfer this is the geometric centre of the can, although the geometric centre may not always be the slowest cooling point. For a liquid product where convection is dominant the cold spot should be determined experimentally but is usually taken to be a point on the can axis, either one fifth or one third from the bottom. For both pure conduction and pure convection the heating curve is linear. With conduction f_H is usually between 30 and 200 min and j_H has a value of approximately 2. The low thermal conductivity of most solid foods is a limiting factor. In contrast liquids are heated by convection, they heat more rapidly, and therefore values of f_H are lower. For convection f_H is usually between 5 and 20 min and j_H is approximately unity.

However not all heat penetration curves are linear. Figure 10.14 is an example of a broken heating curve. These are characterised by two or more linear sections and occur when the product undergoes a physical change due to the effect of heat processing. A common example is the gelation of starch which is responsible for a change from conductive to convective heating. The relationship between the dominant mode of heat transfer and the characteristics of the food is summarised in Table 10.5.

The heat penetration factor f_H depends upon the thermal diffusivity of the contents of the can and on the can dimensions; heat penetration to the centre is more rapid in a smaller

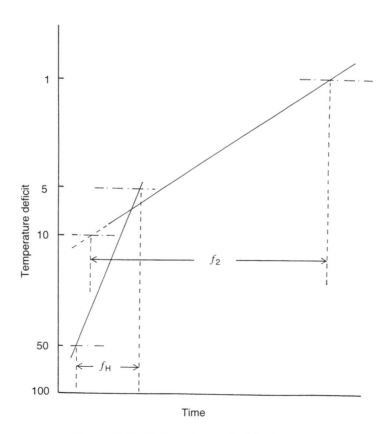

Figure 10.14. Mathematical method: broken curve.

TABLE 10.5
Heat Penetration: Dominant Mode of Heat Transfer

Group	Rate of heat transfer controlled by:	Examples
Conduction	Predominantly conduction	Ham, corned beef
Convection	Predominantly convection	Milk, fruit juice, thin soup
Mixed conduction	Conduction through continuous solid phase; convection through dispersed liquid phase. Conduction is dominant	Chopped ham with pork above the melting point of fat
Mixed convection	Convection through continuous liquid phase; conduction through dispersed solid phase. Convection is dominant	Vegetables in brine, fruit pieces in syrup
Conduction/convection broken curve	Predominantly conduction until solid changes to liquid then predominantly convection	Gels, fats
Convection /conduction broken curve	Predominantly convection until liquid changes to solid then predominantly conduction	Egg, starch-based sauces and soups, meat emulsions

can. For cases where heating occurs within the can solely by convection, and therefore where temperatures gradients are uniform, f_H is proportional to the volume to surface ratio of the can. Therefore to find f_H for a second container

$$\frac{f_{H_1}}{f_{H_2}} = \frac{V_1 A_2}{A_1 V_2}$$
(10.32)

where V and A are the volume and surface area, respectively. If the can radius and height are r and h, respectively, then

$$\frac{f_{H_1}}{f_{H_2}} = \frac{\pi r_1^2 h_1 (2\pi r_2^2 + 2\pi r_2 h_2)}{(2\pi r_1^2 + 2\pi r_1 h_1)\pi r_2^2 h_2}$$
(10.33)

and

$$\frac{f_{H_1}}{f_{H_2}} = \frac{r_1 h_1 (r_2 + h_2)}{(r_1 + h_1)r_2 h_2}$$
(10.34)

Thus f_H for the first can is obtained from the heat penetration curve and a second curve, corresponding to f_{H_2}, is drawn from the adjusted $t = 0$. This procedure assumes that pid and j_H are the same in each case; a reasonable assumption since the driving force (the temperature deficit) is large at the start of the process and is therefore dominant.

Example 10.11

For a liquid food heated in a metal can of diameter 65 mm and height 124 mm the heat penetration curve yields a heat penetration factor of 15 min. What will be the value of f_H if the same food is heated in a can 99 mm in diameter and 166 mm high?

The liquid contents of the can will heat by convection and therefore f_H for the larger can may be found from Equation (10.34). Using the subscript 1 for the smaller can, $r_1 = 32.5$ mm, $h_1 = 62$ mm, $r_2 = 49.5$ mm, and $h_2 = 83$ mm. Thus

$$f_{H_2} = 15 \times \frac{(32.5 + 62)\, 49.5 \times 83}{32.5 \times 62\, (49.5 + 83)} \text{ min} \qquad f_{H_2} = 21.8 \text{ min}$$

Agitation of the container during processing increases the magnitude of the convection currents and reduces the value of f_H. In addition convection currents tend to be increased in cans with a greater height/diameter ratio. For conduction, the value of f_H must be calculated using unsteady-state conduction theory which involves a series solution. This result can be summarised as

$$f_H = \frac{6.63 \times 10^{-7}\alpha}{[1/r^2 + 1.708/h^2]} \tag{10.35}$$

where f_H is in minutes, the radius and height of the can are in cm and the thermal diffusivity is in $m^2\, s^{-1}$. However for comparative purposes, with the same food material, it is not necessary to know the diffusivity and

$$\frac{f_{H_1}}{f_{H_2}} = \frac{[1/r_2^2 + 1.708/h_2^2]}{[1/r_1^2 + 1.708/h_1^2]} \tag{10.36}$$

10.6.4. Integrated F Value

During convective heating it is reasonable to assume that all points in the container receive an approximately equivalent heat treatment and that the F value at the can centre is equal to the mean value throughout the can. However in a conduction 'pack' there is a radial variation in the severity of heating to which the food is subjected. Stumbo describes a model which divides the contents of a cylindrical can into a series of concentric cylindrical portions and considers the heat treatment received by each, that is, an individual F value can be assigned to each hollow cylindrical portion.

Each of the cylindrical surfaces presented in this way is considered to have a uniform F value and, by implication, a uniform j value. Thus Equation (10.21) can be applied to each surface, giving a series of F values which are designated F_λ. The thermal treatment received by the can centre is now designated as F_C and the fraction of the total can volume which is enclosed by a given region of constant and uniform j is v. The model is based upon the observation that a plot of $F_\lambda - F_C$ against v is linear if the fraction v is less than 0.4, that is

$$F_\lambda - F_C = mv \tag{10.37}$$

Now, let F_S be the F value which represents the heat received by the entire container with respect to a given reduction in the microbial population of the whole container. Consequently

$$F_S = D \log\left(\frac{n_1}{n_2}\right) \tag{10.38}$$

where D is the decimal reduction time at the relevant reference temperature. Based upon these definitions and observations Stumbo's model takes the form

$$F_S = F_C - D\{\log(Dv) - \log[2.303(F_\lambda - F_C)]\} \tag{10.39}$$

Thus choosing a suitable value of v, for which the F value received by the cylindrical body defined by v is F_λ, allows F_S, the integrated value of F, to be calculated from F_λ, v, and F_C. The latter quantity is determined using the mathematical method, as described in section 10.6.2, from data taken at the can centre. The relationship between j_λ, j_{centre} and v is such that

$$\frac{j_\lambda}{j_{centre}} = 0.5 \tag{10.40}$$

and also

$$\frac{g_\lambda}{g_{centre}} = 0.5 \tag{10.41}$$

when v has a value of 0.19. Therefore taking $v = 0.19$ the model becomes

$$F_S = F_C + D\left[1.084 + \log\left(\frac{F_\lambda - F_C}{D}\right)\right] \tag{10.42}$$

The first step in the procedure to evaluate an integrated F value is to determine F_C, g and j_H in the usual way. Second j_λ and g_λ are calculated from Equations (10.40) and (10.41), respectively. Next, the value of f_H/U_λ must be found from Table 10.4 using j_λ and g_λ; this may well require interpolation in reverse. Fourth, F_λ is obtained from a modified version of Equation (10.31)

$$U_\lambda = F_\lambda 10^{(121.1 - T_R)/z} \tag{10.43}$$

Finally each of these quantities is substituted into Equation (10.42).

Example 10.12

A foodstuff is to be canned at an F_0 value of 5 using steam at 120°C. The heat penetration data are as follows:

Time (min)	Temperature (°C)
0	88.0
10	92.5
20	98.0
25	101.0
30	104.5
35	108.0
40	111.0
45	113.0
50	115.0
55	116.0
60	117.0
65	118.0

The retort came up to temperature after 10 min and the cooling curve is characterised by $j_C = 1.2$. Estimate the integrated F value, for this process if $z = 10$ K for the target organism.

The gradient of the heat penetration curve, using an adjusted zero time of 5.8 min, gives the heat penetration factor as $f_H = 43.0$ min and the value of U, based on an F_0 value of 5, is then

$$U = 5 \times 10^{\left(\frac{121.1-120}{10}\right)} \text{ min}$$

or $U = 6.44$ min. Consequently $f_H/U = 6.67$ and from Table 10.4 the final temperature deficit is $g = 3.97$ K. The values of j_λ and g_λ are now found from Equations (10.41) and (10.42), respectively. Thus

$$j_\lambda = \frac{1.20}{2} \text{ or } j_\lambda = 0.60$$

and

$$g_\lambda = \frac{3.97}{2} \text{ K or } g_\lambda = 1.99 \text{ K}$$

Again interpolating from Table 10.4, but this time in reverse, gives

$$\frac{f_H}{U} = \frac{0.41}{0.57}(4 - 3) + 3.0, \qquad \frac{f_H}{U} = 3.72$$

from which $U_\lambda = 11.56$ min. F_λ is now found from Equation (10.44), that is,

$$11.56 = F_\lambda 10^{\left(\frac{121.1-120}{10}\right)}$$

and therefore $F_\lambda = 8.97$ min. The integrated process lethality is given by Equation (10.43)

$$F_S = 5.0 + 1.0 \left[1.084 + \log\left(\frac{8.97 - 5.0}{1.0}\right) \right]$$

and $F_S = 6.68$ min. Thus the total integrated process lethality is somewhat greater than that found from the general method implying that the product has been subjected to a greater heat treatment than is actually necessary.

10.7. RETORTS FOR THERMAL PROCESSING

10.7.1. The Batch Retort

A batch operated retort consists of a pressure vessel, usually cylindrical, and a basket or grate into which individual cans are placed. The various layers of cans are separated by grids which allows the through circulation of steam. The retort is supplied with steam, cold water and compressed air. The basic operating procedure is as follows: the cans are placed in a basket which is lowered into the retort and then steam is introduced. Time is allowed for any condensed water to drain away before a drain valve at the base of the retort is closed. Whilst acidic products, for example canned fruit, which require a milder heat treatment may be able to use steam at atmospheric temperature, the higher temperatures required for less acidic foods need heat to be supplied by steam at pressures greater than atmospheric.

It is very important to vent air from the retort in order that the correct temperature is achieved; the presence of air reduces the temperature and tends to insulate the cans giving lower rates of heat transfer. After a short period the vent valve is closed and the retort is brought up to temperature by allowing the steam pressure to rise to a pre-set level. After processing for the required time the steam flow is stopped and the cooling water flow is started. As the temperature of the can contents rises the vapour pressure inside the can increases. During the heating cycle this increase in pressure is balanced by the steam pressure outside the can, however during the cooling cycle, particularly with larger cans, this internal pressure must be counter-balanced in order not to distort the can or cause a rupture of the can wall or seam. This is achieved by admitting compressed air at the same time as the cooling water. Cooling continues in the retort until the temperature falls to about $40°C$ after which the retort is opened and the cans removed.

10.7.2. Design Variations

(a) *Heating medium* A number of variations on the basic principle are possible. For example the batch retort may be aligned vertically or horizontally, it may be static or rotary and the heat source may be either steam, a steam/air mixture or pressurised hot water. Some food packaging, for example, plastic pouches, have less mechanical strength than the traditional steel can and needs an over pressure in the retort in order to balance the rise in vapour pressure within the container and maintain the integrity of the pack. Increasing the steam pressure is not possible without a further increase in temperature and therefore use is made of either steam/air mixtures or pressurised hot water. Steam as a heat source gives the highest film heat transfer coefficients. It is necessary to agitate hot water in order to provide adequate heat transfer coefficients and sufficiently high rates of heat transfer and to avoid localised overheating. Both pressurised hot water and steam/air mixtures are used for sterilising glass bottles. With the former the retort is filled with water, steam is injected, and the water is circulated through the retort by a pump. This results in slower rates of heating which minimises thermal shock to the glass. The available over-pressure in the retort prevents the removal of metal tops and lids from the bottle.

(b) *Rotary retorts* Some foods, viscous liquids in particular, may need agitation in order to induce sufficiently large convection currents and give adequate heating and acceptable processing times. Large cans may require agitation to avoid overheating of the outside layers. Two forms of rotary retort are in use. In end-over-end agitation the retort revolves, at between 6 and 15 rpm, about a horizontal axis. More commonly rotation is arranged about the axis of a cylindrical retort. In each case agitation is brought about by displacement through the food of the bubble which forms the headspace of the can and therefore the size of the head space effects the degree of agitation which is possible.

10.7.3. Continuous Retorts

Batch retorts have the advantage that they are adaptable to different can sizes but suffer from the usual disadvantages of batch operation; the overall production rate is slow and the process is labour intensive. The hydrostatic cooker (Figure 10.15) is a high capacity device consisting of three columns through which cans are transported continuously by a chain mechanism. The cans enter at the top of a column of hot water and are preheated as they travel down the column. The height of the water column is sufficient to maintain the correct steam pressure in the central section. The cans then move through the steam chamber where the residence time

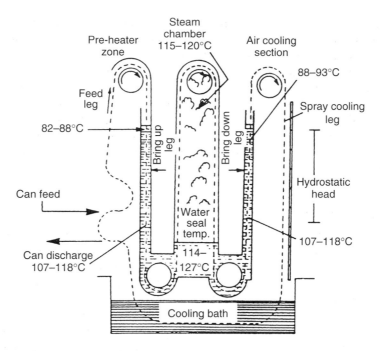

Figure 10.15. Hydrostatic cooker (from S.D. Holdsworth, Thermal Processing of Packaged Foods, Blackie Academic & Professional, 1997, with permission).

is governed by the rate at which they are conveyed and the total length of the path travelled. On leaving the steam chamber the cans pass through a column of slightly cooler water, of the same height so as to maintain the steam chamber pressure, and leave at the top of the column ready to be cooled by a series of water sprays.

10.8. CONTINUOUS FLOW STERILISATION

10.8.1. Principles of UHT Processing

Ultra high temperature (or UHT) processing describes processes in which temperatures up to 140°C are used for periods of the order of 2–4 s to achieve commercial sterility in a continuous flow of liquid or semi-liquid food or a particulate/liquid food mixture. The sterile product is then placed into pre-sterilised packaging under aseptic conditions, hence the term aseptic processing. This method gives products with both superior organoleptic properties and greater retention of nutrients than those produced by in-container sterilisation. It is used for a wide range of products including milk, custard, fruit and tomato purees, concentrated fruit juices, pizza sauce, soups, and baked beans. Conventional sterilised milk (as opposed to pasteurised) is heated to 100°C for between 20 and 30 min and gives a product with a creamy brown colour and a slightly cooked flavour. The effect of UHT processing on the flavour and colour of milk is far less pronounced, although UHT milk does retain a distinctive flavour compared to pasteurised milk. The process dates to the 1940s with the introduction of the Dole process for filling pre-sterilised metal cans. Subsequent developments of the basic idea have included the use of Tetrapacks and plastic packaging.

UHT processing has a number of advantages. There is better retention of flavour and textual properties and less destruction of heat sensitive vitamins such as thiamine. The rates of browning and caramelisation reactions are lower which results in less darkening of the product. Also there are overall savings in processing costs because it is essentially a continuous process in contrast to traditional batch retorting. The disadvantage of UHT processing is that the reduction of enzyme activity is less than in conventional sterilisation.

10.8.2. Process Description

Aseptic processing consists of heating, holding, and cooling steps after which liquid flows continuously to an aseptic balance tank followed by continuous aseptic filling and sealing of the pack. Either direct or indirect heating can be used in UHT processing. With direct heating culinary grade steam is injected into the product at pressure of about 900 kPa and a temperature of 175°C bringing the temperature up to the required level in about 0.1 s. However control of the temperature is difficult and the method is generally suitable only for low-viscosity liquids. Alternatively, the liquid food can be sprayed into a vessel containing steam at the required pressure and temperature. This is known as steam infusion and heating takes about 0.25 s and consequently control of the temperature is easier. Steam infusion can cope with higher viscosities than steam injection. In each case there is dilution of the liquid product and the extra water from the condensed steam must be removed. This is achieved by flash cooling. After holding at the required temperature for the required time the liquid is passed to a chamber under vacuum where the water vaporises. The enthalpy of the water vapour may be partially recovered and reused but heat recovery is not as effective as in direct heating systems. There may also be loss of volatiles during the cooling process.

Ohmic heating involves the direct electrical heating of particulate/liquid food mixtures by the passage of a low frequency (50 Hz) alternating electrical current through the food stream. This is an example of volumetric heating, similar to microwave heating, and has the advantage of rapid and more even heating without large temperature gradients. However there is a much greater depth of penetration than with microwave heating, because of the considerably lower frequency. In contrast to conventional processing techniques there is no need to over-process the liquid phase in order to ensure adequate heating at the centre of particles and therefore ohmic heating avoids thermal damage to the liquid. Applications of this heating technique include fruit pie fillings, ratatouille, cottage cheese, pasta in sauce, and beef bourgignone.

Indirect heating is achieved using either a tubular, scraped surface or plate heat exchanger. Rates of heat transfer are lower than in direct heating but a greater proportion of heat can be recovered in the regenerator section of the exchanger. A plate heat exchanger used for UHT processing is very similar to that used in pasteurisation but with two important differences. After the regeneration section there is an homogenisation step to reduce the size of fat droplets and to prevent the separation of cream during product storage. In addition the sterilised product is cooled rapidly to 85°C after holding because using, for example, milk at the holding temperature for regenerative heating is likely to induce thermal damage in the incoming milk.

In the sterilisation of liquid/particulate mixtures the residence time in the holding tube is determined by the velocity of the liquid. Calculations of process lethality must be based on the fluid of highest velocity, which has the lowest residence time, and therefore knowledge of the velocity profile is essential. The flatter velocity profile in turbulent flow is advantageous because the maximum fluid velocity is approximately 20% greater than the mean velocity whereas in laminar flow the maximum is about twice the mean. The presence of particulate matter complicates the flow pattern and this is a area of considerable current research interest.

NOMENCLATURE

a	Parameter in Equation (10.12)
A	Area; constant
B	Constant; process time from the adjusted zero
Bi	Biot number
c_p	Heat capacity
d	Diameter
D	Decimal reduction time
D_0	Decimal reduction time at $121.1°C$
f_H	Heat penetration factor
F	Total process lethality
F_C	Process lethality at can centre
F_S	Integrated process lethality
F_0	Total process lethality at $121.1°C$
F_λ	Process lethality at a given surface
Fo	Fourier number
g	Temperature deficit after time B
g_λ	Value of g at a given surface
h	Film heat transfer coefficient; can height
I	Initial temperature deficit lag factor or intercept index
j_C	Cooling lag factor
j_{centre}	Thermal lag factor at can centre
j_H	Thermal lag factor
j_λ	Value of j at a given surface
k	Thermal conductivity; rate constant
L	Characteristic length; half slab thickness; lethal rate
m	Mass; reciprocal of Biot number; gradient
n	Dimensionless position; number of micro-organisms
pid	Pseudo initial temperature deficit
Q	Rate of heat transfer
r	Can radius; number of containers in batch
t	Time
T	Temperature
T_0	Initial surface temperature
T_R	Retort temperature
U	Overall heat transfer coefficient; sterilisation time at retort temperature
V	Volume
x	Distance in the x direction
y	Distance in the y direction
Y	Dimensionless temperature
z	Distance in the z direction; thermal resistance constant (z value)

GREEK SYMBOLS

α Thermal diffusivity
ρ Density
υ Volume fraction

SUBSCRIPTS

S Steam

PROBLEMS

10.1. A batch of 1,200 kg of tomato serum, initially at 20°C, is heated in an agitated jacketed vessel. The vessel has a heating area of 4.7 m^2 and uses steam at a pressure of 140 kPa. The temperature of the serum, which has a mean heat capacity of 3,800 J kg^{-1} K^{-1}, increased with time as follows:

Time (min)	Temperature (°C)
20	55.1
30	67.0
40	76.1
50	83.5

Determine the overall heat transfer coefficient.

10.2. A large block of frozen beef 8 cm thick is heated from −25°C by placing it in air at a temperature of −3°C. The thermal conductivity, density and mean heat capacity of the block are 1.40 W m^{-1} K^{-1}, 1,300 kg m^{-3}, and 1,500 J kg^{-1} K^{-1}, respectively. How long will elapse before the centre of block reaches −10°C?

10.3. Slices of meat 4 mm thick and originally at 50°C are cooled in a chiller operating at 5°C. It takes 20 min to reduce the temperature at the centre of the slice to 10°C. How long would it take to achieve this temperature at the centre of a 6 mm thick slice and what will be the temperature 2 mm from the surface of the 6 mm slice when the centre reaches 10°C?

10.4. A liver sausage of diameter 8 mm is cooked in air at 130°C. Its thermal properties are: thermal conductivity, 0.40 W m^{-1} K^{-1}; mean heat capacity, 3,430 J kg^{-1} K^{-1}; density, 800 kg m^{-3}. If the film heat transfer coefficient is 20 W m^{-2} K^{-1}, estimate the time for the centre temperature to reach 110°C from an initial temperature of 20°C.

10.5. Estimate the temperature after 30 min at the centre of the can in Example 10.5 if heat is transferred principally by convection and the heat transfer coefficient is 50 W m^{-2} K^{-1}.

10.6. The initial concentration of a spoilage organism is 100 spores per can. Calculate the F_0 value needed to ensure a concentration after processing of 10^{-10} spores per can. For the target organism, $D_0 = 1.0$ min.

10.7. The initial population of *clostridium botulinum* spores in a batch of cans is 10^3 per can. Estimate the number of surviving spores after a process for which $F_0 = 3$ min. $D_0 = 0.25$ min for *clostridium botulinum*.

10.8. Calculate the process lethality required to reduce the population of a micro-organism from 5×10^4 per container to one organism per container in a batch of 1 million cans processed at 115°C. For the organism concerned $D_{115} = 1.2$ min.

10.9. Calculate the lethal rate at temperatures of 111°C and 125°C, respectively for a organism with a z value of 10 K.

10.10. Heat penetration data was collected from the thermal centre of a retortable pouch containing fish. Determine the value of F_0 for the process.

Time (min)	Temperature (°C)
0	12
2	66
4	99
6	110
8	114
10	118
12	122
14	122
16	122
18	118
20	117
22	116
24	108
26	104

10.11. The following temperature history was recorded for a canned food at a retort temperature of 120°C:

Time (min)	Temperature (°C)
2	70
4	100
6	109
8	113
10	115
12	118
14	119
16	119.5
18	119.5
20	120
24	120
28	120

Estimate the point at which cooling should begin if an F_0 of 10 min is required for the process. Ignore the contribution of cooling to the total lethality.

10.12. Metal cans, 6 cm in diameter and 10 cm in height, were filled with a liquid food material and placed in a retort with a steam temperature of 116°C. This yielded the following penetration data:

Time (min)	Temperature (°C)
0	86
15	88
20	96
25	101.5
30	105.5
35	108.5
40	110.6
50	113.2

The retort came up to temperature after 8.6 min. Determine the initial temperature deficit, the pseudo initial temperature deficit, the heat penetration factor and the thermal lag factor.

10.13. The same product is to be manufactured in larger cans (diameter 10 cm and height 15 cm) under the same retorting conditions. Determine the new f_H and calculate the required total process time for the larger cans from 'steam on' if the required value of F_0 is 15 min and $z = 10$ K for the target organism.

FURTHER READING

S. D. Holdsworth, *Thermal Processing of Packaged Foods*, Blackie Academic & Professional (1997).
C. R. Stumbo, *Thermobacteriology in Food Processing*, Academic Press, New York (1973).

11

Low-Temperature Preservation

11.1. PRINCIPLES OF LOW TEMPERATURE PRESERVATION

Freezing has long been used to preserve high value food products such as meat; fish; particular foods where the quality of the frozen product is significantly better than the alternative, such as peas; and increasingly for other convenience foods ranging from chipped potatoes to complete ready meals. Lowering the temperature of foodstuffs reduces microbiological and biochemical spoilage by decreasing microbial growth rates and by removing liquid water which then becomes unavailable to support microbial growth. Freezing refers to the storage of food at temperatures between $-18°C$ and $-30°C$. In general lower storage temperatures give a longer shelf-life. For example soft fruits may be stored for between 3 and 6 months at $-12°C$ but for 24 months and beyond at $-24°C$. Most meat has a shelf-life of 6 to 9 months at $-12°C$ and this increases to between 15 and 24 months at temperatures down to $-24°C$. In contrast chilling is defined by a storage temperature range between $-1°C$ and $8°C$ and is used for meat, fish, dairy products and chilled recipe dishes prior to consumption. It has little or no effect on the nutritional content or organoleptic properties of food. Similarly, the freezing process itself has little or no effect on the nutritional value of frozen foods. Conversely the quality of the initial raw material cannot be improved by freezing and only high quality raw materials should be selected for freezing. Thus the quality and nutrient content of any food at the point of consumption is dependent upon the quality of the original raw material; the length of storage and the storage conditions; and the extent and nature of the freezing process.

The properly controlled freezing and thawing of foods allows considerable retention of organoleptic properties: taste, colour, texture, and nutritional content. However there are some potential disadvantages and problems associated with freezing. It may have an adverse affect on texture especially where poor temperature control leads to partial thawing and recrystallisation; high rates of vitamin loss are possible during lengthy storage; and oxidative rancidity may occur during the frozen storage of foods with a high fat content.

It is important to remember that some bacteria grow well at low temperatures, for example *Pseudomonas* (a psychrophilic spoilage bacteria which grows on the surface of meat and fish), that the slow growth of some moulds occurs at chill temperatures and that yeast activity is known at temperatures down to $-2°C$. It is essential therefore to cool cooked foods as rapidly as possible before pathogenic bacteria have an opportunity to grow; British Government advice recommends that cook-chill and cook-freeze products should be cooled to below $3°C$ within 90 min.

The advantages to be gained by freezing food in terms of quality must be balanced against the considerable costs involved in removing latent heat from the food, in other words the power requirement of the refrigeration circuit used to drive the freezer, and the costs of maintaining low temperatures in the cold store and during refrigerated transport. The use of appropriate preliminary processing prior to freezing can help to maximise the final product quality. Fruit and vegetables are normally blanched before freezing; this involves holding food at temperatures of about 95°C for 1 or 2 min. The effect of blanching is to destroy the enzymes which result in the loss of nutrients, colour, texture, and flavour. Blanching also reduces the microbiological load on the food. Blanching itself may result in some initial loss of nutrients, but this is offset by improved retention of nutrients over the period of frozen storage by the correct application of blanching.

11.2. FREEZING RATE AND FREEZING POINT

As the temperature of food is lowered a typical freezing curve such as that shown in Figure 11.1 results. Three stages in this process may be identified; the removal of sensible heat from the food between the initial temperature and the freezing temperature, the removal of the latent heat of fusion leading to a change of state and the formation of ice crystals and, after the formation of the majority of the ice, further sensible heat removal down to required storage temperature. However it is important to realise that there is no clear freezing point of a food material unlike that of a pure compound such as water for which a distinctive freezing point (0°C) can be measured. Instead there is a freezing range which is normally between 0°C and about −2°C or −3°C. The freezing point is usually defined as the highest temperature at which ice crystals are found to be stable.

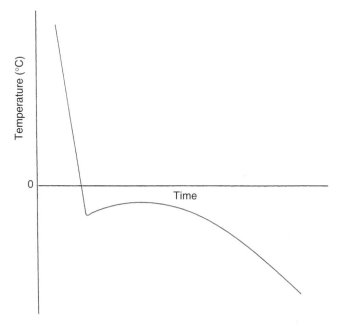

Figure 11.1. Freezing curve.

The formation of ice crystals may be considered to be the start of the freezing period. Immediately before this there is a degree of supercooling which initiates the formation of ice nuclei. Nucleation is where small numbers of water molecules aggregate and begin to form a lattice structure. It may be either homogeneous in which the molecules aggregate spontaneously or heterogeneous in which nucleation begins because of the presence of a very small particle or an impurity in the liquid. The latter is more common in foods because the formation of ice crystals is centred on cell walls.

The existence of a freezing temperature range rather than a clear freezing point is explained by the presence of soluble components in the food. During the removal of latent heat the temperature remains approximately constant (and would be truly constant for pure water). However as the water freezes the remaining liquid water has an increased concentration of these solutes and the freezing point is further depressed. Lower molecular weight solutes also give a greater depression of the freezing point.

Liquid water may be present even at temperatures well below the freezing point (and may constitute up to 10% of the food mass) and therefore there is not necessarily a clear end to the freezing period. Nevertheless it is necessary to calculate the total enthalpy change Δh, or heat load, during the freezing process. This is required in order to calculate the size of both freezing equipment and the refrigeration circuit. The heat load is simply the addition of the two sensible heat components and the latent heat, therefore

$$\Delta h = c_{p_u}(T_i - T_f) + \lambda + c_{p_f}(T_f - T_{final}) \tag{11,1}$$

where T_i, T_f, and T_{final} are the initial, freezing, and final temperatures, respectively, c_{p_u} and c_{p_f} are the heat capacities of the unfrozen and frozen food respectively and λ is the latent heat of fusion. However there is no clear definition of the latent heat of fusion for food materials. Two methods can be used to deal with this problem. The first is to assume that the latent heat is equal to the product of the latent heat of water and the moisture content of the food. The second is to use instead the change in enthalpy between the freezing temperature and an arbitrary temperature at which freezing is considered to be complete. The variation of enthalpy with temperature then takes the form of Figure 11.2. Amongst other sources, Rao and Rizvi give comprehensive tables of the enthalpies of various foods as a function of temperature.

Example 11.1

Calculate the heat load involved in freezing a fish block, initially at a temperature of 10°C, to a final temperature of −25°C. The freezing temperature, the latent heat of fusion, the heat capacities of the frozen and unfrozen fish may be assumed to be −2°C, 280 kJ kg^{-1}, 1.9 kJ kg^{-1} K^{-1} and 3.6 kJ kg^{-1} K^{-1}, respectively.

The total heat load is found by summing the two sensible heat components and the latent heat of fusion. Therefore the heat load per unit mass of fish is

$$\Delta h = 3.6(10 - (-2)) + 280 + 1.9(-2 - (-25)) \text{ kJ kg}^{-1}$$

and

$$\Delta h = 366.9 \text{ kJ kg}^{-1}$$

Note that the latent heat component represents 76% of the total heat load.

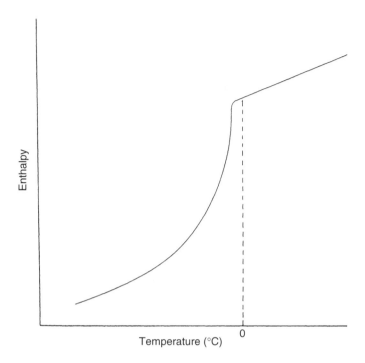

Figure 11.2. Enthalpy of frozen food as a function of temperature.

The rate at which freezing takes place determines the size of ice crystals produced and can have a very significant effect on the quality of the frozen product. Slow freezing leads to the formation of a limited number of larger nuclei and a tendency for these nuclei to grow into large crystals which form in the spaces between cells and result in structural damage. The effect of large ice crystal formation is to cause the disruption of cellular integrity. For example, vegetables have a rigid cellular structure with semi-permeable cell walls composed mainly of cellulose. On thawing the cells will be unable to maintain rigidity due to disruption of the cell walls by the ice crystals and a soft, flaccid texture will result when the vegetable is cooked. High drip loss occurs where cellular damage has occurred in slow freezing leading to the inability of the food to retain water.

In contrast rapid freezing, that is a high rate of heat removal, results in a high nucleation rate, the formation of a large number of small nuclei and the subsequent growth of many small ice crystals. These small crystals grow both within and outside cells and consequently the cells maintain their integrity which in turn minimises drip loss during thawing. The cellular structure of fish and meat differs from that of vegetable material. The disruptive effect of ice crystal formation is minimised due to the elasticity of the cellular structure in muscle. The loss of quality in fish and meat products is associated largely with a loss of functionality of proteins. As the water in fish and meat freezes, two effects are observed which have a destructive effect on proteins. First, as water begins to form ice, there is an increased concentration of enzymes in the remaining water. These enzymes denature the proteins and therefore effect protein functionality. Second, as water forms ice, the build up of salt concentration in the remaining water causes protein denaturation. These destructive effects on protein due to slow freezing occur during the formation of ice crystals in the temperature range $0°C$ to $-5°C$. The associated loss of texture and water holding capacity results in a poor quality thawed product. It is important to minimise the loss of texture in foods by rapidly freezing the product and ensuring that the time taken for a food to pass through the temperature range $0°C$ to $-5°C$, the zone of ice crystal formation, is as short as possible.

11.3. THE FROZEN STATE

11.3.1. Physical Properties of Frozen Food

There are significant changes in the physical properties of food as it enters the frozen state. The important properties are those which are combined in thermal diffusivity, namely density, heat capacity, and thermal conductivity. The density of frozen water is less than that of liquid water; at $0°C$ water has a density of $999.8\,\mathrm{kg\,m^{-3}}$ and this falls to about $920\,\mathrm{kg\,m^{-3}}$ for ice over the temperature range encountered in food freezing. The density of unfrozen foods are of course greater than that of water and therefore, as the proportion of frozen water increases during freezing, the density of frozen food falls as indicated in Figure 11.3. However these changes in density are relatively small; the reduction is of the order of 10–15%.

There is a sudden decrease in the heat capacity of water as it enters the frozen state and the heat capacity of ice continues to fall, although much less dramatically, as the temperature is reduced further (Table 11.1). This is reflected in the lower heat capacities of frozen foods which are listed in Table 11.2.

In contrast the thermal conductivity of foods increases on freezing; Table 11.3 lists some typical data and Figure 11.4 indicates the relationship between thermal conductivity and temperature at and below the freezing point. Thermal conductivity falls sharply as the temperature increases from frozen storage temperatures but remains approximately constant at temperatures just above freezing. The dependence of thermal diffusivity on temperature combines

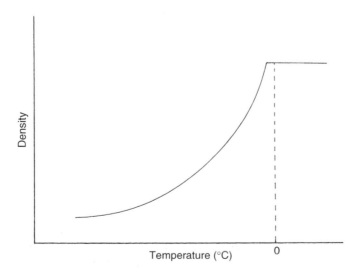

Figure 11.3. Density of frozen food as a function of temperature.

TABLE 11.1

Heat Capacity of Water and Ice

Temperature (°C)	Heat capacity (kJ kg^{-1} K^{-1})	
	Water	Ice
5	4.202	
4	4.205	
3	4.208	
2	4.211	
1	4.214	
0	4.218	
−1	4.226	
−2	4.228	
−3	4.230	
−4	4.231	
−5	4.234	
−2.2		2.101
−2.6		2.095
−4.9		2.065
−8.1		2.050
−11		2.035
−14.6		2.000
−20.8		1.954
−24.5		1.928
−31.8		1.865

TABLE 11.2

Approximate Heat Capacities of Frozen
Foods

Material	Heat capacity (kJ kg^{-1} K^{-1})	
	Above freezing	Below freezing
Fish	3,5–3.7	1.8–2.0
Beef	2.5	1.5
Pork	2.85	1.6
Poultry	3.3	1.55
Bread	2.7–2.9	1.4
Carrots	3.7	1.8
Strawberries	3.9	1.1
Peas	3.3	1.8
Potatoes	3.4	1.8
Tomatoes	4.0	2.0

the effects on density, heat capacity and thermal conductivity. Above the freezing temperature thermal diffusivity remains approximately independent of temperature but at the freezing point there is a clear discontinuity and as the temperature continues to fall thermal diffusivity increases steeply (Figure 11.5).

TABLE 11.3
Thermal Conductivities of Frozen Foods

Material	Temperature (°C)	Thermal conductivity $(W\,m^{-1}\,K^{-1})$
Water	0	0.569
Ice	0	2.25
	−23	2.41
Fish	0	0.43–0.54
	−10	1.2–1.5
Salmon	−29	1.30
	4	0.502
Cod	−10	1.66
Beef	−10	1.35
Lamb	5	0.41–0.48
Pork	2	0.44–0.54
Pork	−15	1.11
Turkey	2.8	0.50
	−10	1.46
Strawberries	13	0.675
	−12	1.10
Beans	−12	0.80
Cauliflower	−6.6	0.80

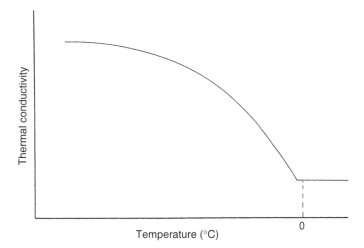

Figure 11.4. Thermal conductivity of frozen food as a function of temperature.

11.3.2. Food Quality During Frozen Storage

It is important to ensure that a constant low temperature is maintained within an industrial cold store, or indeed within a domestic freezer, during the storage of frozen food. Fluctuations in freezer temperature give rise to the thawing and subsequent recrystallisation of ice and can result in a severe loss of quality; problems similar to those associated with slow freezing can

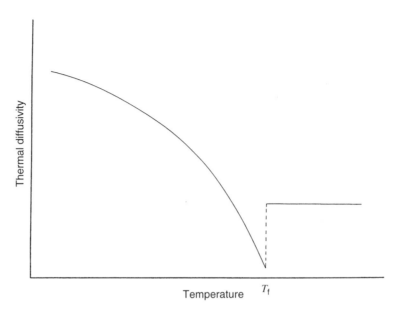

Figure 11.5. Thermal diffusivity of frozen food as a function of temperature.

occur. As the temperature rises small ice crystals will melt but on re-freezing the crystal size increases resulting in disruption to cell walls and a loss of texture.

Commercial cold stores are held at about $-25°C$ and domestic freezers normally operate at $-18°C$. Maintenance of these temperatures depends not only on the design of the equipment and the thickness of insulation but on how frequently the doors are opened and closed. If the temperature of incoming material is too high then this may affect the performance of the refrigeration circuit. In addition it is important to maintain a high humidity within a cold store or freezer. A decrease in humidity creates a difference in the partial pressure of water vapour between the air immediately above the surface of frozen food and the bulk air within the storage compartment. This concentration difference gives rise to the sublimation of ice on the food surface and a loss of weight. Sublimation tends to dry out the food surface which results in discoloration, the denaturation of proteins and has an adverse effect on texture. This phenomenon is usually referred to as freezer burn and is more likely to occur with food pieces which have a high surface area to volume ratio. Food is also more prone to oxidative rancidity as a result of the sublimation of ice because the removal of water will increase the exposure of macromolecules to oxygen.

The loss of quality during frozen storage can be minimised by adopting a number of good practices. First the relative humidity should be kept above 90% and measures taken to reduce the movement of air across the stored product by ensuring that the opening and closing of cold store doors is kept to a minimum. Second, frozen food should be glazed. A glaze is a layer of water or aqueous solution applied to the food surface which turns to ice in the initial stages of freezing. It is applied by a spray or by immersion, prior to freezing, and protects the food during storage by subliming preferentially from the surface. It may be necessary to re-apply glazes during long-term storage. Third, the use of appropriate packaging has a beneficial effect by preventing freezer burn and reducing the risk of oxidative rancidity.

11.4. FREEZING EQUIPMENT

11.4.1. Plate Freezer

A plate freezer consists of a series of parallel, hollow extruded aluminium plates through which a refrigerant circulates at temperatures down to -40°C. Food blocks are placed between the plates which are then moved together hydraulically and a slight pressure is exerted on the food to be frozen. This is illustrated in Figure 11.6. Spacers, fractionally smaller than the food, may be inserted to prevent the food being crushed. The principal heat transfer mechanism is conduction and the overall thermal resistance includes both that of the plate and the packaging material. Consequently it is important to achieve good contact between the food and the plate and to exclude air. The plate freezer is a very efficient method of freezing food with relatively high rates of heat transfer but the technique is inevitably limited to flat foods and packs of relatively shallow dimension. Typically, plate freezers are used for fish blocks and convenience foods in rectangular shaped packs, for example, fish fingers and hamburgers. The freezing time is dependant upon the pack thickness. The plates are usually horizontal and commonly are operated batchwise with several parallel plates contained in a insulated cabinet; however continuous systems are available and are used for high throughputs. Vertical plate freezers are used on board fishing vessels to facilitate easier loading and operation and to reduce plant volume. Whole fish are packed into the spaces between the plates and the gaps between the fish are filled with water. This results in an ice block containing tightly packed fish which is delivered to the fish processing plant. Many plate freezer systems have a facility for circulating water through the plates at the end of the freezing process in order to thaw the surface of the product slightly and assist in its release.

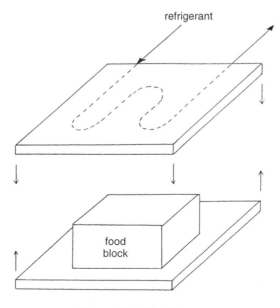

Figure 11.6. Plate freezer.

11.4.2. Blast Freezer

In the blast freezer cold air, which has been cooled by indirect contact with a refrigerant in a heat exchanger, is blown over the surface of the food. The simplest form of blast freezer is a batch operated cabinet freezer in which the food to be frozen is placed on trays which are then wheeled into the cabinet on trolleys. The cabinet is designed to move air evenly at equal velocity over all the surfaces to be frozen and trolleys and trays are designed to offer the same resistance to air flow no matter which path the air takes through the freezer. For this reason the freezer must always be used at full capacity. The great advantage of the blast freezer is its ability to accommodate foods of all shapes and sizes; in this respect it is far more versatile than the plate freezer. However the freezing time will be relatively long because the temperature difference between the food and the cold air is rather smaller than that between food and plate in a plate freezer and it is difficult to achieve intimate contact between the air and food pieces.

In a continuous air blast freezer the food travels on a conveyor belt through a tunnel in which there is a counter-current flow of cold air. The residence time in the freezer, which is a function of the speed of the conveyor, is made equal to the required freezing time. This in turn is dependant on the air velocity and the heat transfer coefficient can be obtained from an expression of the form

$$Nu \propto Re^{0.8} \qquad \qquad (11.2)$$

In the spiral freezer the conveyor belt is wound around, and in contact with, a vertical rotating drum and moves slowly 'up' the drum to exit at the top. Air is blown up through the drum before being directed downwards over the food, via a series of baffles. The food is loaded into the freezer at the bottom and exits at the top. By passing the belt to a second drum it is possible for the food to both enter and leave at the base of the freezer. This arrangement is a major advantage when freezers are incorporated into process lines. The major advantages of the spiral freezer is the flexibility of operation (it can be used for more than one product at a time) and the high throughput of up to about $8 \, t \, h^{-1}$.

11.4.3. Fluidised Bed Freezer

The principle of the fluidised bed freezer is covered in section 13.4.6. Suffice it to say here that it may be seen as a modification of the air blast freezer in which there is far better contact between the air and the food pieces being frozen. The gas-particle heat transfer coefficient can be found from the Kunii and Levenspiel correlation for particle Reynolds numbers below 100. This takes the form

$$Nu = 0.03 \, Re^{1.3} \qquad \qquad (11.3)$$

However the Reynolds number of, say, a pea in a fluidised bed freezer is of the order of 1,000 and thus Equation (11.3) tends to over estimate the heat transfer coefficient somewhat.

11.4.4. Scraped Surface Freezer

The principal example of the use of scraped surface heat exchangers as freezers is in the production of ice-cream and margarine where a uniform product with a consistently smooth texture is required. The desirable texture is achieved by controlling crystal growth by scraping the newly formed solid crystals from the refrigerated internal surface with the scraper blade. In addition the blades function as beaters and help to incorporate air into the product giving the desired light texture. Ice cream usually leaves a scraped surface freezer at about $-5°C$ and is then removed to a blast or plate freezer, depending upon the geometry of the pack, for further temperature reduction.

11.4.5. Cryogenic and Immersion Freezing

In cryogenic freezing the food to be frozen is placed in direct contact with a liquid refrigerant; the food releases heat to provide the latent heat of vaporisation of the refrigerant which then undergoes a phase change. The most commonly used refrigerant is liquid nitrogen (boiling point $-196°C$) but solid carbon dioxide which sublimes at a temperature $-78°C$ can be used and its use is widespread in the U.S.A. Cryogenic freezing with nitrogen is carried out by first passing the food through nitrogen vapour at about $-50°C$ and then freezing the food by spraying the refrigerant directly onto the food. Immersion of the food product in the liquid refrigerant is undesirable because of the risk of thermal shock which produces internal pressures within the food causing deformation and a resultant loss of texture and quality. The cryogenic technique avoids this problem.

A variety of equipment configurations is available. The simplest is the batch cabinet freezer in which liquid nitrogen is injected into a high velocity gas stream circulating in the insulated cabinet. The food to be frozen is placed in trays which fit onto a trolley rather in the manner of a tray drier. In a tunnel freezer food is conveyed on a mesh belt through an insulated tunnel. The food passes through a region in which cold nitrogen gas circulates at a high velocity before being conveyed to the spray heads, where liquid nitrogen is sprayed directly onto the food surface. Fish, meat, poultry, fruit, vegetables, and bakery products can all be frozen in this way. Particulate foods such as diced meat or vegetables are best processed in a rotary cryogenic freezer. This consists of a long hollow drum, which is inclined slightly to the horizontal to aid the flow of material through the device. Food falls through a curtain of nitrogen vapour and liquid nitrogen resulting in an IQF product. Such equipment may be up to 1.5 m in diameter and 10 m in length.

The equipment for cryogenic freezing is comparatively simple and consequently the capital cost is relatively low. Very rapid freezing of the product is obtained with short freezing times and in comparison with air blast freezers a much shorter start-up and defrost cycle is required. In addition there is a lower mass loss from the food due to dehydration. However the major disadvantage of this method of freezing is the high cost of the refrigerant; a mass of liquid nitrogen up to three times the mass of food to be frozen may be required. Consequently cryogenic freezing is used for premium high-value products, such as trout or strawberries, where the high cost of freezing may be recovered. Often it is used in periods of glut as a back up to conventional freezers.

In immersion freezing the food is immersed directly in a liquid refrigerant such as liquid nitrogen. The refrigerant is selected on the basis of its thermodynamic properties and safety; other refrigerants can be used but there is a risk of contamination. Supersaturated salt brines, with temperatures down to $-21°C$, have been widely used but packaging of the product is necessary to prevent salt contamination and the materials of construction of the freezer require anti-corrosive treatment. More recently liquids such as glycerol, glycol, and calcium chloride solution have been used. A specific application is the immersion freezing of tightly packaged poultry carcasses in glycol. The process causes surface freezing, giving a desirable white surface due to ice crystal formation, prior to blast freezing of the whole carcass. The major advantage of this type of freezing is that very high heat transfer coefficients (up to

$1,000\ \mathrm{W\,m^{-2}\,K^{-1}}$) are possible and freezing is very rapid. Furthermore there is no mass loss on freezing. However foods suffer a high degree of thermal shock and immersion freezing should be used only for foods which can withstand this treatment or for foods where the consequent changes are acceptable, for example in the freezing of diced meat.

11.5. PREDICTION OF FREEZING TIME

11.5.1. Plank's Equation

Plank's model is one of the most widely used equations to predict freezing time and is one of the most simple to use. The derivation makes a number of assumptions which may be summarised as follows:

i   initially all food is at a distinct freezing point, which remains constant for both the frozen and unfrozen layers;

ii   the thermal conductivity of the food is constant and the frozen and unfrozen layers have equal density;

iii there is a distinct interface, or ice front, which moves from the freezing medium into the block at a uniform rate;

iv   the heat capacity of the frozen layer is negligible, that is, the latent heat changes are far greater than the sensible heat changes to the frozen layer; and

v   heat transfer is sufficiently slow that it approximates to a steady-state process.

Figure 11.7 represents a food block of cross-sectional area A and thickness a which is frozen by contact with cold air on both sides of the block. The frozen layer on each side of the block

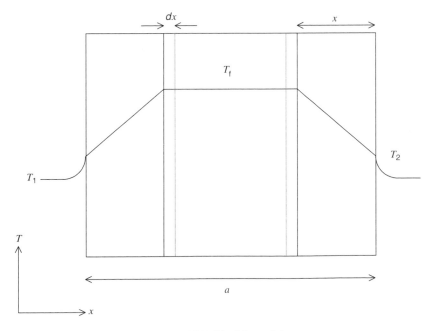

Figure 11.7. Plank's model.

then has a thickness x and the interface between the frozen and unfrozen layers advances into the block until $x = a/2$. Heat is removed from the surface of the block to the surrounding air by convection and therefore the rate of removal of heat is

$$Q = hA(T_2 - T_1) \tag{11.4}$$

where T_2 is the surface temperature and T_1 is the temperature of the surrounding air. The rate of heat transfer by conduction through the frozen layer is given by Fourier's equation, thus

$$Q = \frac{kA(T_f - T_2)}{x} \tag{11.5}$$

where k is the thermal conductivity of the frozen food. Consider now a thin layer of thickness dx which freezes in a time dt. The volume of this frozen layer is then $A\,dx$ and the mass of the layer is $\rho A\,dx$ if ρ is the food density. Now, if the latent heat of fusion is λ then the heat removed in time dt is $\rho A \lambda\,dx$ and the rate of removal of this heat is

$$Q = \rho A \lambda \frac{dx}{dt} \tag{11.6}$$

The surface temperature T_2 is not necessarily known, and in any case is not of any great interest, and it may be eliminated from Equations (11.4) and (11.5). Consequently

$$\frac{Q}{hA} + T_1 = T_f - \frac{Qx}{kA} \tag{11.7}$$

Rearranging this expression gives

$$Q = \frac{A(T_f - T_1)}{(1/h + x/k)} \tag{11.8}$$

The rate of removal of latent heat from the unfrozen layer must be equal to the rate at which heat is transferred through the frozen layer to the surrounding air and therefore

$$\frac{A(T_f - T_1)}{(1/h + x/k)} = \rho A \lambda \frac{dx}{dt} \tag{11.9}$$

This expression can now be integrated between the boundary conditions $t = 0$ at $x = 0$ and $t = t$ (the total freezing time) at $x = a/2$ (the centre of the block)

$$\int_0^t dt = \frac{\rho \lambda}{(T_f - T_1)} \int_0^{a/2} \left(\frac{1}{h} + \frac{x}{k} \right) dx \tag{11.10}$$

which gives the freezing time to be

$$t = \frac{\rho \lambda}{(T_f - T_1)} \left[\frac{a}{2h} + \frac{a^2}{8k} \right] \tag{11.11}$$

A more generalised form of Plank's equation can be written as

$$t = \frac{\rho \lambda}{(T_f - T_1)} \left[\frac{Pa}{h} + \frac{Ra^2}{k} \right] \tag{11.12}$$

where the parameters P and R assume different values for different geometries. These are summarised in Table 11.4.

TABLE 11.4

Values of Parameters in Plank's Equation

	P	R	a
Infinite slab	$\frac{1}{2}$	$\frac{1}{8}$	Thickness^a
Infinite cylinder	$\frac{1}{4}$	$\frac{1}{16}$	Diameter
Sphere	$\frac{1}{6}$	$\frac{1}{24}$	Diameter

^{a}When frozen from both surfaces; a = half thickness for freezing from one surface.

Example 11.2

Fish is to be frozen in a flat pack, 2 cm thick, using a plate freezer. Estimate the time to just freeze the pack if the fish is already at the freezing temperature. The plates are at $-30°C$ and the heat transfer coefficient between the blocks and the freezer plates can be assumed to be 80 W m^{-2} K^{-1}. The following data are available:

freezing temperature $-2°C$
latent heat of fusion   280 kJ kg^{-1}
density   880 kg m^{-3}
thermal conductivity   1.5 W m^{-1} K^{-1}

The fish is already at its freezing temperature and therefore the heat load is equal to the latent heat of fusion. The freezing time using Plank's equation is then

$$t = \frac{880 \times 280 \times 10^3}{(-2 - (-30))} \left[\frac{0.02}{2 \times 80} + \frac{(0.02)^2}{8 \times 1.5} \right] \text{ s}, \qquad t = 1,393 \text{ s}$$

Plank's equation suffers from some serious limitations in addition to the simplicity of the assumptions made in the course of its derivation. The first of these is the lack of accurate physical property data. For example the latent heat of fusion is usually assumed to be equal to the product of moisture content and the latent heat of water. Second, it takes no account of the sensible heat changes and therefore does not predict the total processing time required. Although it is simple to use, Plank's model is accurate only to within about 20% at best. However it does indicate the dependence of freezing time upon a number of variables.

First, freezing time is inversely proportional to the temperature driving force. For example in Example 11.2 if the plate temperature is lowered by 10°C to $-40°C$ the freezing time decreases by 26% to 1,026 s. Second, if the conductive thermal resistance is dominant, and consequently the term Pa/h in Equation (11.12) is negligible, then freezing time is proportional to the square of the size of the object being frozen. Changing the value of a will have a dramatic effect on t. However increasing the air velocity in a blast freezer for example, although increasing the heat transfer coefficient, will have only a relatively small effect on the freezing time. On the other hand if convection is dominant then freezing time is directly proportional to a and changing the thickness of the slab or cylinder diameter has less effect. However the influence of the heat transfer coefficient, which characterises heat transfer to the freezing medium, will be greater.

The thermal resistance of any packaging should be taken into account when determining an appropriate value for the heat transfer coefficient in Plank's equation.

Example 11.3

If the fish in Example 11.2 is packed using 1 mm thick cardboard, of thermal conductivity $0.07\ \mathrm{W\ m^{-1}\ K^{-1}}$, estimate the freezing time taking into account the extra thermal resistance provided by the packaging.

The packaging has the effect of reducing the overall heat transfer coefficient which is then given by

$$\frac{1}{U} = \frac{1}{h} + \frac{x}{k}$$

where x and k are the thickness and thermal conductivity of the packaging respectively. Hence

$$\frac{1}{U} = \frac{1}{80} + \frac{0.001}{0.07}$$

and $U = 37.3\ \mathrm{W\ m^{-2}\ K^{-1}}$. Substituting this value into Plank's equation now gives the freezing time to be 2653 s.

11.5.2. Nagaoka's Equation

Nagaoka's model takes into account the time required to reduce the temperature from an initial temperature T_i above the freezing point. It contains an empirical correction factor and the latent heat of fusion in Equation (11.12) is replaced by the total enthalpy change Δh which includes the sensible heat which must be removed in reducing the temperature from T_i.

$$t = \frac{\rho \Delta h}{(T_f - T_1)} [1 + 0.008(T_i - T_f)] \left[\frac{Pa}{h} + \frac{Ra^2}{k} \right] \tag{11.13}$$

Example 11.4

A factory which works two eight-hour shifts produces burgers at a rate of 5 tonnes per day. The burgers enter a high velocity blast freezer at a temperature of $18°C$ and are frozen down to $-15°C$. The following physical property data are available:

freezing temperature	$-2°C$
frozen density	$950\ \mathrm{kg\ m^{-3}}$
thermal conductivity (below freezing)	$1.1\ \mathrm{W\ m^{-1}\ K^{-1}}$
heat capacity (above freezing)	$2{,}700\ \mathrm{J\ kg^{-1}\ K^{-1}}$
heat capacity (below freezing)	$1{,}500\ \mathrm{J\ kg^{-1}\ K^{-1}}$
latent heat of fusion	$200\ \mathrm{kJ\ kg^{-1}}$

i Calculate the necessary rate of heat removal during production.

ii The burgers, which are 20 mm thick, are to be frozen using air at $-32°C$. If the heat transfer coefficient is $200\ \mathrm{W\ m^{-2}\ K^{-1}}$, estimate the freezing time.

The total heat load per unit mass is

$$\Delta h = 2.7(18 - (-2)) + 200 + 1.5(-2 - (-15)) \, \text{kJ kg}^{-1}$$

or

$$\Delta h = 273.5 \, \text{kJ kg}^{-1}$$

This heat is removed from 5,000 kg over a sixteen-hour period and therefore the rate of heat removal is

$$Q = \frac{5000}{16 \times 3600} \times 273.5 \, \text{kW}$$

and $Q = 23.7 \, \text{kW}$. Now using Nagaoka's equation the freezing time is

$$t = \frac{950 \times 273.5 \times 10^3}{(-2 - (-32))}[1 + 0.008(18 - (-2))]\left[\frac{0.02}{2 \times 200} + \frac{(0.02)^2}{8 \times 1.1}\right] \, \text{s}$$

and

$$t = 959 \, \text{s}$$

11.5.3. Stefan's Model

Stefan's method of estimating freezing time is based upon a simplification of the solution of the unsteady-state conduction equation. It assumes a negligible resistance at the surface of the object being frozen, in other words that the film heat transfer coefficient is infinite. Stefan's equation takes the form

$$Fo = \frac{K}{2}\left(1 + \frac{1}{6Fo}\right) \tag{11.14}$$

where the Fourier number is defined by Equation (10.15) and the group K by

$$K = \frac{\lambda}{c_p(T_f - T_1)} \tag{11.15}$$

Example 11.5

Use Stefan's method to estimate the freezing time for the fish block in Example 11.2 if the heat capacity of the frozen fish is $2,000 \, \text{J kg}^{-1} \, \text{K}^{-1}$.

The thermal diffusivity of the fish block is

$$\alpha = \frac{1.5}{880 \times 2000} \, \text{m}^2 \, \text{s}^{-1} \quad \text{or} \quad \alpha = 8.52 \times 10^{-7} \, \text{m}^2 \, \text{s}^{-1}$$

Substituting this into the Fourier number gives

$$Fo = \frac{8.52 \times 10^{-7}}{(0.01)^2}t, \qquad Fo = 8.52 \times 10^{-3}t$$

where t is the unknown freezing time. The dimensionless group K is now

$$K = \frac{280 \times 10^3}{2000(-2 - (-30))} \quad \text{or} \quad K = 5$$

and therefore Stefan's equation becomes

$$8.52 \times 10^{-3} t = \frac{5}{2} \left[1 + \frac{1}{6 \times 8.52 \times 10^{-3} t} \right]$$

which simplifies to the quadratic expression

$$4.356 \times 10^{-4} t^2 - 0.1278 - 2.5 = 0$$

Solving this gives the positive root as $t = 312\,\text{s}$. Stefan's equation considers only conduction and therefore this result can be compared with that from Plank's equation which suggests a freezing time of 293 s with an infinite heat transfer coefficient.

11.5.4. Plank's Equation for Brick-shaped Objects

A modified version of Plank's equation can be used for a cuboid, or brick-shaped object, of dimensions $a \times E_1 a \times E_2 a$, where a is the shortest dimension. Here the Plank equation is written as

$$t = \frac{\rho \Delta h D}{(T_f - T_1)} \left[\frac{a}{2h} + \frac{Ga^2}{4k} \right] \tag{11.16}$$

where D and G are shape factors. The factor D is defined by

$$D = \frac{2 \times \text{volume}}{\text{cooled surface area} \times \text{shortest dimension}} \tag{11.17}$$

and for a cuboid which is cooled on all six faces this becomes

$$D = \frac{2 E_1 E_2 a^3}{2a^3 (E_1 E_2 + E_1 + E_2)} \tag{11.18}$$

which in turn simplifies to

$$D = \frac{E_1 E_2}{E_1 E_2 + E_1 + E_2} \tag{11.19}$$

The shape factor G is obtained from Figure 11.8 as a function of E_1 and E_2. In Equation (11.16) the term Δh may be either simply the latent heat of fusion or may include sensible heat changes as in Equation (11.1).

Example 11.6

A food product is packaged in cartons measuring 8 cm $\times$ 8 cm $\times$ 2 cm and frozen using an air temperature of $-30°C$ and with a surface heat transfer coefficient of $200\,\text{W}\,\text{m}^{-2}\,\text{K}^{-1}$.

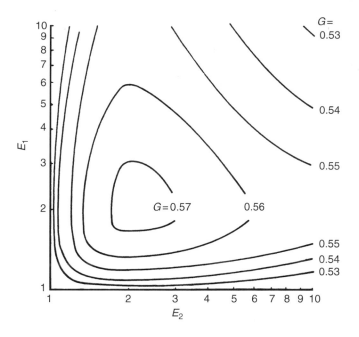

Figure 11.8. Plank's model: shape factor for cuboids (from A.T. Jackson and J. Lamb, Calculations in Food and Chemical Engineering, Macmillan, 1981, reprinted by permission of Macmillan Press Ltd).

The product enters the freezer at $12°C$, freezes at $-3°C$ and leaves with a centre temperature of $-18°C$. If all six surfaces are available for heat transfer, then determine the freezing time. The physical properties for the food are as follows:

frozen density	$850\,\mathrm{kg\,m^{-3}}$
thermal conductivity (below freezing)	$1.5\,\mathrm{W\,m^{-1}\,K^{-1}}$
heat capacity (above freezing)	$2{,}800\,\mathrm{J\,kg^{-1}\,K^{-1}}$
heat capacity (below freezing)	$1{,}650\,\mathrm{J\,kg^{-1}\,K^{-1}}$
latent heat of fusion	$260\,\mathrm{kJ\,kg^{-1}}$

The heat load per unit mass of product is

$$\Delta h = 2.8(12 - (-3)) + 260 + 1.65(-3 - (-18))\,\mathrm{kJ\,kg^{-1}}$$

or

$$\Delta h = 326.75\,\mathrm{kJ\,kg^{-1}}$$

The shortest dimension of the carton a is $0.02\,\mathrm{m}$ and therefore $E_1 = E_2 = 4$ and Equation (11.19) gives the shape factor D as

$$D = \frac{4 \times 4}{((4 \times 4) + 4 + 4)}, \qquad D = 0.67$$

Now from Figure 11.8 the shape factor G equals 0.557 and the freezing time is then

$$t = \frac{850 \times 326.75 \times 10^3 \times 0.67}{(-3 - (-30))}[1 + 0.008(12 - (-3))]\left[\frac{0.02}{2 \times 200} + \frac{0.557(0.02)^2}{4 \times 1.5}\right] \text{s}$$

and

$$t = 673\,\text{s}$$

11.6. THAWING

The controlled thawing of frozen foods is an important aspect of low-temperature preservation which is often overlooked. Frozen foods can be thawed either by applying heat to the surface, for example by immersion in water or exposure to warm air, or by applying heat evenly throughout the food mass using microwave or dielectric methods. Whilst thawing involves melting ice crystals within the frozen food, tempering is intended to raise the temperature of the food from frozen storage temperatures to between $-5°C$ and $-1°C$ in order to facilitate further processing such as cutting or slicing whilst still in the frozen state. The thawing process is critical to the final quality and safety of foods which have been preserved by freezing. It is important that the surface temperature during thawing is maintained at levels which do not promote microbial growth. The time required for thawing must be considered if the thawed product is to be further processed. Generally, thawing is a slower process than freezing for two reasons. First, the temperature difference between the food and the thawing medium is likely to be smaller. Second, as the ice melts progressively from the surface, the rate of heat transfer to the thermal centre decreases because unfrozen food has a lower thermal diffusivity than frozen food. However few models exist to predict thawing times and therefore it is common to use the freezing time as a first estimate.

Air thawing is achieved by holding frozen food in a chilled room at temperatures below $8°C$. More rapid thawing can be achieved by allowing a flow of air to be circulated around the food product however there is a danger that drying of the surface may result from this technique and therefore humid air is used to minimise the effect. Another possibility is to thaw food wrapped in packaging which will minimise the loss of surface quality. However, there will be an increase in thawing time because of the insulating effect of the packaging. Water thawing uses temperatures up to $20°C$. Normally the product to be thawed is packaged to minimise leaching of important constituents and to avoid disintegration of the food in the water. Thawing times can be reduced by recirculating the water across the frozen food but a balance must be struck between the cost of pumping the water and the reduced processing time. If the food is unpackaged contaminants can build up in the water and chlorination and filtration of the water becomes necessary.

Vacuum thawing is an alternative to the use of air or water. Frozen food is placed in a vacuum chamber at absolute pressures of the order of 2.5 kPa. Steam at this pressure is then admitted to the chamber and condenses on the surface of the food releasing its latent heat. Surface temperatures of the order of $20°C$ can be used to give rapid and controlled thawing. Volumetric heating involves placing frozen food in a high-frequency electric field. In dielectric thawing the food is placed between parallel capacitor plates but this is relevant only to foods which have a standard pack shape and uniform thickness and composition. Typically a voltage of 5 kV and a frequency of 80 MHz is applied across the plates and gives uniform heating throughout the frozen block. Thawing times are much reduced compared to water and air thawing. The details of microwave thawing are covered in chapter seven.

The re-freezing of thawed food presents considerable risks to food safety. First, thawing can increase any microbiological load on the food particularly if the thawing is carried out at ambient temperatures. Refreezing of this food and subsequent thawing will result in a greatly increased initial microbiological load after the second thawing and a greater risk of food poisoning and food spoilage problems. Second, even with fast-freezing processes, there will be some deterioration in texture. A series of freeze–thaw cycles will inevitably result in poorer quality food products.

11.7. PRINCIPLES OF VAPOUR COMPRESSION REFRIGERATION

11.7.1. Introduction

As we saw in chapter three the refrigerator is an example of a heat pump in which a net work input W is used to transfer heat from a low temperature to a higher temperature; heat is taken from a cold reservoir at a rate Q_2 and is rejected at a rate Q_1 to a heat sink. The flow of heat removed from the cold chamber is the important quantity. This is often compared to the mechanical work needed to produce it and thus a coefficient of performance (COP) can be defined as

$$COP = \frac{Q_2}{W} \qquad (11.20)$$

The refrigerator is a simple example of a closed-cycle thermodynamic system. The principle is to use a fluid, the refrigerant, to extract energy from the cold compartment. This energy is used to vaporise the refrigerant and then later in the cycle the resulting vapour is condensed back into a liquid and releases its energy. The refrigerant can then be returned to the cold compartment and the cycle is repeated. Thermal energy is rejected continuously from the system in the condenser. It is important to remember that the latent heat of vaporisation of a volatile liquid (even water) is very large; the enthalpy of a dry saturated vapour at low and moderate pressures is many times greater than that of the same substance in its liquid form. It is this property of a refrigerant which is exploited in the freezer or refrigerator. It would be possible to pass a cold gas or liquid through the cold compartment and allow it to extract energy without a change of phase but the low heat capacity would necessitate very large flow rates of refrigerant and consequently very bulky equipment. Vaporisation and condensation of a volatile liquid enable energy flows to be absorbed from the cold compartment whilst using the minimum quantity of fluid and the smallest equipment possible. A schematic diagram of a simple vapour compression refrigeration cycle is shown in Figure 11.9.

11.7.2. The Refrigerant

There are particular chemical, physical and thermodynamic properties that make a fluid suitable for use as a refrigerant. First, the fluid must have suitable freezing and boiling points, relative to the temperature of its surroundings, at convenient working pressures. Second, the other thermodynamic properties of the refrigerant should be chosen to give the minimum power input to the cycle. These are: the latent heat of vaporisation in the appropriate pressure range; the specific volume of the vapour; the heat capacity in both liquid and vapour phases.

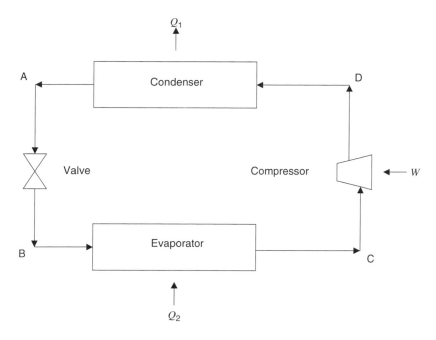

Figure 11.9. Schematic diagram of a vapour compression refrigerator.

Ammonia, sulphur dioxide, carbon dioxide, methyl chloride and methylene chloride all have suitable characteristics and have been used as industrial refrigerants in the past; ammonia is still widely used. However, the most extensively used refrigerants in recent years have been the chlorofluorocarbons or CFCs. The ideal refrigerant, widely used in domestic refrigerators until very recently, is dichlorodifluoromethane or R12. R12 is completely safe in that it is non-toxic, non-flammable and non-explosive. It is highly stable and is difficult to decompose even under extreme operating conditions. Along with these safe properties, the fact that it has a saturation temperature of about 243 K at atmospheric pressure makes it a suitable refrigerant for all types of vapour compression system. However CFCs are now known to cause depletion of the ozone layer and since the Montreal Protocol of 1987 have been phased out of use. Dichlorotrifluoroethane or R123 (an HCFC) was identified as a suitable replacement, and has a very much lower ozone depletion potential than R12, but this is now thought to be a carcinogen. The longer term replacement appears to be tetrafluoroethane or R134a. However ammonia could well be retained as a refrigerant but its toxicity means that a secondary refrigerant is often employed. A secondary refrigerant exchanges heat with the primary refrigerant in a conventional heat exchanger and then extracts heat from the freezer. In this way any toxic or explosive risk associated with the primary refrigerant can be confined to a small area and the safer secondary refrigerant allowed into food processing areas. Suitable secondary refrigerants might be brine or ethylene glycol.

11.7.3. The Evaporator

The evaporator forms the freezing cabinet in a plate or blast freezer and the ice-box or freezing compartment of a domestic refrigerator. Refrigerant circulates through the pipework on the surface of the compartment walls or within the hollow walls and evaporates, thereby absorbing heat at a rate Q_2 from within the freezer cabinet. Evaporation will continue at a constant temperature as long as the evaporator is never allowed to run dry; in practice the system is arranged so that there is always a continuous supply of liquid refrigerant being pumped into it at a controlled rate.

11.7.4. The Compressor

The compressor has two functions. It pumps refrigerant around the circuit by drawing the vapour out of the evaporator, compressing it and discharging it at a higher pressure into the next component of the system. Second, by running at constant speed and capacity, the compressor helps us to maintain a constant pressure, and therefore a constant temperature, in the evaporator.

11.7.5. The Condenser

To allow the refrigerant to be used over and over again it must be condensed back into a liquid. Furthermore, if the system is sealed against the ingress of air and contaminants, it will never need replacing. Therefore the condenser enables the warm refrigerant vapour to reject a quantity of heat Q_1 to a heat sink at a temperature lower than itself. For the domestic refrigerator the ambient air in a kitchen is cool enough to achieve this condensation and the condenser consists simply of an exposed length of tubing. In an industrial freezer, however, a water cooled condenser is used.

11.7.6. The Valve or Nozzle

The purpose of this valve is to control or constrict the flow of liquid refrigerant between the condenser and the evaporator and, by maintaining a pressure difference between the two components, control the evaporating pressure and temperature. The precise design of this device depends on the particular application, but in the domestic refrigerator the constriction is often a fixed length of capillary tubing between the condenser and the evaporator. It thus forms a non-adjustable valve with a high resistance to flow because of its length and small bore. Since the capillary tube and compressor are connected in series the flow capacity of the tube and system must equal the pumping capacity of the compressor. Once the compressor, capillary and other system equipment have been selected, the system will settle down to run at *one* particular condition. That is, the system will operate at a given mass flow rate, depending upon the choice of refrigerant, the compressor and the capillary tube, but also on the working temperatures (and design) of the evaporator and condenser.

11.7.7. The Refrigeration Cycle

The thermodynamic cycle can be followed in a number of ways but one of the most convenient is to use a pressure–enthalpy chart for the refrigerant. Figure 11.10 illustrates an ideal cycle which assumes that the refrigerant vapour leaves the evaporator and enters the compressor as a dry saturated vapour at the evaporation temperature and pressure, and that the

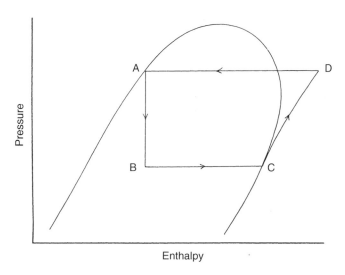

Figure 11.10. Vapour compression refrigeration cycle.

liquid refrigerant leaves the condenser and enters the capillary tube as a just-saturated liquid at the condensing temperature and pressure.

Starting with the relatively high pressure, warm liquid at the outlet of the condenser (point A) the refrigerant flows at a controlled rate through the valve. A fairly large pressure difference exists across the valve in order to maintain flow through it and consequently the refrigerant pressure on the downstream side of this throttling device is very low. The saturation temperature of the refrigerant corresponding to this low pressure must also be much less than it was before the fluid entered the valve or capillary, and in fact the saturation temperature falls below the temperature of the refrigerated space in the freezer cabinet.

Liquid (point B) now enters the evaporator and vaporises at a constant pressure and temperature as the latent heat of vaporisation is transferred from the refrigerated space through the walls of the evaporator to the refrigerant. Vapour then flows from the evaporator (point C) through the suction line and into the inlet of the compressor. Ideally, the vapour leaving the evaporator is dry and saturated (and its pressure and temperature are the same as that of the vaporising liquid) but while flowing through the suction line it usually absorbs energy from the surrounding air and may become slightly superheated. Although this tends to increase the temperature of the vapour, the pressure remains substantially constant. In the compressor the vapour is compressed and consequently undergoes a temperature increase; the hot, high-pressure superheated vapour is discharged into the hot gas line which leads back to the condenser. It is the pressure difference across the compressor which maintains flow through the system.

Now the refrigeration effect, the heat removed in the freezer compartment, is given by

$$Q = h_C - h_B \qquad\qquad (11.21)$$

and the work input is the enthalpy difference across the compressor, that is

$$W = h_D - h_C \qquad\qquad (11.22)$$

Therefore the coefficient of performance becomes

$$COP = \frac{h_C - h_B}{h_D - h_C} \tag{11.23}$$

In all of this no account is taken of the power required to overcome mechanical friction in the compressor. Furthermore, both cooling and heating are assumed to take place at constant pressure. In fact, of course, there will be some pressure losses in the pipelines.

NOMENCLATURE

a	Characteristic length
A	Cross-sectional area
c_{p_f}	Heat capacity of frozen food
c_{p_u}	Heat capacity of unfrozen food
COP	Coefficient of performance
D	Shape factor
E_1	Dimensionless dimension
E_2	Dimensionless dimension
Fo	Fourier number
G	Shape factor
h	Heat transfer coefficient; enthalpy
k	Thermal conductivity of frozen food
K	Group defined by Equation 11.15
Nu	Nusselt number
P	Parameter in Plank's equation
Q	Rate of heat transfer in freezing
Q_1	Rate of heat rejection in refrigeration condenser
Q_2	Rate of heat extraction from refrigeration evaporator
R	Parameter in Plank's equation
Re	Reynolds number
t	Time; freezing time
T_1	Temperature of freezing medium
T_2	Surface temperature
T_f	Freezing temperature
T_{final}	Final temperature
T_i	Initial temperature
W	Work input in refrigeration cycle
x	Thickness of frozen layer

GREEK SYMBOLS

Δh	Total enthalpy change
λ	Latent heat of fusion
ρ	Food density

PROBLEMS

11.1. The latent heat of fusion of a vegetable product is $256 \, \text{kJ kg}^{-1}$. At what rate must heat be removed to freeze $500 \, \text{kg h}^{-1}$ of this product down to $-30°\text{C}$ from $15°\text{C}$? The freezing temperature is $-1.5°\text{C}$ and the frozen and unfrozen heat capacities are $1.8 \, \text{kJ kg}^{-1} \, \text{K}^{-1}$ and $3.3 \, \text{kJ kg}^{-1} \, \text{K}^{-1}$, respectively.

11.2. A joint of beef approximates to a cylinder $0.15 \, \text{m}$ in diameter. Calculate the time to just freeze the centre of the beef in a blast freezer operating at $-25°\text{C}$ with a surface heat transfer coefficient of $170 \, \text{W m}^{-2} \, \text{K}^{-1}$ if the freezing temperature is $-1°\text{C}$ and the latent heat of fusion is $240 \, \text{kJ kg}^{-1}$. The density and thermal conductivity of the frozen beef are $1,100 \, \text{kg m}^{-3}$ and $1.35 \, \text{W m}^{-1} \, \text{K}^{-1}$, respectively.

11.3. What increase in freezing time can be expected if the beef in problem 11.2 is wrapped in a plastic film $0.2 \, \text{mm}$ thick and of thermal conductivity $0.04 \, \text{W m}^{-1} \, \text{K}^{-1}$?

11.4. Mashed potato is formed into blocks $9 \, \text{mm}$ deep and frozen in a plate freezer. The plates are at $-28°\text{C}$ and the blocks enters the freezer at $12°\text{C}$. A surface heat transfer coefficient of $50 \, \text{W m}^{-2} \, \text{K}^{-1}$ may be assumed. Each block weighs $1.5 \, \text{kg}$ and has a frozen volume of $1.3 \times 10^{-3} \, \text{m}^3$. Other data available for the potato blocks is as follows: thermal conductivity (frozen), $1.3 \, \text{W m}^{-1} \, \text{K}^{-1}$; latent heat of fusion $276 \, \text{kJ kg}^{-1}$; heat capacity (unfrozen) $3,350 \, \text{J kg}^{-1} \, \text{K}^{-1}$; freezing temperature $-2°\text{C}$. Calculate the time required to just freeze the centre of the potato blocks.

11.5. A food product is packaged in cuboidal packs measuring $15 \, \text{cm} \times 15 \, \text{cm} \times 5 \, \text{cm}$ which enter a blast freezer at $10°\text{C}$ and leave when the centre of the packs reach a temperature of $-18°\text{C}$. The air temperature is $-30°\text{C}$ and the heat transfer coefficient is $150 \, \text{W m}^{-2} \, \text{K}^{-1}$. Assuming that the packs lie on one of the $15 \, \text{cm} \times 15 \, \text{cm}$ faces, and that freezing takes place from the other five faces, determine the freezing time using the following properties:

initial freezing temperature	$-1.5°\text{C}$
frozen density	$900 \, \text{kg m}^{-3}$
thermal conductivity (below freezing)	$1.0 \, \text{W m}^{-1} \, \text{K}^{-1}$
heat capacity (above freezing)	$2,700 \, \text{J kg}^{-1} \, \text{K}^{-1}$
heat capacity (below freezing)	$1,500 \, \text{J kg}^{-1} \, \text{K}^{-1}$
latent heat of fusion	$300 \, \text{kJ kg}^{-1}$

FURTHER READING

A. C. Cleland, *Food Refrigeration Processes*, Elsevier, Amsterdam (1990).
W. B. Gosney, *Principles of Refrigeration*, Cambridge University Press, Cambridge (1982).
D. R. Heldman and R. W. Hartel, *Principles of Food Processing*, Chapman and Hall, London (1997).
A. T. Jackson and J. Lamb, *Calculations in Food and Chemical Engineering*, Macmillan, London (1981).
M. A. Rao and S. S. H. Rizvi, *Engineering Properties of Foods*, Dekker, New York (1995).

12

Evaporation and Drying

12.1. INTRODUCTION TO EVAPORATION

Evaporation is a food preservation technique in which dilute liquid foods and solutions are concentrated by the evaporation of water, with the aim of increasing microbiological stability and shelf life. A second major reason for the concentration of liquids is the reduction in transport and storage costs which can be achieved by reducing the product bulk volume. In this way concentrated liquids can be transported at relatively low cost and water added later, closer to the point of sale. In addition, evaporation can be used to increase the concentration of solutions prior to the removal of the remaining water by drying, particular by spray drying. This is an attractive option because high efficiency evaporation is significantly less costly than drying and other methods of removing water as the data in Table 12.1 shows.

Evaporation is used extensively for the concentration of milk, fruit and vegetable juices and sugar solutions. The principle of evaporation is very simple; the liquid is boiled in a suitable vessel and the vapour is removed leaving behind a residue of concentrated liquor in the evaporator. The concentrated liquor is the desired product except in the case of desalination where the vapour is condensed into drinking water. However, even in most other evaporation operations the vapour is either reused as lower grade steam or is condensed.

A number of problems arise when food liquids are evaporated. The most obvious of these is that lengthy exposure to heat and to high temperatures must be avoided to prevent thermal degradation of the food. Consequently methods have been devised to reduce residence times in evaporators and to maximise the rate of heat transfer by using thin liquid films rather than add heat to liquids in bulk. Other difficulties arise because of the concentration increase

TABLE 12.1
Comparative Costs of Water Removal (after
Timmins, 1975)

	Separation costs per unit volume of water removed (arbitrary units)
Spray drying	17–50
Drum drying	10–25
Centrifugation	0.1–10
UF/RO	0.2–7
Evaporation	0.2–5

303

itself. The viscosity of the liquor usually increases substantially with concentration and this brings problems associated with pumping and with poor heat transfer. The boiling point of the solution may also rise as the concentration increases and this in turn reduces the temperature driving force between the steam and the feed liquid. Evaporators are usually operated under vacuum because the boiling point of a solution is reduced at reduced pressure and thus thermal degradation can be minimised. This also has the effect of increasing the temperature driving force between the steam and the boiling liquor, however the lower temperature may lead to a higher viscosity and therefore to lower heat transfer coefficients.

As in any other heat transfer operation involving foods there is the likelihood of fouling layers being deposited on heat transfer surfaces which reduce the heat transfer coefficient as well as presenting a microbiological hazard; fouling can be particularly severe in the evaporation of milk. Finally, great care should be taken to avoid crystallisation of the solute. Evaporation is one means used to bring about crystallisation and this is usually undesirable in a conventional concentration process because the presence of crystals can cause serious blockages in process equipment and lead to the shut-down of the evaporator.

12.2. EQUIPMENT FOR EVAPORATION

12.2.1. Natural Circulation Evaporators

Natural circulation evaporators consist of a tubular heat exchanger placed within a cylindrical vessel. The heat for vaporisation is provided by process steam which enters the tubes. Such evaporators rely entirely on free convection currents to circulate the liquid to be concentrated. In horizontal tube evaporators (Figure 12.1) the tube bundle, known as a calandria, consists of short tubes fixed to tube plates on either side of the evaporator vessel. The tube

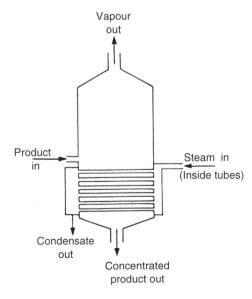

Figure 12.1. Horizontal tube natural circulation evaporator (from C.W. Hall, A.W. Farrall and A.L. Rippen, Encyclopaedia of Food Engineering, AVI Publishing, 2nd ed., 1986, reprinted with permission).

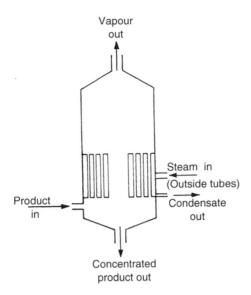

Figure 12.2. Vertical tube natural circulation evaporator (from C.W. Hall, A.W. Farrall and A.L. Rippen, Encyclopaedia of Food Engineering, AVI Publishing, 2nd ed., 1986, reprinted with permission).

diameter is normally 25 mm and the evaporator diameter between 1 and 3.5 m. Vapour leaves through a de-entrainment device, often consisting of a series of baffles, to prevent the carry over of droplets. Horizontal tube evaporators are suitable for non-crystallising solutions and low viscosity fluids and either batch or continuous operation is possible.

In vertical tube evaporators (Figure 12.2) tubes between 1 and 2 m long are held between horizontal tube plates in the lower part of the vessel. In this case steam normally condenses on the outside of the tubes and the liquid to be concentrated rises in the tubes which are usually larger in diameter than those in horizontal evaporators, between 35 and 75 mm. The circulation of liquid is aided by a central downcomer, that is, the calandria is annular in plan with the downcomer equal to between 40 and 100% of the area of the tubes.

12.2.2. Forced Circulation Evaporators

Increasing the rate of circulation of liquid increases the heat transfer coefficient which can be obtained. This is achieved either by installing an impeller in the central downcomer of a vertical tube evaporator or by employing an external pump which circulates fluid between the calandria and the evaporating chamber. This kind of evaporator is more suitable for viscous fluids; with gear pumps being used for particularly high viscosities. In general a greater degree of concentration is obtained but with increased operating costs.

12.2.3. Thin Film Evaporators

Natural and forced circulation evaporators have very high residence times, often of the order of several hours, whereas heat sensitive materials need very short residence times. In a thin film evaporator the liquid feed is formed into a thin film across the heat transfer surface

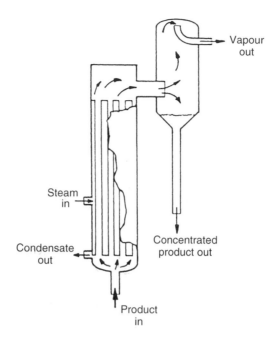

Vapour
out

Steam
in

Condensate
out

Concentrated
product out

Product
in

Figure 12.3. Climbing film evaporator (from C.W. Hall, A.W. Farrall and A.L. Rippen, Encyclopaedia of Food Engineering, AVI Publishing, 2nd ed., 1986, reprinted with permission).

in order the maximise the rate of heat transfer to the liquid and minimise the residence time. In a climbing film evaporator (Figure 12.3) the feed enters the bottom of a series of vertical tubes about 50 mm in diameter and forms an annular film of liquid on the inside of the tube. This film is maintained by the drag which is exerted by the rising vapour on the liquid surface. Thus the vapour and any entrained liquid leave at the top whilst the concentrated product is withdrawn at the bottom. Falling film evaporators are used for more viscous feeds which are unsuitable for the climbing film technique. The general principle is similar but the feed enters at the top and both vapour and liquor leave at the bottom. The distribution of liquid at the top of the tube often presents a difficulty and sprays are used to obtain an annular film.

The principle of both the scraped surface heat exchanger and the plate heat exchanger can be exploited in evaporation. Scraped surface evaporators are used for highly viscous concentrates where a thin film is created by a rotating blade. In the plate evaporator steam is admitted to alternate spaces between the plates. Feed liquid enters the remaining spaces and it is possible to have either climbing or falling film arrangements or indeed a mixture of the two.

12.3. SIZING OF A SINGLE EFFECT EVAPORATOR

The primary quantities required to size an evaporator are the flow rates of the major streams: feed, vapour and concentrated liquor; the steam flow rate and the area of the heat transfer surface across which heat is transferred from steam to liquid in the evaporator. This requires a material balance, an enthalpy balance and a heat transfer rate equation to be solved simultaneously.

12.3.1. Material and Energy Balances

Referring to Figure 12.4, an overall material balance across a single effect evaporator yields

$$F = V + L \tag{12.1}$$

where F, V, and L are the mass flow rates of feed, vapour, and liquor, respectively. The component material balance is now

$$x_F F = yV + x_L L \tag{12.2}$$

where x_F, y, and x_L are the mass fractions of solute or non-soluble solids in the feed, vapour and liquor, respectively. However, because there is no solute in the vapour stream, $y = 0$ and the component balance reduces to

$$x_F F = x_L L \tag{12.3}$$

An enthalpy balance over the evaporator involves two further streams, the inlet steam S and the condensate C. The combined enthalpy of the feed and the steam must balance that of the vapour, liquor and condensate. Thus, if h_F is the specific enthalpy of the feed, h_S that of the steam and so on,

$$F h_F + S h_S = V h_V + L h_L + S h_C \tag{12.4}$$

where each term in Equation (12.4) represents a flow of heat associated with that particular stream and has units of either W or kW if h has units of J kg^{-1} or kJ kg^{-1}, respectively. Strictly, the steam and condensate should be included in the material balances. However they may be omitted because, of course, the flow rate and composition of the steam remains unchanged as it gives up heat and condenses. Thus, in Equation (12.4), $S = C$. For the purpose of calculation, the enthalpies of liquid streams are obtained from steam tables and care must be taken not to confuse the subscripts F and f.

Rearranging Equation (12.4) to give the steam flow rate produces

$$S(h_S - h_C) = V h_V + L h_L - F h_F \tag{12.5}$$

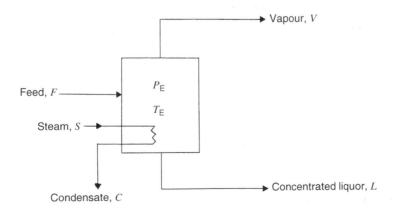

Figure 12.4. Single effect evaporator: material and enthalpy balance.

Now the left-hand side of Equation (12.5), the difference between the enthalpy of the steam and that of the condensate, must be equal to the rate Q at which heat is transferred from the steam to the feed, that is,

$$Q = S(h_S - h_C) \tag{12.6}$$

The enthalpy given up by the steam is transferred across the tube walls of the calandria, across which the temperature difference is ΔT, and therefore

$$S(h_S - h_C) = U A \, \Delta T \tag{12.7}$$

where A is the heat transfer surface area; this area must be determined in order for the calandria and the evaporator to be sized. An evaporator of course is a kind of heat exchanger and the overall heat transfer coefficient can be obtained by summing the various thermal resistances as in the examples of chapter seven. The temperature driving force is that between the steam and the boiling liquor in the evaporator. Hence

$$\Delta T = T_S - T_E \tag{12.8}$$

Any evaporator problem now requires the simultaneous solution of Equations (12.3), (12.5), and (12.7). However it is very likely that there will be sufficient information available to solve the material balance independently. Equally, if the working pressures of the evaporator are specified then the steam and condensate enthalpies can be determined from steam tables. The major difficulty may be in finding values of enthalpy and boiling point for food solutions, but a reasonable first estimate is to assume that the properties of food solutions approximate to those of water.

Example 12.1

A single effect evaporator is to be used to concentrate a food solution containing 15% (by mass) dissolved solids to 50% solids. The feed stream enters the evaporator at 291 K with a feed rate of 1.0 kg s^{-1}. Steam is available at a pressure of 2.4 bar and an absolute pressure of 0.07 bar is maintained in the evaporator. Assuming that the properties of the solution are the same as those of water, and taking the overall heat transfer coefficient to be $2,300 \text{ W m}^{-2} \text{ K}^{-1}$, calculate the rate of steam consumption and the necessary heat transfer surface area.

Working in units of kg s^{-1}, the overall material balance becomes

$$1.0 = V + L$$

Substituting into the component material balance for $x_F = 0.15$ and $x_L = 0.50$ gives

$$0.15 \times 1.0 = 0.50 \, L$$

from which the unknown liquor flow rate is

$$L = 0.3 \text{ kg s}^{-1}$$

Hence from the overall balance the flow rate of vapour is

$$V = 0.7 \text{ kg s}^{-1}$$

To proceed with an enthalpy balance, specific enthalpies must be obtained from steam tables. If the steam and condensate remain saturated at 2.40 bar then h_s is equal to h_g at 2.40 bar and h_c is equal to h_f at 2.40 bar. Thus $h_S = 2{,}715 \text{ kJ kg}^{-1}$ and $h_C = 530 \text{ kJ kg}^{-1}$. The feed enthalpy is determined by its temperature. Assuming the feed to be pure water, h_F is equal to h_f at 291 K and therefore $h_F = 75.5 \text{ kJ kg}^{-1}$. The enthalpies of the vapour and liquor streams are a function of the pressure within the evaporator: $h_V = 2{,}572 \text{ kJ kg}^{-1}$ (h_g at 0.07 bar) and $h_L = 163 \text{ kJ kg}^{-1}$ (h_f at 0.07 bar). The enthalpy balance [Equation (12.5)] now becomes

$$S(2715 - 530) = (0.70 \times 2572) + (0.30 \times 163) - (1.0 \times 75.5)$$

from which $S = 0.812 \text{ kg s}^{-1}$. The rate of heat transfer, from Equation (12.6), is now

$$Q = 0.812(2715 - 530) \text{ kW} \quad \text{or} \quad Q = 1774 \text{ kW}$$

The temperature of steam at 2.4 bar is $T_S = 126.1°\text{C}$ and the temperature of saturated liquid water at the evaporator pressure of 0.07 bar is $T_E = 39.0°\text{C}$. Thus to find the heat transfer area from the rate equation,

$$A = \frac{Q}{U(T_S - T_E)}, \qquad A = \frac{1774}{2.30(126.1 - 39.0)} \text{ m}^2$$

or

$$A = 8.86 \text{ m}^2$$

12.3.2. Evaporator Efficiency

A common measure of the efficiency of an evaporator is the mass of vapour generated per unit mass of steam admitted to the calandria. This quantity is known as the economy. Thus

$$\text{economy} = \frac{V}{S} \tag{12.9}$$

Clearly it is impossible, in a single effect evaporator, for 1 kg of steam to generate more than 1 kg of vapour. In practice, because of energy losses, the economy will be below unity and values of 0.8 or slightly greater may be expected for industrial units.

Example 12.2

Calculate the economy of the evaporator in Example 12.1.

From the definition in Equation (12.9)

$$\text{economy} = \frac{0.70}{0.812}, \qquad \text{economy} = 0.862$$

In other words 0.862 kg of vapour are produced for every kg of steam used.

Although this quantity is a kind of efficiency it is not appropriate to express it as a percentage because, as will seen in section 12.4, it is possible to increase this figure beyond unity.

Example 12.3

An aqueous food solution at a temperature of 18°C contains 6% solids by mass and is to be concentrated to 24% solids in a single effect evaporator. The evaporator has a total heat transfer surface area of 30 m², uses steam at 300 kPa and operates under a vacuum of 79.3 kPa. Previous operating experience with these conditions suggests an overall heat transfer coefficient of 2,200 W m^{-2} K^{-1}. Determine the mass flow rate of steam required and the evaporator economy.

The various specific enthalpies and temperatures are obtained from steam tables as follows:

For the steam and condensate at 300 kPa, $h_S = 2{,}725$ kJ kg^{-1}, $h_C = 561$ kJ kg^{-1} and the steam temperature is 133.5°C. The enthalpy of feed at a temperature of 18°C is $h_F = 75.5$ kJ kg^{-1}. Taking atmospheric pressure to be 101.3 kPa, the pressure within the evaporator is 22.0 kPa and therefore the evaporator temperature (assuming no boiling point elevation) is 62.2°C. Consequently $h_V = 2{,}613$ kJ kg^{-1} (h_g at 0.22 bar) and $h_L = 260$ kJ kg^{-1} (h_f at 0.22 bar).

From Equation (12.7) the steam flow rate is

$$S = \frac{U A \Delta T}{(h_S - h_C)}, \qquad S = \frac{2.20 \times 30 \times (133.5 - 62.2)}{(2725 - 561)}$$

or

$$S = 2.175 \text{ kg s}^{-1}$$

Because neither the feed flow rate nor the product flow rate is specified, the material and energy balances must be solved simultaneously. The component balance, with $x_F = 0.06$ and $x_L = 0.24$, is

$$0.06F = 0.24L$$

from which

$$F = 4L$$

Substituting this into the enthalpy balance gives

$$S(h_S - h_C) = (4L - L)h_V + Lh_L - 4Lh_F$$

and

$$2.175(2725 - 561) = (3L \times 2613) + 260L - 302L$$

This can be solved to give $L = 0.604$ kg s^{-1} and from the overall material balance therefore $V = 1.811$ kg s^{-1}. Consequently the economy becomes

$$\text{economy} = \tfrac{1.811}{2.175}, \qquad \text{economy} = 0.833$$

12.3.3. Boiling Point Elevation

The vapour pressure of an aqueous solution is less than that of pure water. Consequently the boiling point of the solution is higher than that of pure water and this difference must be taken into account in the enthalpy balance. The boiling point rise or boiling point elevation is defined as the difference between the boiling point of the solution and that of pure water, at the same pressure.

Example 12.4

An aqueous solution at 15.5°C, and containing 4% solids, is concentrated to 20% solids. A single effect evaporator with a heat transfer surface area of 37.2 m² and an overall heat transfer coefficient of 2,000 W m⁻² K⁻¹ is to be used. The calandria contains dry saturated steam at a pressure of 200 kPa and the evaporator operates under a vacuum of 81.3 kPa. If the boiling point rise is 5°C, then calculate the evaporator capacity.

As in Example 12.3 the steam flow rate is obtained from

$$S = \frac{U A \, \Delta T}{(h_S - h_C)}$$

At 200 kPa the steam and condensate enthalpies are $h_S = 2,707 \, \text{kJ kg}^{-1}$ and $h_C = 505 \, \text{kJ kg}^{-1}$, respectively, and the steam temperature is 120.2°C. The pressure within the evaporator is $101.3 - 81.3 = 20.0 \, \text{kPa}$ at which the boiling point of water is 60.1°C. The evaporator temperature is now 60.1°C plus the boiling point elevation and therefore $T_E = 65.1°C$. Hence

$$S = \frac{2.0 \times 37.2 \times (120.2 - 65.1)}{(2707 - 505)} \, \text{kg s}^{-1}, \qquad S = 1.862 \, \text{kg s}^{-1}$$

From steam tables, the feed enthalpy at 15.5°C is $h_F = 65 \, \text{kJ kg}^{-1}$. The vapour enthalpy can be estimated by adding the enthalpy of water at the evaporator pressure and the sensible heat added to the vapour in raising its temperature from the boiling point of water to the actual evaporator temperature. Consequently h_V is equal to 2,609 kJ kg⁻¹ (h_g at 0.20 bar) plus $(1.91 \times 5) \, \text{kJ kg}^{-1}$ where the heat capacity of water vapour at 60.1°C is 1.91 kJ kg⁻¹ K⁻¹. Thus $h_V = 2,618.6 \, \text{kJ kg}^{-1}$. The enthalpy of the concentrated liquor stream at the evaporator temperature is $h_L = 272 \, \text{kJ kg}^{-1}$ (h_f at 65.1°C). The component balance becomes

$$0.04F = 0.20L \quad \text{or} \quad F = 5L$$

and therefore

$$V = 4L$$

Substituting this into the enthalpy balance yields

$$S(h_S - h_C) = 4Lh_V + Lh_L - 5Lh_F$$

and thus

$$1.862(2707 - 505) = (4L \times 2618.55) + 272L - 325L$$

from which $L = 0.393 \, \text{kg s}^{-1}$ and the evaporator capacity is $F = 1.97 \, \text{kg s}^{-1}$.

12.4. METHODS OF IMPROVING EVAPORATOR EFFICIENCY

In single stage evaporation the enthalpy of the vapour is wasted because the vapour is either vented to atmosphere or condensed. This poor use of steam results in low thermal efficiency and a low steam economy. Reusing the vapour, either by recycling it to the calandria or by passing it to the calandria of a second evaporator, means that 1 kg of original steam can be used to generate more than 1 kg of vapour giving economies greater than unity.

12.4.1. Vapour Recompression

(*a*) *Mechanical recompression*   Mechanical recompression (Figure 12.5a) of the exhaust vapour from an evaporator allows the enthalpy of the vapour to be reused. In compressing the vapour its enthalpy is increased to that of the original steam. Because of inevitable heat losses in the system some make-up steam will be required but a large increase in economy can be expected. A major disadvantage of this technique is that a large volume of vapour must be handled which in turn requires a large compressor; positive displacement compressors are normally used. The increase in steam economy must be balanced against the running costs of the compressor.

(*b*) *Steam jet ejector*   An alternative method of reusing the vapour is to inject high pressure steam via a nozzle, or steam jet ejector (Figure 12.5b). This creates a vacuum which entrains the low pressure vapour from the evaporator at right angles. The combined stream is then recycled to the calandria. Again, the fresh steam requirement is reduced but there are several advantages over mechanical recompression. The steam jet ejector has the ability to handle very large volumes of vapour and can operate at lower pressures. There are no moving parts, no power requirement and corrosion resistant materials can easily be used. The major disadvantage is that optimum operation of such a device occurs at a specific pressure and temperature; variation of the conditions in the evaporator may well lead to a reduction in the economy which can be achieved.

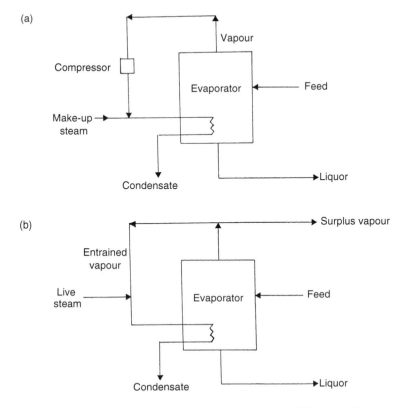

Figure 12.5. Vapour recompression: (a) mechanical, and (b) steam ejector.

12.4.2. Multiple Effect Evaporation

In multiple effect evaporation a series of identical evaporators are used and the individual units are known as effects. The concentrated liquor from the first effect is passed to the second effect and the vapour from the first effect is used as the heat source for the second. The overall economy is increased because the original unit mass of steam produces a greatly increased total mass of vapour. Any number of effects can be used in series but in practice four or five is the maximum. A vacuum is applied to second and subsequent effects to lower the boiling point and therefore maintain an adequate temperature difference because the temperature of the vapour (steam) produced in successive effects will decrease. However multiple effect evaporation has the disadvantage that the capacity is reduced if the boiling point rise is significant. In the descriptions that follow, and in the accompanying diagrams, three effects have been assumed but the principles apply equally to a larger number of stages. The first effect is defined as that to which steam is fed, and where the pressure in the evaporator is highest.

(*a*) *Forward feed*   Figure 12.6 shows the outline of a forward feed arrangement. Steam is fed to the first effect and the resultant vapour is passed to the calandria of the second effect; the vapour from the second effect is passed directly the calandria of the third effect. The concentrated liquor from effect 1 is passed to effect 2 and to effect 3 in turn. A vacuum is applied to the final effect giving an absolute pressure of about 13 kPa. The pressure in the second effect will be higher and will be higher still in the first effect. Forward feed is the simplest of the multiple effect arrangements which are possible. It requires a pump for the feed (because the first effect may well be at atmospheric pressure) and a pump to remove the concentrated and viscous liquor from the final effect. However no pumps are required between each effect because liquid flows under the pressure gradient. With forward feed the steam economy tends to increase strongly as a function of feed temperature.

(*b*) *Backward*   An alternative arrangement is backward feed (Figure 12.7). Here, the feed solution is introduced to effect 3, and pumped successively to effect 2 and effect 1. Steam is fed in the forwards direction as before. The lowest liquid concentration is now at the lowest temperature (in effect 3) and the most concentrated liquid, and therefore the most viscous,

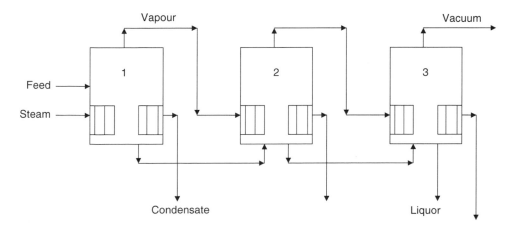

Figure 12.6. Triple effect evaporator: forward feed.

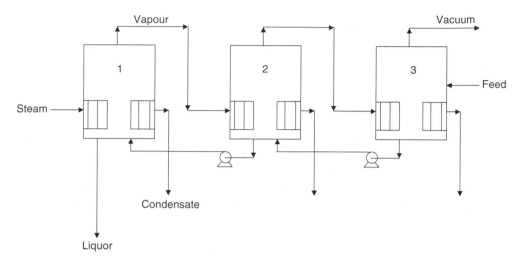

Figure 12.7. Triple effect evaporator: backward feed.

is at the highest temperature (in effect 1). This has the advantage that higher heat transfer coefficients can be maintained with particularly viscous liquids and consequently a higher capacity results. Even so, the overall heat transfer coefficient in effect 1 will be smaller than that obtained in forward feed and the coefficient for effect 3 will be higher than for the forward feed case. The disadvantage to backward feed is the increased cost associated with the pumps required to move liquid between each effect and the less strong dependence of economy on the feed temperature.

(c) *Mixed*   It is possible to combine some of the elements of both forward and backward feed. In mixed feed mode the steam flows in the forward direction but the feed is introduced to some intermediate effect and flows forwards to the last effect (at the lowest pressure) and is then pumped in a backward direction through the first few effects for final concentration. This arrangement has the advantage of eliminating some of the pumps that are required in backward feed and also allows the most concentrated liquor to encounter the highest temperatures.

12.4.3. An Example of Multiple Effect Evaporation: The Concentration of Tomato Juice

Tomato juices and purees are examples of high volume products in which evaporation is a major production step. Tomato juice is concentrated by evaporation under partial vacuum either in a batch or continuous process. In the traditional batch process the evaporation may be entirely carried out in steam jacketed vacuum pans (known as 'boules') fitted with agitators, or the juice may be preconcentrated in a tubular evaporator to about 12% solids before transfer to the boules. Evaporation at low pressure reduces the boiling point of the juice so that the resulting paste retains most its colour and flavour. Continuous processes tend to produce a more consistent paste than batch processes. Multiple-effect evaporators with backward feed are used to obtain improved steam economy because energy consumption for evaporation is the second highest production cost after raw materials. The concentration of tomato juice for paste results in a large increase in viscosity. Consequently early continuous double effect evaporators used a rotary steam coil or forced circulation in the first effect (which experiences the highest temperature and deals with the most concentrated juice) to ensure adequate circulation of the concentrate. Such circulation prevents burn-on fouling and improves heat transfer. More recent evaporators use up to four effects with forced circulation or double or triple effects with scraped surface evaporation. An alternative method for improving energy efficiency is the use of a single forced circulation evaporator with mechanical vapour recompression. A single-stage system with vapour recompression has generally been found to be more energy efficient than a triple-effect evaporator.

12.5. SIZING OF MULTIPLE EFFECT EVAPORATORS

The rate equation can be written for each effect in turn

$$\left. \begin{array}{l} Q_1 = U_1 A_1 \, \Delta T_1 \\ Q_2 = U_2 A_2 \, \Delta T_2 \\ Q_3 = U_3 A_3 \, \Delta T_3 \end{array} \right\} \tag{12.10}$$

where the temperature differences are defined by

$$\left. \begin{array}{l} \Delta T_1 = T_S - T_1 \\ \Delta T_2 = T_1 - T_2 \\ \Delta T_3 = T_2 - T_3 \end{array} \right\} \tag{12.11}$$

and the subscripts 1, 2, and 3 refer to the first, second and third effects, respectively.

If now it is assumed that there is no boiling point rise, that the enthalpy required to raise the feed to the temperature T_1 can be neglected and that the enthalpy carried by the concentrated liquor to subsequent effects is negligible, then the heat flux Q_1 appears as the latent heat of the vapour in the calandria of effect 2. Therefore

$$Q_1 = Q_2 = Q_3 \tag{12.12}$$

and

$$U_1 A_1 \, \Delta T_1 = U_2 A_2 \, \Delta T_2 = U_3 A_3 \, \Delta T_3 \tag{12.13}$$

If each unit is geometrically similar each will have the same area and

$$U_1 \, \Delta T_1 = U_2 \, \Delta T_2 = U_3 \, \Delta T_3 \tag{12.14}$$

The total capacity of the evaporator Q is then given by

$$Q = Q_1 + Q_2 + Q_3 \tag{12.15}$$

and

$$Q = U_{av} A (\Delta T_1 + \Delta T_2 + \Delta T_3) \tag{12.16}$$

where A is the area of a *single* effect and U_{av} is an average overall heat transfer coefficient.

It is important to understand that a single effect evaporator will have approximately the same capacity Q as the multiple effect evaporator if the temperature difference is the same as the *total* temperature difference of the multiple effect unit, the area is the same as the area of *one* effect and the overall heat transfer coefficient is the same. The advantage of multiple effect evaporation is not an increased capacity but an increased steam economy. In an n effect evaporator 1 kg of steam evaporates approximately n kg of vapour. Thus the economy of the multiple effect system is greater but the capital cost is greatly increased. For n effects the capital cost will be approximately n times that of a single effect and the optimum number of effects is a balance between the capital cost on the one hand and the improved economy and therefore lower operating costs on the other.

In order to determine the area of a multiple effect evaporator an iterative calculation is required. If the likely values of the overall heat transfer coefficient for each effect are known then, together with the temperature in the final effect (which is a function of the degree of vacuum applied), a first approximation of the temperature differences ΔT_1, ΔT_2 and ΔT_3 can be obtained from Equation (12.13). This will give the temperature in each effect from which the enthalpies of vaporisation can be found and hence the material and energy balances can be solved to give the steam and vapour flow rates. A first approximation of the area of each effect can now be made from the rate equations [Equation (12.9)]. Because of the assumptions made in this procedure, it is very likely that this first iteration will give unequal areas. A new approximation of each temperature difference is now obtained from an equation of the form

$$\text{new } \Delta T_1 = \frac{\Delta T_1 A_1}{A_{\text{mean}}} \tag{12.17}$$

where A_{mean} is the average of each effect area. This calculation is repeated until the areas of each effect are sufficiently close together.

Example 12.5

A 4% aqueous food solution is fed at a rate of $2.0 \, \text{kg s}^{-1}$ and a temperature of $70°C$ to a forward feed double-effect evaporator with equal surface areas. The solution is concentrated to 20% by mass. The second effect is maintained at a pressure of 20 kPa with a boiling point elevation of 8 K. Steam at 240 kPa is available. The heat transfer coefficients in the first and second effects are 2.20 and $1.50 \, \text{kW m}^{-2} \, \text{K}^{-1}$, respectively, and the heat capacity of each liquid stream may be assumed to be $4.18 \, \text{kJ kg}^{-1} \, \text{K}^{-1}$. Calculate the heat transfer surface area of each effect and the economy.

A component material balance gives

$$0.04 \times 2.0 = 0.20 L_2$$

and therefore the flow rate of concentrate from the second effect L_2 equals $0.40 \, \text{kg s}^{-1}$. From an overall balance the total vapour generated V, equal to $V_1 + V_2$, is then $(2.0 - 0.40) = 1.60 \, \text{kg s}^{-1}$. The steam temperature at 240 kPa is $T_S = 126.1°C$. The temperature in the second effect is equal to the boiling point of water at 20 kPa plus the boiling point elevation, thus

$$T_2 = 60.1 + 8.0 = 68.1°C$$

and the overall temperature difference, equal to $\Delta T_1 + \Delta T_2$, is $(126.1 - 68.1) = 58\,\mathrm{K}$. Now from Equation (12.14)

$$2.20\,\Delta T_1 = 1.50\,\Delta T_2$$

and consequently $\Delta T_1 = 23.5\,\mathrm{K}$ and $\Delta T_2 = 34.5\,\mathrm{K}$. From Equation 12.11 the temperature in the first effect T_1 is $(126.1 - 23.5) = 102.6°\mathrm{C}$. Ignoring the enthalpy of the stream L_1, an enthalpy balance over the first effect can now be written as

$$S(h_S - h_C) + Fc_p(T_F - T_1) = V_1 h_{v_1}$$

where h_{v_1} represents the enthalpy of vapour at T_1 and T_F is the feed temperature. Thus

$$2185S + (2.0 \times 4.18(70 - 102.6)) = 2680.4V_1$$

and

$$2185S = 2680.4V_1 + 272.5 \tag{i}$$

The enthalpy balance over the second effect, ignoring the stream L_2, is

$$V_1 h_{fg_1} + (F - V_1)c_p(T_2 - T_1) = V_2 h_{v_2}$$

where h_{fg_1} represents the latent heat of vaporisation at T_1. Thus

$$2250.2V_1 + (2.0 - V_1)4.18(102.6 - 68.1) = 2623V_2$$

which simplifies to

$$2106.0V_1 + 288.4 = 2623V_2 \tag{ii}$$

From the material balance

$$V_1 + V_2 = 1.60 \tag{iii}$$

Now solving (i)–(iii) simultaneously yields

$$V_1 = 0.826\,\mathrm{kg\,s}^{-1}$$

$$V_2 = 0.774\,\mathrm{kg\,s}^{-1}, \qquad S = 1.138\,\mathrm{kg\,s}^{-1}$$

Hence the first estimate of the area of each effect can be obtained from the rate equation [Equation (12.10)] and

$$A_1 = \frac{1.138 \times 2185.0}{2.20 \times 23.50}\,\mathrm{m}^2, \quad \text{or } A_1 = 48.1\,\mathrm{m}^2$$

Similarly

$$A_2 = \frac{0.826 \times 2250.2}{1.50 \times 34.5}\,\mathrm{m}^2, \qquad A_2 = 35.9\,\mathrm{m}^2$$

The values of A_1 and A_2 are not equal and a second iteration is necessary. The temperature difference ΔT_1 is now amended using Equation (12.17).

$$\text{new } \Delta T_1 = \frac{\Delta T_1 A_1}{(A_1 + A_2)/2}$$

therefore

$$\Delta T_1 = \frac{23.5 \times 48.1}{(48.1 + 35.9)/2} \, K, \qquad \Delta T_1 = 26.9 \, K$$

By difference the amended temperature driving force for the second effect is $\Delta T_2 = 31.1 \, K$.

The procedure is now to recalculate T_1, which becomes $(126.1 - 26.9) = 99.2°C$, determine new values of h_{v_1} and h_{fg_1} from steam tables, repeat the calculation of V_1, V_2, and S and so on. This produces

$$V_1 = 0.825 \, kg \, s^{-1}, \qquad V_2 = 0.775 \, kg \, s^{-1}, \qquad S = 1.122 \, kg \, s^{-1}$$

The area of the first effect is now

$$A_1 = \frac{1.122 \times 2185.0}{2.20 \times 26.9} \, m^2 \quad or \quad A_1 = 41.4 \, m^2$$

and that of the second effect is

$$A_2 = \frac{0.825 \times 2280.8}{1.50 \times 31.1} \, m^2, \qquad A_2 = 40.3 \, m^2$$

The estimates of the two areas are now sufficiently close together giving a mean heat transfer surface area of $40.9 \, m^2$. Two iterations are therefore sufficient. The economy is now equal to the total vapour produced $(1.60 \, kg \, s^{-1})$ divided by the steam flow rate of $1.122 \, kg \, s^{-1}$. Thus the economy at 1.426 is greater than unity.

12.6. DRYING

12.6.1. Introduction

Drying, or dehydration, is a food preservation technique which relies upon reducing the moisture content of a food to a level where microbial growth is inhibited or where the rate of an adverse chemical reaction is minimised. The minimum moisture content for the activity of bacteria, fungi or moulds is relatively clear cut, with a rapid increase in the rate of growth above the minimum; the form of this relationship is shown in Figure 12.8. However the rate of a chemical reaction is likely to decrease much more slowly with reduced moisture content. Drying requires the removal of water from a food solid or food solution by vaporisation and therefore requires thermal energy; this is often supplied in the form of steam or hot air. Consequently drying is both a heat transfer and a mass transfer operation.

Other than as a preservation process, drying is used because of the need to provide better physical properties, to make material handling easier (for example improving the flow of powders) or to reduce weight and therefore reduce transport costs. Alternatively drying may be considered solely as a processing operation, in which case it may be fall into one of two categories. First, those applications where drying forms the major process stage, for example the spray drying of milk, and a very different product form results. Thus a solid is produced from a liquid with the ability to be reconstituted as a liquid on the addition of water. Second, drying may form a much smaller stage in the overall process with smaller amounts of water removed (for some of the reasons outlined above). This may follow other steps such as filtration or crystallisation and may well precede mixing or size reduction.

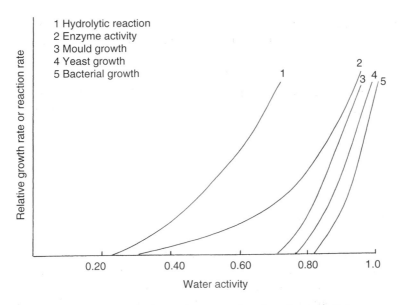

Figure 12.8. Microbial growth rate as a function of water activity.

In the case where water is removed from a solid food the ideal is to be able to reproduce the original structure, appearance and taste after reconstitution. However the disadvantages of drying include the change in appearance of foods and the difficulties in regaining the original properties on rehydration. The removal of water from fruit and vegetable tissue may damage the structure of the food; the rate of drying and the temperature to which food is exposed are both important in determining the quality of the dried product. Equally, the structure of the tissue may influence the mechanism by which water is removed and therefore influence the rate of drying. In any case a further major disadvantage of drying is the operating cost associated with the necessary thermal energy input.

12.6.2. Water Activity

The moisture content of foods is often expressed in terms of water activity. Water activity a_w is defined as the partial pressure of water vapour (p_w) above the food surface divided by the pure component vapour pressure of water (p_w') at the same temperature as the sample, thus

$$a_w = \frac{p_w}{p_w'} \tag{12.18}$$

Consider an aqueous solution held at a constant temperature. Raoult's law states that the partial pressure above the liquid surface is the product of the mole fraction of water and the pure component vapour pressure of water, that is,

$$p_w = x_w p_w' \tag{3.25}$$

Clearly when the liquid phase is pure water, and $x_w = 1$, then $p_w = p_w'$ and the partial pressure is equal to the vapour pressure. Equally, when the water content is zero ($x_w = 0$) there can be no partial pressure of water in the vapour phase. However Raoult's law applies to ideal systems and water is not an ideal material. It is therefore necessary to introduce an activity coefficient γ such that

$$p_w = \gamma x_w p_w'$$

(12.19)

However, activity coefficients are complex functions of both temperature and moisture content and cannot be readily determined. Substituting into the definition of water activity gives

$$a_w = \frac{\gamma x_w p_w'}{p_w'}$$

(12.20)

and therefore

$$a_w = \gamma x_w$$

(12.21)

which for an ideal system reduces to

$$a_w = x_w$$

(12.22)

Raoult's law generally holds only at relatively high values of x_w. Thus for both an ideal system, and for high moisture contents, water activity effectively is nothing more than the mole fraction of water in the liquid phase within the food. It is important to realise that x_w is defined in relation only to the soluble constituents within a food and ignores the insoluble components.

Clearly there are difficulties in using Equation (12.21) to find a_w. However if the food is in contact with, and at thermal equilibrium with, the surrounding air then, from the definition of relative humidity [Equation (9.6)], water activity is equal to the fractional relative humidity. That is

$$a_w = \frac{\%RH}{100}$$

(12.23)

The right-hand side of Equation (12.23) is sometimes designated ERH or equilibrium relative humidity. This now forms the basis of the measurement of water activity.

12.6.3. Effect of Water Activity on Microbial Growth

The ability of micro-organisms to grow is reduced as water activity is reduced. In general, bacteria require a greater water activity or moisture content for growth than do fungi. However there is a range of water activity for growth of all micro-organisms and this range is widest at the optimum growth temperature for a given bacterium, yeast or mould. Also, yeasts and moulds tend to grow over a wider range of water activity than do bacteria. Thus the lower limit to water activity for spoilage bacteria, yeasts and moulds are approximately 0.90, 0.85–0.88, and 0.80, respectively. However these limits may be as low as 0.75 for halophilic bacteria, that is, those which can tolerate a high salt content, 0.60 for xerophilic moulds (which tolerate low moisture contents) and about 0.60 for osmophilic yeasts which thrive with high osmotic pressures. These figures should be considered in the context of the water activity of fresh food which is very often greater than 0.99. Conversely, enzymatic activity in foods may be significant at water activities as low as 0.30.

12.6.4. Moisture Content

For the purposes of process analysis, the moisture content of a solid food is usually expressed either on a wet basis, in the form of a mass fraction, or on a dry basis as a mass ratio.

Example 12.6

A 50 kg mass of food has a moisture content of 33.3% (dry basis). How much water is present? What is the moisture content on a wet basis?

On a wet basis (i.e., mass fraction) the moisture content is $0.333/1.333 = 0.25$ or 25%. Therefore the mass of water present is $0.25 \times 50 = 12.5$ kg.

12.6.5. Isotherms and Equilibrium

It is important to understand that there are limits to the amount of water which can be removed from a food material under any given conditions. The equilibrium moisture content is the moisture content of a material when in equilibrium with the partial pressure of water vapour in the surroundings and inevitably this quantity varies widely for different materials. If now the equilibrium moisture content is measured at different humidities and temperatures, in other words as a function of water activity, a sorption isotherm is generated. This takes the form of the sigmoidal curve shown in Figure 12.9; the characteristic hysteresis is produced by differences in the adsorption and desorption of water. Figure 12.10, which represents a plot of %RH against the moisture content of a food, is a different way of presenting the same information.

Figure 12.10 highlights the important point that, for a given condition of the surrounding air, that is, for a given relative humidity, it is not possible to dry a material below the equilibrium

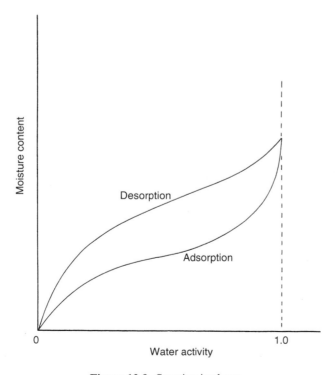

Figure 12.9. Sorption isotherm.

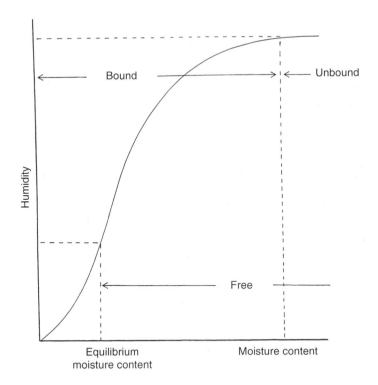

Figure 12.10. Relationship between moisture content and surrounding humidity.

moisture content. As the relative humidity falls with temperature, there is a corresponding decrease in the equilibrium moisture content and a change to the theoretical end point of any drying operation. In addition, Figure 12.10 illustrates a number of other terms which need to be defined. Bound moisture is water which exerts a vapour pressure less than that of free water of the same temperature, in other words it has a water activity less than unity. An example of this would be water held in small capillaries. Unbound moisture is defined as any water in excess of that which is bound. Finally, free moisture is any water in excess of equilibrium moisture, that is, where the water activity is equal to unity. It is only free moisture that may be evaporated.

12.7. BATCH DRYING

12.7.1. Rate of Drying

For a batch process the rate of drying varies with the moisture content of the material. Figure 12.11 is a somewhat idealised curve but it serves to illustrate the different mechanisms of drying which can be identified. This type of curve can be obtained experimentally by measuring the mass of a drying sample as a function of time. The change in mass of the sample in a given interval divided by that interval equals the rate of drying. The moisture content of the food can be obtained from the mass at any given time and from a knowledge of the final dry mass of the sample.

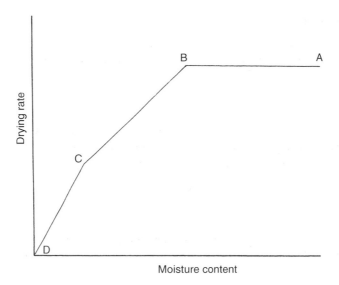

Figure 12.11. Rate of drying as a function of moisture content.

In the constant rate period (A to B in Figure 12.11), the rate of drying is constant as the moisture content is reduced. During this period drying takes place from a saturated surface and the vaporised water molecules diffuse through a thin stagnant film of air close to the surface of the material before being transported into the bulk of the air stream. This air film contains all the resistance to mass transfer. Consequently, because water vapour is leaving what is effectively a free water surface, the drying rates for different materials are remarkably similar under similar conditions (i.e., similar temperature, air velocity and humidity). The drying rate therefore is not a function of the material being dried but of the mass transfer characteristics in the surrounding air stream.

However, below point B in Figure 12.11 (which is known as the critical moisture content) the rate of drying decreases in the falling rate period (B to D). Two distinct zones may exist although often it will be possible to distinguish only one. In the first falling rate period (B to C) the food surface is no longer capable of supplying sufficient free moisture to saturate the air above it. This means that the rate of drying is then influenced by the mechanism of transport of moisture from within the food mass to the surface. In the second falling rate period (C to D) the surface is completely dry and the plane of separation, or water interface, moves into the solid. Evaporation now depends upon the diffusion of vapour through the material and is therefore increasingly slow. The forces controlling vapour diffusion determine the rate of drying and these are largely independent of air conditions at the surface. Consequently the nature and structure of the material being dried influences the rate of drying.

The analysis of this stage of drying is much more complex and two possible mechanisms may account for the transport of water to the surface. First, it is often proposed that capillary forces may control the movement of water in the pore spaces between particles within a granular solid. However it is more likely that diffusion of water vapour through the porous structure of the food governs the drying rate. Mass transfer through the solid phase is very slow, particularly at low-moisture contents, and difficult to predict because of the uncertainty of determining effective diffusivities in the solid phase.

12.7.2. Batch Drying Time

It is possible to make an estimate of the batch drying time t if both the equilibrium (X_e) and critical (X_c) moisture contents are known. By definition, the rate of drying R is given by:

$$R = -\frac{W}{A}\frac{dX}{dt} \tag{12.24}$$

where W is the mass of dry solid, A is the area over which drying takes place and dX/dt represents the rate of change of the moisture content on a dry basis. The negative sign indicates a decrease in moisture content. Separating the variables in Equation (12.24) gives

$$t = \frac{W}{A}\int_{X_2}^{X_1}\frac{dX}{R} \tag{12.25}$$

where X_1 and X_2 are the initial and final moisture contents respectively. In the constant rate period, where $X_1 > X_c$ and $X_2 > X_c$, Equation (12.24) can be solved readily to give

$$t = \frac{W}{AR_c}(X_1 - X_2) \tag{12.26}$$

where R_c is the rate of drying in the constant rate period.

Example 12.7

A 100 kg batch of a food powder contains 28% moisture on a wet basis. It is dried down to 16% moisture at a constant rate of 0.006 kg m^{-2} s^{-1}. The critical moisture content is 15%. Calculate the batch drying time if the drying surface is 0.03 m^2 per kg of dry weight.

The initial and final moisture contents are greater than the critical moisture content and therefore drying takes place in the constant rate period and the drying time can be found from Equation (12.26). Converting the initial moisture content to a dry weight basis gives

$$X_1 = \tfrac{28}{72}, \qquad X_1 = 0.389.$$

Similarly the final moisture content is $X_2 = 0.190$. The available surface area for drying is $0.03W$ m^2 and hence the drying time t is

$$t = \frac{W(0.389 - 0.190)}{0.03W\,6\times 10^{-3}}\,\text{s}, \qquad t = 1106\,\text{s} \ \text{or} \ 18.4\,\text{min}$$

For the falling rate period Equation (12.25) must be solved graphically. In other words the area under the curve in Figure 12.12 must be evaluated from experimental data. However, if there is a known relationship between R and X an analytical solution for drying time can be obtained. It is reasonable to assume a linear relationship as in the simplified drying curve shown in Figure 12.13 and therefore

$$R = mX + b \tag{12.27}$$

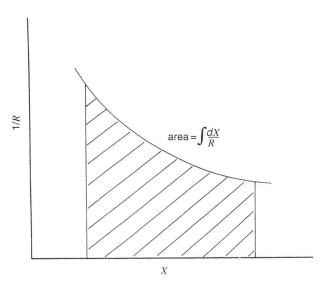

Figure 12.12. Graphical integration to find batch drying time.

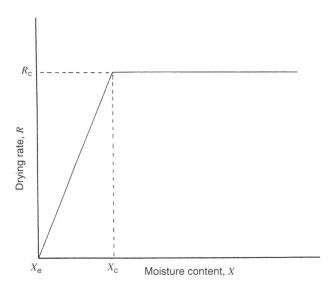

Figure 12.13. Batch drying curve.

where m and b are constants. Substituting from Equation (12.27) to Equation (12.25) produces

$$t = \frac{W}{A} \int_{X_2}^{X_1} \frac{dX}{mX + b} \tag{12.28}$$

where X_1 and X_2 are now both smaller than X_c and which on integration gives

$$t = \frac{W}{mA} \ln \left\{ \frac{mX_1 + b}{mX_2 + b} \right\} \tag{12.29}$$

However it must be the case that the drying rates at X_1 and X_2 are

$$\left.\begin{array}{l} R_1 = mX_1 + b \\ R_2 = mX_2 + b \end{array}\right\} \tag{12.30}$$

from which the gradient m can be obtained as

$$m = \frac{R_1 - R_2}{X_1 - X_2} \tag{12.31}$$

Thus, substituting back into Equation (12.29), the batch drying time becomes

$$t = \frac{W}{A} \frac{(X_1 - X_2)}{(R_1 - R_2)} \ln\left(\frac{R_1}{R_2}\right) \tag{12.32}$$

As the falling rate period can be represented by a straight line from X_c to X_e, and where $X_1 < X_c$ and $X_2 < X_c$, then the drying rate at any moisture content X is

$$R = m(X - X_e) \tag{12.33}$$

and consequently the gradient becomes

$$m = \frac{R_c}{X_c - X_e} \tag{12.34}$$

Finally, substituting into Equation (12.33) yields

$$R = \frac{R_c(X - X_e)}{(X_c - X_e)} \tag{12.35}$$

and combining this with Equation (12.32) results in

$$t = \frac{W}{AR_c}(X_c - X_e) \ln\left(\frac{X_1 - X_e}{X_2 - X_e}\right) \tag{12.36}$$

In other words the drying time in the falling rate period can be determined if the rate of drying in the constant rate period is known.

Example 12.8

A food solid was dried from 40% to 10% moisture content in two hours in a batch drier with constant air conditions. The drying rate remained constant down to a moisture content of 15%. If the equilibrium moisture content is 2% calculate the total time required to dry from 40% to 4% moisture content. All moisture contents are given on a dry basis.

For the constant rate period only, the initial moisture content is $X_1 = 0.40$ and the final moisture content is $X_2 = 0.15$ and therefore, using Equation (12.26), the drying time in the constant rate period is given by

$$t = \frac{W}{AR_c}(0.40 - 0.15) \text{ or } t = \frac{0.25W}{AR_c}$$

where W, A, and R_c are all unknown. For the falling rate period between 15% and 10% the values of X_1 and X_2 become 0.15 and 0.10, respectively. From Equation (12.36) the drying time over this range of moisture content is given by

$$t = \frac{W}{AR_c}(0.15 - 0.02) \ln \left\{ \frac{(0.15 - 0.02)}{(0.10 - 0.02)} \right\}, \qquad t = 0.0631 \frac{W}{AR_c}$$

However, the total drying time from a moisture content of 40% to a moisture content of 10% is 120 min. Thus

$$\frac{W}{AR_c}(0.25 + 0.0631) = 120, \qquad \frac{W}{AR_c} = 383.2 \text{ min}$$

For the falling rate period between moisture contents of 10% and 4%, respectively, the drying time is

$$t = 383.2(0.15 - 0.02) \ln \left\{ \frac{(0.10 - 0.02)}{(0.04 - 0.02)} \right\}$$

or

$$t = 69.1 \text{ min}$$

Hence the total drying time down to a moisture content of 4% is $(120 + 69.1) = 189.1$ min

12.8. TYPES OF DRIER

The classification of industrial driers is difficult because of the huge variety of drier types available and the large number of variables to be considered. However some general groupings can be identified.

12.8.1. Batch and Continuous Operation

The usual disadvantages associated with batch operation apply to batch driers: higher labour costs and a variation in product quality from batch to batch. However as the throughput, or capacity, of the drier falls the capital cost becomes rather more significant and batch operation may well become cheaper overall. The upper limit to throughput for batch operation is about 5 t per day. This is well within the required throughput of many food processes and this, coupled with the general ease of operation, means that batch driers are very popular in food industry applications. On the other hand continuous driers allow the drying step to be integrated into a continuous process and result in lower unit costs at high tonnages. It should be noted that many types of drier can be adapted for either batch or continuous operation.

12.8.2. Direct and Indirect Driers

A direct drier is one in which wet solids are directly exposed to the hot gases which provide the latent heat of vaporisation; a simple tray drier is an example. In indirect driers there is a physical barrier between the food being dried and the heat source. For example heat may be transferred from condensing steam through a metal surface. This technique is useful in reducing material losses considerably because small dried particles are not carried away in a high velocity stream of gas. It also prevents exposure to higher temperatures than are necessary and thus avoids thermal damage to the food.

12.8.3. Cross-circulation and Through-circulation

This is a way of classifying driers according to the nature of contact of solids and hot gas. In cross-circulation the drying medium passes over the surface of the solids, as in a tray drier. There is limited contact between solids and gas which limits the surface area available for heat and mass transfer. In through-circulation driers the hot gas passes in some way through the solid, for example, a granular material or a solid which is perforated deliberately, and which is supported on a grid allowing the free passage of gas. The fluidised bed drier is the most obvious example here.

A further possible classification would be based upon the form of the wet feed to the drier, whether it is a wet granular solid, a paste or a solution. However, once again, many driers can be adapted for a number of applications and this approach has limited value.

12.8.4. Tray Drier

A tray drier consists of shallow trays mounted on racks which can be wheeled into a drying cabinet. The cabinet contains a heater and a fan to circulate the hot air at velocities up to about $5 \, \text{m s}^{-1}$. Tray driers are batch operated with residence times of several hours. They are used generally for low throughputs and therefore for high value products. Operating costs tend to be high because of the intensive use of labour. If the risk of thermal damage imposes a maximum temperature on the inlet air there is an advantage in reheating the air between stages in a tray drier rather than heat the gas to a higher initial temperature. This is illustrated in Figure 12.14 and by Example 9.8. Reheating of the air in this way has the advantage that unit mass of air removes more water than in a single stage and there is a reduction in the volume of air flow through the drier. In turn this simplifies the heating system and, for granular materials, reduces the carry over of small particles. The overall thermal efficiency of the drier is increased because the smaller air flow reduces heat losses.

Example 12.9

Using the data of Example 9.8, determine the temperature to which the inlet air must be heated to obtain the same result without inter-shelf reheating.

Drying on the second shelf takes place at a wet bulb temperature of 28.3°C. Extending this line to meet a humidity of 0.007 gives a temperature of 69°C. Thus heating the inlet air to 69°C and humidifying at a constant wet bulb temperature of 28.3°C achieves the same overall drying rate.

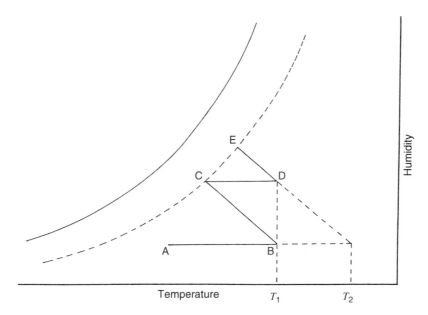

Figure 12.14. Reheating of air.

The air which is discharged from a drier has a greatly increased humidity but also a considerable enthalpy. This enthalpy is wasted, and thermal efficiencies are lower than necessary, unless the air is recycled. Figure 12.15a illustrates a simple a simple recycle system in which some of the exhaust air (B) from the drier is recycled and mixed with fresh air (A) to give the inlet air (C). This process is shown on a psychrometric chart in Figure 12.15b. The air at C is heated to a temperature T (point D) and then picks up water adiabatically before leaving the drier (B). Recycling of air has the advantage that the heat required per unit mass of water evaporated is reduced by effectively reusing some of the heat which would be rejected in the exhaust stream. However the drying rate is reduced because the humidity of the air entering the drier is greater. This can be an advantage when it is necessary to control the rate of drying or when an increase in humidity is needed, for example, to control the risk of a dust explosion in spray drying.

12.8.5. Tunnel Drier

Essentially this is a continuous version of the tray drier. Trucks containing the wet solid move slowly through a tunnel in which air is circulated either co- or counter-currently. It is used where large throughputs necessitate a reduction in bulk handling of the product. It may be likened to a belt-drier where the wet material forms a 'continuous' phase.

12.8.6. Rotary Drier

A horizontal drum or cylinder is operated continuously with material entering and discharging at opposite ends. The air flows through the drum either co-currently or counter-currently to the solids. The drum rotates slowly (typically at 25 rpm for 1 m diameter and more

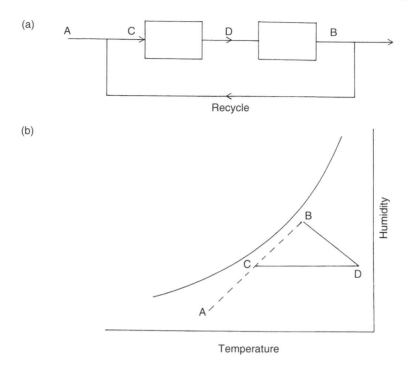

Figure 12.15. Recirculation of air.

slowly for larger driers) and baffles are placed along the length of the interior to lift material
away from the surface and thus improve contact between the warm air and the solids. Transport
of solids through the drier is achieved by inclining it slightly, of the order of 5°. The residence
time can be varied by changing either the angle of inclination or the rim height. Throughput
can be very high, of the order of several tens of tonnes per hour. Rotary driers are suitable for
most granular materials but solids handling difficulties will increase as the moisture content
of the feed increases. It is possible to arrange indirect heating by using condensing steam in a
double shell.

12.8.7. Fluidised Bed Drier

A gas–solid fluidised bed can be used as a drier with the heat for evaporation of the
solvent being supplied in the inlet gas. The efficient heat and mass transfer characteristics of
fluidisation are exploited to give uniform bed temperatures and high drying rates. The details
of fluidised bed driers are covered in chapter thirteen.

12.8.8. Drum Drier

A dry flaky solid can be formed from a solution or wet paste by running the solution
onto the hot surface of a slowly rotating drum which is heated internally by steam. The solid
remains on the drum surface after evaporation of the solvent and is scraped off by a knife or
wire. The residence times obtainable are of the order of 10 s and throughputs up to 50 kg h^{-1} per

square metre of drum surface area are possible. Drums may be sized up to 2–3 m in diameter and 3–4 m in length.

12.8.9. Spray Drier

In spray drying a solution or slurry is pumped to the top of a cylindrical chamber, atomised into small droplets and contacted with a hot air stream either co-currently or counter-currently. In counter-current operation, which is used most frequently, the hot air stream enters the drying chamber at a number of points from a circular ring main. A large drying chamber is needed because the droplets must dry before hitting the walls. Consequently, for very high throughputs (up to 50 t h^{-1}), the chamber diameter may be as large as 10 m although spray driers as small as 1 m diameter are available. Atomisation of the feed solution is necessary to produce droplets of approximately 10–100 μm which will dry in the time available. Three methods of atomisation are used:

   i   the pressure nozzle, in which the solution is forced through a small orifice at pressures up to 5,000 kPa,

   ii   the spinning disc where liquid impinges on a disc rotating at a velocity of several thousand rpm, and

   iii the twin-fluid nozzle in which a liquid stream is broken up by a high velocity air stream.

The choice of atomisation method depends upon the rheology and flow rate of the feed but method (ii) is the most common in the food industry.

The correct mixing of droplets/particles and air in the drying chamber and the consequent droplet drying time are crucial for the success of spray drying. Residence times are much shorter (about 5 s in a large diameter drier) for co-current drying than for counter-current drying, where comparable residence times are of the order of 30 s. Counter-current operation is the most thermally efficient of the two but great care is needed because dry and dusty particles are in contact with the highest gas temperatures and thus there is a considerable explosion risk.

Spray drying has the advantage that a particulate solid can be formed from a solution in a one-stage process at very high throughputs and capacities of 0.5 t m^2 h^{-1} are possible. Spray dried particles tend to be hollow, of low density and uniformly spherical which gives spray dried products an attractive appearance. Thus it is used extensively to produce dried milk powders. Some control of product properties is possible, for example by adjusting drying temperature and droplet size. However in the production of dried milk powders a subsequent agglomeration stage is necessary to produce particles which can be easily dispersed in water in order to aid reconstitution. The disadvantages of spray drying are the relative complexity of the process and the associated high capital cost. Considerable ancillary equipment is required such as high-pressure pumps, cyclones for the separation of air and solids and a heater control system. A further disadvantage is the risk of dust explosion.

12.8.10. Freeze Drying

Freeze drying is a technique that finds wide application to heat sensitive materials, especially foods, and is capable of removing water without impairing product quality. It operates on a very different principle to other drying methods. The material to be dried is placed on hollow shelves inside a cabinet and frozen at temperatures down to $-30°C$. A vacuum is applied to the cabinet (absolute pressures are below 600 Pa and often much lower) and the food is then heated by circulating a hot fluid in the hollow shelves. This results in sublimation of the water which is removed by the vacuum pump. Sublimation takes place at a receding ice surface within the food and water vapour then diffuses out through the porous structure of the dried solid.

NOMENCLATURE

a_w	Water activity
A	Area
A_{mean}	Average heat transfer surface area in a multiple effect evaporator
b	Constant
C	Mass flow rate of condensate
ERH	Equilibrium relative humidity
F	Mass flow rate of feed
h	Enthalpy
L	Mass flow rate of concentrated liquor
m	Constant
n	Number of evaporator effects
p_w	Partial pressure of water vapour
p_w'	Pure component vapour pressure of water
Q	Rate of heat transfer
R	Rate of drying
R_c	Rate of drying in the constant rate period
S	Mass flow rate of steam
t	Time
T	Temperature
T_E	Evaporator temperature
U	Overall heat transfer coefficient
U_{av}	Average overall heat transfer coefficient
V	Mass flow rate of vapour
W	Mass of dry solid
x	Mass fraction of solute or non-soluble solids
X	Moisture content (dry basis)
X_c	Critical moisture content
X_e	Equilibrium moisture content
X_1	Initial moisture content
X_2	Final moisture content
y	Mass fraction (vapour phase)

GREEK SYMBOLS

γ	Activity coefficient
ΔT	Temperature difference

SUBSCRIPTS

C   Condensate
f   Saturated liquid
F   Feed
g   Saturated vapour
L   Concentrated liquor
S   Steam
V   Vapour
w   Water

PROBLEMS

12.1. A 5% solids by mass aqueous food solution, flowing at $1.0 \, \text{kg s}^{-1}$ and 288 K, is to be concentrated to 30% solids by evaporation in a single effect evaporator. Dry saturated steam at 2 bar is available and the overall heat transfer coefficient is $2.50 \, \text{kW m}^{-2} \, \text{K}^{-1}$. Assuming that the properties of the solution are those of pure water and that the evaporator operates at 0.2 bar, calculate the rate of steam consumption and the heat transfer area required.

12.2. Calculate the economy for the evaporator in Problem 12.1.

12.3. Fresh milk containing 11.5% solids is fed to an evaporator, operating under vacuum at 0.30 bar, at the rate of $0.50 \, \text{kg s}^{-1}$ after being pre-heated to the evaporator temperature. The evaporated milk leaves with a solids content of 31%. The steam in the calandria is at a pressure of 3 bar. Calculate the flow rate of vapour leaving the evaporator and the steam consumption.

12.4. An overall heat transfer coefficient of $2{,}100 \, \text{W m}^{-2} \, \text{K}^{-1}$ may be expected in the evaporator in Problem 12.3. Calculate the heat transfer surface area.

12.5. A 5% sugar solution is concentrated, at a rate of $1{,}000 \, \text{kg h}^{-1}$ to 25% in a single effect evaporator operating at a pressure of 14 kPa with steam supplied at 170 kPa. The feed temperature is 293 K and the product leaves at its boiling point of 335 K. For an overall heat transfer coefficient of $2{,}250 \, \text{W m}^{-2} \text{K}^{-1}$, calculate the boiling point elevation, the heat transfer area and the economy.

12.6. A double-effect, forward feed evaporator with identical surface areas is used to concentrate a vegetable juice from 10% to 40% solids using 2.7 bar steam. A boiling point elevation of 5 K is expected in the second effect where a vacuum of 89.3 kPa is maintained. The juice is fed to the evaporator at 80°C. Assuming the heat capacity of each liquid stream to be $4.18 \, \text{kJ kg}^{-1} \, \text{K}^{-1}$ and the overall heat transfer coefficients in the first and second effects to be 2,500 and $1{,}500 \, \text{W m}^{-2} \, \text{K}^{-1}$, respectively, calculate the economy and the heat transfer surface area of each effect if the product rate is $3{,}600 \, \text{kg h}^{-1}$.

12.7. A 15 kg batch of dried carrot has an overall moisture content of 20% on a dry basis. What mass of water does this represent? What is the moisture content on a wet basis?

12.8. How much water must be added to potato granules containing 10% moisture (dry basis) to give a final mass of 25 kg with a moisture content of 80% (wet basis)?

12.9. A wet solid leaves a mixing step containing 25% moisture (dry basis) and is then batch dried to a final moisture content of 7%. Under constant drying conditions 3 h are required to reduce the moisture content to 12% and the drying rate is constant down to 18% moisture. If the equilibrium moisture content is 5% under the conditions used, calculate the total time for drying.

12.10. The following data was obtained by drying a small sample of food in an air stream at constant humidity and temperature and recording the mass of the sample at intervals of 5 min. The sample was then dried to a constant mass of 70 g. Plot a curve of drying rate against moisture content and hence determine the rate of drying in the constant rate period, the critical moisture content (wet basis) and the equilibrium moisture content.

Time (min)	Mass (g)
0	100.0
5	97.20
10	94.40
15	91.60
20	88.80
25	86.00
30	83.20
35	80.64
40	78.70
45	77.26
50	76.20

12.11. Air with an absolute humidity of 0.005 kg per kg is heated to 325 K and passed over the lower shelf of a tray drier. It leaves at 60% saturation and is then reheated to 325 K. The drier contains four shelves and the above cycle is repeated for each. Determine the temperature to which the air would need to be heated to avoid inter-shelf reheating.

FURTHER READING

C. G. J. Baker, (ed.), *Industrial Drying of Foods*, Chapman and Hall, London (1996).
D. R. Heldman and R. W. Hartel, *Principles of Food Processing*, Chapman and Hall, London (1997).
R. B. Keey, *Introduction to Industrial Drying Operations*, Pergamon, London (1978).
K. Masters, *Spray Drying Handbook*, Godwin, (1979).
D. MacCarthy, (ed.), *Concentration and Drying of Foods*, Elsevier Applied Science, Amsterdam (1986).

13

Solids Processing and Particle Manufacture

13.1. CHARACTERISATION OF PARTICULATE SOLIDS

13.1.1. Particle Size Distributions

Any sample of solids whether natural (such as the grains of sand on a beach) or manufactured (such as particles of spray dried milk) will contain a distribution of particle sizes. The size of particulate material is an important product characteristic, both because of its effect on product appearance, dissolution, flowability and because particle size determines the behaviour of powders in process equipment. It is necessary therefore to be able to measure the particle size distribution and to present the data in an intelligible way.

A frequency distribution curve is a plot of the frequency with which a given particle size occurs (the percentage of total particles either on a number or a mass basis) against that size; an example is shown in Figure 13.1. The frequency distribution shows clearly the overall shape of

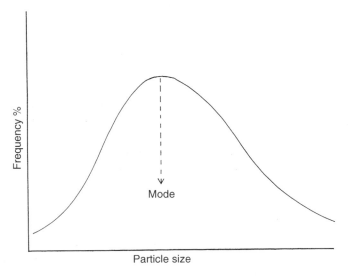

Figure 13.1. Frequency distribution curve.

335

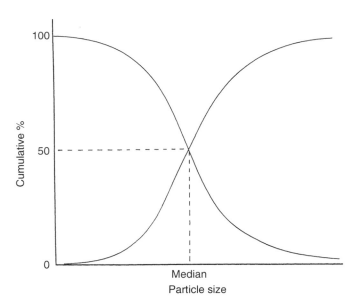

Figure 13.2. Cumulative frequency distribution.

a particle size distribution and the presence of very large or very small particles. Alternatively the data can be plotted as a cumulative distribution in which either the cumulative percentage undersize or cumulative percentage oversize is plotted against size. The two curves are mirror images of each other (Figure 13.2). A graphical representation of a distribution can be very helpful in assessing the presence of oversized particles or of fines and therefore deciding whether or not a product is within specification. Alternatively it may be possible to draw conclusions about the behaviour of a process, for example whether particular size fractions are being lost or if unwanted agglomeration is occurring.

Despite the usefulness of a particle size distribution, it is greatly desirable to use a single value as a measure of the size of the particles in a sample. Two expressions of 'average' size can be obtained from distribution curves. The mode size is the most frequently occurring size (Figure 13.1) whilst the median particle size is that which cuts the area under the frequency distribution curve in half, that is it is the 50% point on the cumulative curve, which in turn must be the intersection of the undersize and oversize curves. Whilst each of these quantities gives an indication of the average size, neither takes account of the spread of the distribution and it is possible to obtain the same mode or median with either a very narrow or a very wide distribution. These difficulties can be overcome by a using a carefully defined mean particle size.

13.1.2. Mean Particle Size

The characteristics of a distribution can be defined as the total number, total length, total surface area, and total volume of the particles. A distribution of particle sizes can be represented by a set of uniformly sized particles which retains two characteristics of the original distribution. The mean particle size of a distribution is then equal to the size of the uniform particles with respect to the two characteristics.

Example 13.1

A sample consists of five spherical particles of size 2, 4, 6, 8, and 10 arbitrary units, respectively. What is the mean particle size if (a) the same total number and the same total length are retained and (b) if the total number and total volume are retained?

The total number of particles is 5 and the total length of all particles is $2+4+6+\cdots = 30$ units. Therefore the mean particle size x with respect to the total number and total length is

$$x = \tfrac{30}{5} \text{ units}, \qquad x = 6.0 \text{ units}$$

The total volume V of particles in the original distribution is

$$V = \frac{\pi}{6}(2^3 + 4^3 + \cdots + 10^3) \text{ cubic units}$$

and

$$V = \frac{\pi}{6}1,800 \text{ cubic units}$$

Thus the mean particle size x with respect to the total number and total volume is obtained from

$$\frac{\pi}{6}1,800 = 5\frac{\pi}{6}x^3, \qquad x = \sqrt[3]{360}$$

or

$$x = 7.11 \text{ units}$$

Any of the possible mean particle sizes can be calculated using the model of Mugele and Evans which is summarised by Equation (13.1)

$$x_{q,p}^{q-p} = \frac{\int_{x_1}^{x_u} x^q (dN/dx)\,dx}{\int_{x_1}^{x_u} x^p (dN/dx)\,dx} \tag{13.1}$$

where N is the number of particles of size x and p and q are parameters representing the characteristics of a distribution; x_1 and x_u are the lower and upper limits of the size distribution, respectively. The parameters p and q have values of zero for number, one for length, two for area and three for volume. Most methods of determining particle size generate data in the form of the number of particles within a given size band. The mean size is then obtained by summing over all the band widths. Thus the model becomes

$$x_{q,p}^{q-p} = \frac{\sum(x^q N)}{\sum(x^p N)} \tag{13.2}$$

Substituting all the possible combinations of characteristics, that is values of p and q, into Equation (13.2) gives rise to seven different definitions of the mean size of a distribution. These are listed in Table 13.1.

Of these the surface-volume mean, also known as the Sauter mean diameter and the harmonic mean diameter, is perhaps the most commonly used and the most useful (see, e.g., section 13.4). The weight–moment mean is based on weight, or volume, ($p = 3$) and the moment, that is, mass × distance, ($q = 4$) and is particularly sensitive to the presence of large particles at the top end of a distribution.

TABLE 13.1
Definitions of Mean Particle Diameter

	Number distribution	Mass distribution
Number–length $p = 0,\ q = 1$	$x_{1,0} = \dfrac{\sum(xN)}{\sum N}$	$x_{1,0} = \dfrac{\sum(\omega/x^2)}{\sum(\omega/x^3)}$
Number–surface $p = 0,\ q = 2$	$x_{2,0} = \sqrt{\dfrac{\sum(x^2 N)}{\sum N}}$	$x_{2,0} = \sqrt{\dfrac{\sum(\omega/x)}{\sum(\omega/x^3)}}$
Number–volume $p = 0,\ q = 3$	$x_{3,0} = \sqrt[3]{\dfrac{\sum(x^3 N)}{\sum N}}$	$x_{3,0} = \dfrac{1}{\sqrt[3]{\sum(\omega/x^3)}}$
Length–surface $p = 1,\ q = 2$	$x_{2,1} = \dfrac{\sum(x^2 N)}{\sum(xN)}$	$x_{2,1} = \dfrac{\sum(\omega/x)}{\sum(\omega/x^2)}$
Length–volume $p = 1,\ q = 3$	$x_{3,1} = \sqrt{\dfrac{\sum(x^3 N)}{\sum(xN)}}$	$x_{3,1} = \dfrac{1}{\sqrt{\sum(\omega/x^2)}}$
Surface–volume $p = 2,\ q = 3$	$x_{3,2} = \dfrac{\sum(x^3 N)}{\sum(x^2 N)}$	$x_{3,2} = \dfrac{1}{\sum(\omega/x)}$
Volume–moment Weight-moment $p = 3,\ q = 4$	$x_{4,3} = \dfrac{\sum(x^4 N)}{\sum(x^3 N)}$	$x_{4,3} = \sum(\omega x)$

Example 13.2

The particle size distribution of a sample was measured as follows:

Size band (μm)	Number of particles
0–4	30
4–8	40
8–12	90
12–16	100
16–20	120
20–24	80
24–28	65
28–32	15

Determine the number–length mean particle diameter and the length–volume mean particle diameter.

The number–length and length–volume mean particle diameters are defined by

$$x_{1,0} = \frac{\sum(xN)}{\sum N}, \qquad x_{3,1} = \sqrt{\frac{\sum(x^3 N)}{\sum(xN)}}$$

respectively. The most appropriate method of solving this problem is to tabulate values of x, N, xN, and $x^3 N$. Assuming the particle size to be the arithmetic average of the size band, then the following table can be constructed:

x (μm)	N	xN	x^3N
2	30	60	240
6	40	240	8,640
10	90	900	90,000
14	100	1,400	274,400
18	120	2,160	699,840
22	80	1,760	851,840
26	65	1,690	1,142,440
30	15	450	405,000
$\sum N = 540$		$\sum xN = 8.66 \times 10^3$	$\sum x^3N = 3.47 \times 10^6$

Thus

$$x_{1,0} = \frac{8.66 \times 10^3}{540} \mu\text{m} \text{ or } x_{1,0} = 16.04 \mu\text{m}$$

and

$$x_{3,1} = \sqrt{\frac{3.47 \times 10^6}{8.66 \times 10^3}} \mu\text{m} \text{ or } x_{3,1} = 20.02 \mu\text{m}$$

13.1.3. Particle Shape

Particle shape is also important in determining the behaviour of particulate solids although little is known about the influence of shape and it is exceedingly difficult to measure or define. A detailed treatment of particle shape would be out of place in a text of this kind but it is necessary to consider the implications for size analysis. A sphere may be characterised uniquely by its diameter and a cube by the length of a side. However few natural or manufactured food particles are spherical or cubic. For irregular particles, or for regular but non-spherical particles, an equivalent spherical diameter d_e can be defined as the diameter of a sphere with the same volume V as the original particle.

$$V = \frac{\pi}{6}d_e^3 \tag{13.3}$$

Example 13.3

A regular cuboid has sides of 1.0, 1.5 and, 3.0 units. What is the equivalent spherical diameter?

The volume of the cuboid $= 1.0 \times 1.5 \times 3.0 = 4.5$ cubic units. Therefore, from the definition of the equivalent spherical diameter,

$$4.5 = \frac{\pi}{6}d_e^3, \qquad d_e = 2.05 \text{ units}$$

A further way of defining a diameter for a non-spherical particle is the diameter of a sphere having the same terminal falling velocity as the particle (see section 13.2). The density of the particle and the density and viscosity of the fluid must be the same and this definition is especially useful if settling behaviour is under investigation. More commonly, shape factors are used to relate particle volume to the cube of diameter. In the case of a sphere [Equation (13.3)], they are related by a factor of $\pi/6$. For non-spherical but regular particles the volume V can be defined in terms of a volume shape factor k and therefore

$$V = kx^3 \tag{13.4}$$

Particle shape may also be characterised by comparing the respective surface areas of the particle and a sphere. Thus sphericity, ϕ is defined as

$$\phi = \frac{\text{surface area of sphere of equal volume to particle}}{\text{surface area of particle}} \tag{13.5}$$

Example 13.4

Determine the sphericity of a cube.

Let the side of the cube be equal to 1 unit. The volume of the cube then equals 1 cubic unit and the total surface area of the six faces equals 6 square units. The equivalent spherical diameter d_e is given by

$$1.0 = \frac{\pi}{6} d_e^3$$

and therefore

$$d_e = \left(\frac{6}{\pi}\right)^{1/3} \text{ units}$$

The surface area of the sphere (of equal volume to the cube) is πd_e^2 and hence

$$\phi = \frac{\pi \left(\frac{6}{\pi}\right)^{2/3}}{6}, \qquad \phi = 0.806$$

13.1.4. Methods of Determining Particle Size

There are very many ways of measuring or deducing the size of particles although very few of these are suitable for assessing samples taken from bulk powders. The most common methods are outlined below.

(a) *Sieving*   Sieving remains one of the easiest and cheapest ways of determining a particle size distribution and is very widely used. It is the only method which gives a mass distribution and the only method which can be used for a reasonably large sample of particles which are not in suspension. A sieve analysis is carried out by placing a sample of about 50 g (depending on the sieve diameter) on the coarsest of a set of standard sieves made from woven wire. Below this sieve the other sieves are arranged in order of decreasing aperture size. The sample is then shaken for a fixed period of time, and the material on each sieve collected and weighed. In a series of sieves, aperture sizes are normally related by a factor of $2^{1/2}$ so that the aperture area doubles between consecutive sieves (Table 13.2). For particularly fine or cohesive powders air swept sieving can be used in which an upward flow of air from a rotating arm underneath the mesh prevents blockage of the sieve apertures.

TABLE 13.2
British Standard Sieve Sizes

Aperture size (μm)	Aperture size (μm)
1,400	
	1,180
1,000	
	850
710	
	605
500	
	425
355	
	300
250	
	212
180	
	150
125	
	100
90	
	75
63	
	53
45	

(*b*) *Microscopy*   This is the only method of measuring particle size directly. Individual particles are placed on a slide and viewed with a microscope which is fitted with a graticule. The method is tedious and is not suitable for generating distribution data. Microscopy is used only where it is not possible to use other techniques or when only individual measurements are required. For sub-microscopic particles scanning electron microscopy can be used.

(*c*) *Image analysis*   Image analysis is an extension of microscopy in that the contrast between the particle and the background is used to measure the projected area of a particle when placed on a slide and viewed with a photo-optic imaging system. From this the particle diameter and volume can be calculated. Photographs of particles, cells, bubbles or droplets can all be used with this technique.

(*d*) *Electrolyte resistivity (Coulter counter)*   The so-called Coulter counter measures the change in resistance as particles suspended in an electrolyte pass through a small orifice between two electrodes. The change in resistance generates a pulsing electromotive force, the amplitude of which is proportional to the particle volume. The particle size distribution can be constructed by counting and measuring the amplitudes. Essentially this technique measures the volume of an envelope around the particle and therefore porous particles give erroneous results. The particles must be in suspension and must not dissolve in the electrolytic solution.

(*e*) *Laser scattering*   The angle at which light is scattered by a particle depends upon the particle size. Commercial instruments now use intense laser light and the intensity of light scattered in a forward direction is measured as a function of radius by a series of photocells.

Other devices use Fraunhofer diffraction where the distance of the maximum intensities from the beam axis is a function of particle size.

13.1.5. Mass Distributions

The definitions of mean particle diameter given by Equation (13.2) must be modified for use with data from a sieve analysis. In transforming a number distribution to a mass distribution assumptions must be made about the shape and density of the particles. For a spherical particle of density ρ the particle mass m is given by

$$m = \frac{\pi}{6}x^3\rho \tag{13.6}$$

but for non-spherical albeit regular particles, using a volume shape factor, the mass is

$$m = kx^3\rho \tag{13.7}$$

If there are now N_1 particles of size x_1 the total mass of these particles is $N_1 kx_1^3\rho$. By setting the total mass of particles in the whole sample arbitrarily to unity then

$$1.0 = \sum Nkx^3\rho \tag{13.8}$$

and the mass fraction ω_1 of particles of size x_1 is now given by

$$\omega_1 = N_1 kx_1^3\rho \tag{13.9}$$

Rearranging Equation (13.9) gives the number of particles as

$$N_1 = \frac{\omega_1}{kx_1^3\rho} \tag{13.10}$$

which can be substituted into the definition of any mean diameter. For example, the surface-volume diameter becomes

$$x_{3,2} = \frac{\sum(\omega/kx^3\rho)x^3}{\sum(\omega/kx^3\rho)x^2} \tag{13.11}$$

This can be simplified, if it is assumed that the shape factor and particle density are constant for all size fractions (a not unreasonable assumption), to

$$x_{3,2} = \frac{\sum\omega}{\sum(\omega/x)} \tag{13.12}$$

The sum of all the mass fractions $\sum\omega$ is of course unity and therefore

$$x_{3,2} = \frac{1}{\sum(\omega/x)} \tag{13.13}$$

Expressions for each mean diameter based on a mass distribution are given in Table 13.1.

Example 13.5

A sample of granulated sugar was subjected to sieve analysis and yielded the following data.

Sieve aperture (μm)	Mass retained (g)
2,000	0
1,400	0.40
1,000	3.1
710	18.4
500	12.7
355	9.8
250	4.1
<250	2.9

Calculate the surface-volume mean particle diameter.

Material passing through the 2,000 μm sieve but retained on the 1,400 μm sieve may be assumed to have a diameter equal to the arithmetic average aperture, that is, 1,700 μm. Applying this reasoning to each sieve fraction and tabulating the mass fractions ω gives the following:

Particle size (μm)	Mass (g)	ω	ω/x
1,700	0.4	0.007782	4.58×10^{-6}
1,200	3.1	0.06031	5.03×10^{-5}
855	18.4	0.3580	4.19×10^{-4}
605	12.7	0.2471	4.08×10^{-4}
427.5	9.8	0.1907	4.46×10^{-4}
302.5	4.1	0.07977	2.64×10^{-4}
125	2.9	0.05642	4.51×10^{-4}
	$\sum = 51.4\,\mathrm{g}$		$\sum = 2.043 \times 10^{-3}$

Thus the surface-volume mean particle diameter is

$$x_{3,2} = \frac{1}{2.043 \times 10^{-3}}\,\mu\mathrm{m}, \qquad x_{3,2} = 489.5\,\mu\mathrm{m}$$

If the relationship between particle size and mass fraction is known precisely then, rather than sum across a number of size bands, this relationship can be integrated directly to give the mean diameter as in the following example.

Example 13.6

Milk is atomised in a spray drier and it is found that the cumulative droplet size distribution on a mass basis can be represented by a straight line from 0% at 50 μm to 100% at 250 μm. Calculate both the weight–moment and the surface-volume mean diameters of the droplets.

For a continuous function the definition of the surface-volume mean diameter becomes

$$x_{3,2} = \frac{1}{\int_0^1 d\omega/x}$$

The equation representing the straight line cumulative distribution is

$$x = 200\omega + 50$$

where x is the particle size in μm. Thus the mean particle diameter is given by

$$x_{3,2} = \frac{1}{\int_0^1 d\omega/(200\omega + 50)}$$

which on integration becomes

$$x_{3,2} = \frac{1}{\frac{1}{200}[\ln(200\omega + 50)]_0^1}$$

On evaluation this gives $x_{3,2}$ as 124.3 μm.

Similarly, the weight–moment mean is defined by

$$x_{4,3} = \int_0^1 x \, d\omega$$

which on substitution for x gives

$$x_{4,3} = \int_0^1 (200\omega + 50)d\omega, \qquad x_{4,3} = [100\omega^2 + 50\omega]_0^1$$

Thus the weight–moment mean diameter is $x_{4,3} = 150 \, \mu$m.

13.1.6. Other Particle Characteristics

The behaviour and appearance of manufactured powdered foods depends not only upon the size of individual particles but also upon the properties of the particles in bulk such as surface area, particle density, porosity and the packing density of powders. Many of these quantities are inter-related. The specific surface S is the external surface area of a particle per unit particle volume and for a sphere this is equal to $6/d$. The total surface area of a porous particle, including the internal pore spaces, can be measured by gas adsorption. Such areas can be of the order of several hundred square metres per gram. The solids density ρ_s is the density of the solid material from which the particle is made. This excludes any pore spaces within the particle. It can be measured using a specific gravity bottle and a liquid in which the particle does not dissolve. The envelope density of a particle is that which would be measured if an envelope covered the external particle surface, that is, it is equal to the particle mass divided by the external volume. The bulk density of a powder ρ_B is the effective density of the particle bed defined by

$$\rho_B = \frac{\text{mass of solids}}{\text{total bed volume}} \tag{13.14}$$

The bulk density will be considerably smaller than the solids density because the bed volume includes the volume of the spaces between particles. Bulk density is particularly important in the packing and appearance of food powders. Although packets are sold by mass they are usually filled by volume and it is crucial to know the bulk density accurately.

Intra-particle porosity refers to the fraction of the particle volume which is occupied by internal pores; most manufactured food particles are porous. This quantity should not be confused with the bed voidage. The inter-particle voidage ε is the fraction of the packed bed occupied by the void spaces between particles. This is defined as

$$\varepsilon = \frac{\text{void volume}}{\text{total bed volume}} \tag{13.15}$$

which can be written as

$$\varepsilon = \frac{\text{total bed volume} - \text{particle volume}}{\text{total bed volume}} \tag{13.16}$$

or

$$\varepsilon = 1 - \frac{\text{particle volume}}{\text{total bed volume}} \tag{13.17}$$

Volume is inversely proportional to density and therefore

$$\varepsilon = 1 - \left(\frac{\rho_B}{\rho_S}\right) \tag{13.18}$$

13.2. THE MOTION OF A PARTICLE IN A FLUID

The motion of a particle in a fluid has important consequences for the separation of particles from gases in a cyclone, from liquids in sedimentation and thickening, for particle size analysis in the sub-sieve range and for a number of unit operations including fluidisation and the drying of droplets.

Figure 13.3 represents the cross-section through a spherical particle over which an ideal non-viscous fluid flows. The fluid is at rest at points 1 and 3 but the fluid velocity is a maximum at points 2 and 4. There is a corresponding decrease in pressure from point 1 to point 2 and from 1 to 4. However the pressure rises to a maximum again at point 3. If now the ideal fluid is

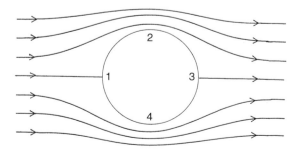

Figure 13.3. Flow of a fluid over a spherical particle.

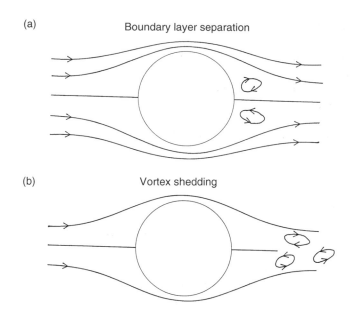

Figure 13.4. Boundary layer separation and vortex shedding.

replaced with a real viscous fluid then, as the pressure increases towards point 3, the boundary layer next to the particle surface becomes thicker and then separates from the surface as in Figure 13.4a. This separation of the boundary layer gives rise to turbulent eddies within which energy is dissipated and which creates a force acting on the particle. This force is called form drag. The total force acting on the particle because of the fluid flow is then the sum of form drag and viscous drag (or skin friction) over the surface.

13.2.1. Terminal Falling Velocity

A particle falling from rest, through a fluid, under gravity, will accelerate until it reaches a constant velocity known as the terminal falling velocity u_t. Four forces act on the particle: the particle weight (mg) which pulls the particle down under gravity, the upthrust due to the fluid displaced as the particle falls $(m'g)$, the drag force (F) acting against the particle weight and finally the product of particle mass and acceleration. These forces must be in balance and therefore

$$mg - m'g - F = m\frac{du}{dt} \tag{13.19}$$

Stokes first showed that the drag force F on a sphere was given by

$$F = 3\pi\, d\mu u \tag{13.20}$$

where u is the relative velocity between the sphere and the fluid. At its terminal falling velocity the particle no longer accelerates and therefore $du/dt = 0$. The mass (m) of a particle of diameter d is $(\pi/6)\rho_s d^3$ and the mass of displaced fluid (m') is $(\pi/6)\rho_f d^3$. Substituting each of these relationships into Equation (13.19) gives

$$3\pi\, d\mu u = \frac{\pi}{6}g(\rho_s - \rho_f)d^3 \tag{13.21}$$

and putting $u = u_t$, the terminal falling velocity, this becomes

$$u_t = \frac{g(\rho_s - \rho_f)d^2}{18\mu} \tag{13.22}$$

This is Stokes' law and gives the terminal falling velocity for a single smooth, rigid sphere falling in an homogeneous fluid as a function of the size and density of the particle and the density and viscosity of the fluid. It assumes that the settling particle is unaffected by the presence of other particles, that the walls of the vessel do not exert a retarding effect on the particle and that the particle size is much larger than the size of the molecules in the fluid. It is valid in the range $10^{-4} < Re < 0.20$ where the particle Reynolds number is defined by

$$Re = \frac{\rho_f u_t d}{\mu} \tag{13.23}$$

Thus the Stokes region is limited to low Reynolds numbers and therefore to relatively small particles.

Example 13.7

Calculate the terminal falling velocity of $80\,\mu$m diameter starch granules (density $1,600\,\text{kg m}^{-3}$) in water at $20°C$.

The density and viscosity of water at $20°C$ are $998.2\,\text{kg m}^{-3}$ and $1.002 \times 10^{-3}\,\text{Pa s}$, respectively. From Stokes' law the terminal falling velocity is

$$u_t = \frac{9.81(1600 - 998.2)(80 \times 10^{-6})^2}{18 \times 1.002 \times 10^{-3}}\,\text{m s}^{-1}$$

and therefore

$$u_t = 2.095 \times 10^{-3}\,\text{m s}^{-1}$$

However it is necessary to check the validity of this calculation. The particle Reynolds number is

$$Re = \frac{998.2 \times 2.095 \times 10^{-3} \times 80 \times 10^{-6}}{1.002 \times 10^{-3}} \quad \text{or} \quad Re = 0.167$$

and therefore the assumption of using Stokes' law is valid.

The measurement of terminal falling velocity can be used to determine the dynamic viscosity of Newtonian fluids if the fluid density is known.

Example 13.8

Coloured glass beads are allowed to fall through an aqueous glycerol solution and take $100\,\text{s}$ to travel a distance of $0.5\,\text{m}$. The beads have a diameter of $250\,\mu$m and a density of $2,500\,\text{kg m}^{-3}$. If the density of glycerol is $1,100\,\text{kg m}^{-3}$ what is its viscosity?

The glass beads fall with a velocity of 0.005 m s^{-1}. From Equation (13.22) the viscosity of the glycerol is

$$\mu = \frac{(2500 - 1100)9.81 \times (250 \times 10^{-6})^2}{18 \times 0.005} \text{ Pa s}$$

giving

$$\mu = 0.00954 \text{ Pa s}$$

This calculation is valid if the Reynolds number is below 0.20. Now

$$Re = \frac{1100 \times 0.005 \times 250 \times 10^{-6}}{0.00954} \quad \text{or} \quad Re = 0.144$$

and consequently the requirements of Stokes' law are met.

Example 13.9

What is the limiting diameter of a starch granule in Example 13.7 for which Stokes' law is valid?

The upper limit for Stokes' law is $Re = 0.20$ and therefore

$$0.20 = \frac{\rho_f u_t d}{\mu}$$

Now substituting for u_t from Stokes' law [Equation (13.22)] gives

$$0.20 = \frac{\rho_f g(\rho_s - \rho_f)d^3}{18\mu^2}, \qquad d = \left(\frac{3.6\mu^2}{g\rho_f(\rho_s - \rho_f)}\right)^{1/3}$$

Thus

$$d = \left(\frac{3.6(1.002 \times 10^{-3})^2}{9.81 \times 998.2(1600 - 998.2)}\right)^{1/3}, \qquad d = 85.0 \times 10^{-6} \text{ m}$$

Thus the limiting particle diameter is 85 μm.

13.2.2. Particle Drag Coefficient

A drag coefficient c_D can be defined as the drag force divided by the product of the dynamic pressure acting on the particle (i.e., the velocity head expressed as an absolute pressure) and the cross-sectional area of the particle. This definition is analogous to that of the friction factor in Equation (6.28). Hence

$$c_D = \frac{F}{(\pi d^2/4)(\rho_f u^2/2)} \tag{13.24}$$

Substitution for the drag force from Stokes' law gives

$$c_D = \frac{3\pi d\mu u}{(\pi d^2/4)(\rho_f u^2/2)} \tag{13.25}$$

which can be simplified to give

$$c_D = \frac{24}{Re} \tag{13.26}$$

for the Stokes region.

13.2.3. Effect of Increasing Reynolds Number

Beyond the Stokes region the boundary layer separates from the particle surface at a point just forward of the centre line of the sphere. A wake is formed containing vortices which results in a large frictional loss and the drag force on the particle increases significantly. This is the transition region where there is no analytical solution for the drag force. As the Reynolds number increases further, vortex shedding takes place (Figure 13.4b); this is known as the Newton region. At still greater Reynolds numbers the boundary layer itself becomes turbulent and separation occurs at the rear of the sphere and closer to the particle; drag is reduced considerably.

Figure 13.5 shows the experimentally determined relationship between the drag coefficient and Reynolds number. The linear part of the curve is described by Equation (13.26) but beyond the Stokes region empirical equations are required to describe the curve in the transition region. The curve levels off in the Newton region where the drag coefficient has a value of approximately 0.44 and as the boundary layer becomes turbulent the drag coefficient falls further to a value of about 0.1.

(*a*) *Transition region* The transition region is described by $0.20 < Re < 500$ and Schiller and Neumann proposed the empirical equation

$$c_D = \frac{24}{Re}(1 + 0.15\,Re^{0.687}) \tag{13.27}$$

for prediction of the drag coefficient. However the calculation of c_D presents some difficulty because both the drag coefficient and the Reynolds number contain both particle diameter and terminal falling velocity. In order to find the velocity for a known diameter it is necessary to use tables or a graph of values of the group $c_D Re^2$ as a function of Re. From Equation (13.19),

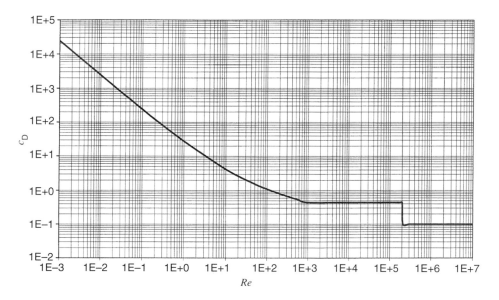

Figure 13.5. Drag coefficient as a function of Reynolds number.

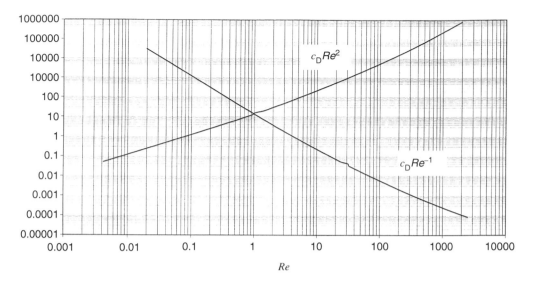

Figure 13.6. Drag coefficient–Reynolds number relationship for transition region.

putting $du/dt = 0$ the drag force becomes

$$F = \frac{\pi}{6} g(\rho_s - \rho_f) d^3 \tag{13.28}$$

Combining this with Equations (13.23) and (13.24) and simplifying gives

$$c_D Re^2 = \frac{4(\rho_s - \rho_f)\rho_f d^3 g}{3\mu^2} \tag{13.29}$$

This expression does not contain the terminal falling velocity. Consequently a knowledge of the size and density of the particle, and of the density and viscosity of the fluid, allows $c_D Re^2$ to be calculated and the particle Reynolds number to be found from Figure 13.6. Hence the terminal falling velocity can be determined.

Example 13.10

Calculate the terminal falling velocity of liquid milk droplets of density $1,350 \, \text{kg m}^{-3}$ falling in air through the chamber of a spray drier at $227°C$. The droplet diameter is $150 \, \mu m$.

At $227°C$ the density and viscosity of air are $0.706 \, \text{kg m}^{-3}$ and $2.67 \times 10^{-5} \, \text{Pa s}$, respectively. Stokes' law gives the terminal falling velocity as

$$u_t = \frac{9.81(1350 - 0.706)(150 \times 10^{-6})^2}{18 \times 2.67 \times 10^{-5}} \, \text{m s}^{-1}$$

and therefore

$$u_t = 0.620 \, \text{m s}^{-1}$$

However the particle Reynolds number is

$$Re = \frac{0.706 \times 0.620 \times 150 \times 10^{-6}}{2.67 \times 10^{-5}} \quad \text{or} \quad Re = 2.46$$

which is beyond the limit for Stokes' law and suggests that the drag coefficient lies in the transition region. Accordingly

$$c_D Re^2 = \frac{4(1350 - 0.706)0.706(150 \times 10^{-6})^3 9.81}{3(2.67 \times 10^{-5})^2} \quad \text{or} \quad c_D Re^2 = 59.0$$

From Figure 13.6 $Re = 2$, which lies in the transition range, and thus

$$2 = \frac{0.706 \times u_t \times 150 \times 10^{-6}}{2.67 \times 10^{-5}}, \quad u_t = 0.504 \, \text{m s}^{-1}$$

In order to find a particle diameter from a known velocity it is necessary to use tables or a graph of c_D/Re as a function of Re. In a similar manner to the derivation of Equation (13.29), it can be shown that

$$\frac{c_D}{Re} = \frac{4(\rho_s - \rho_f)g\mu}{3u^3 \rho_f^2} \tag{13.30}$$

This expression does not contain the particle diameter; knowledge of the terminal velocity gives c_D/Re and the Reynolds number can again be determined from Figure 13.6.

Example 13.11

Stones and gravel are to be separated from peas by sedimentation in water. Find the diameter of gravel, density $2,000 \, \text{kg m}^{-3}$, which falls in water at a constant velocity of $0.23 \, \text{m s}^{-1}$. The density and viscosity of water may be assumed to be $1,000 \, \text{kg m}^{-3}$ and $0.001 \, \text{Pa s}$, respectively.

Assume that the particle Reynolds number lies between 0.20 and 500. The particle diameter is unknown and therefore it is necessary to determine the value of c_D/Re. From the given data

$$\frac{c_D}{Re} = \frac{4(2000 - 1000)9.81 \times 10^{-3}}{3(0.23)^3 (1000)^2}, \quad \frac{c_D}{Re} = 1.075 \times 10^{-3}$$

Thus, from Figure 13.6, $Re = 450$ which is within the transition region and justifies the original assumption. The particle diameter can now be found from the Reynolds number.

$$450 = \frac{1000 \times 0.23 \, d}{10^{-3}}$$

and

$$d = 1.956 \times 10^{-3} \, \text{m or about 2 mm.}$$

(b) *Newton region*   In the Newton region, $500 < Re < 2 \times 10^5$, the drag coefficient c_D is approximately 0.44 and therefore Equation (13.24) gives the drag force as

$$F = 0.055\pi d^2 \rho_f u^2 \tag{13.31}$$

Thus to find the terminal falling velocity the drag force on the particle is set equal to the net weight and

$$0.055\pi d^2 \rho_f u_t^2 = \frac{\pi}{6} g(\rho_s - \rho_f)d^3 \tag{13.32}$$

from which

$$u_t = 1.74 \sqrt{\frac{dg(\rho_s - \rho_f)}{\rho_f}} \tag{13.33}$$

Example 13.12

At what velocity will 10 mm potato cubes (density 1080 kg m^{-3}) settle in water assuming that they have reached their terminal falling velocity?

It is first necessary to find a suitable particle diameter. From the definition of equivalent spherical diameter, Equation (13.3),

$$(0.01)^3 = \frac{\pi}{6} d_e^3, \qquad d_e = 12.4 \times 10^{-3} \text{ m}$$

Now, assuming that the potato cubes are sufficiently large to give a Reynolds number greater than 500, the terminal falling velocity an be found from Equation (13.33). Taking the density of water to be $1,000 \text{ kg m}^{-3}$

$$u_t = 1.74 \sqrt{\frac{12.4 \times 10^{-3} \times 9.81(1080 - 1000)}{1000}}, \qquad u_t = 0.172 \text{ m s}^{-1}$$

The Reynolds number is

$$Re = \frac{1000 \times 0.172 \times 12.4 \times 10^{-3}}{10^{-3}} \quad \text{or} \quad Re = 2128$$

which is in the range for the Newton region.

(c) *Turbulent region*   In the fully turbulent region, beyond $Re = 2 \times 10^5$, the drag coefficient has a value of about 0.10 and therefore the drag force is

$$F = 0.0125\pi d^2 \rho_f u^2 \tag{13.34}$$

and the expression for terminal falling velocity becomes

$$u_t = 3.65 \sqrt{\frac{dg(\rho_s - \rho_f)}{\rho_f}} \tag{13.35}$$

13.3. PACKED BEDS: THE BEHAVIOUR OF PARTICLES IN BULK

The relationship between the velocity of a fluid passing through a packed bed of particles and the consequent pressure drop across the bed must be known in order to understand, inter alia, fluidisation and filtration.

Darcy observed that the superficial velocity u of water flowing through a packed bed of sand was directly proportional to the pressure drop across the bed and inversely proportional to the bed depth L, and proposed an equation of the form

$$u = \frac{K' \Delta P}{L} \tag{13.36}$$

where K' is a constant. The resistance to flow in a packed bed is due to the viscous drag between the fluid and the particles, the flow is laminar and consequently the velocity is inversely proportional to viscosity. Darcy's relationship can be expressed therefore as

$$u = \frac{\beta \Delta P}{\mu L} \tag{13.37}$$

where β is the permeability of the bed, a measure of the ease with which a fluid can pass through a packed bed of particles, and which subsumes the constant K'. Modelling the relationship between pressure drop and fluid velocity is a difficult problem because of the irregular nature of the void spaces in a bed of irregular non-uniformly sized particles; an exact solution is not possible because of the tortuous flow paths followed by the fluid (Figure 13.7). Kozeny proposed that a packed bed could be modelled by a series of capillaries (Figure 13.8), by applying the Hagan–Poiseuille relationship for laminar flow in a tube [Equation (6.25)] and by using an equivalent diameter d_B and length L_B to represent the void spaces. Thus

$$u' = \frac{\Delta P d^2}{32 L_B \mu} \tag{13.38}$$

where the velocity u' represents the interstitial velocity, that is, the velocity between the particles in a packed bed. Kozeny, and later Carman, proposed that the interstitial and superficial

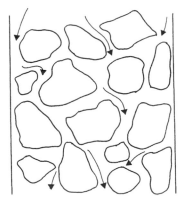

Figure 13.7. Flow through a packed bed.

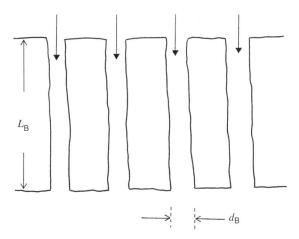

Figure 13.8. Kozeny's model of a packed bed.

velocities are related by the bed voidage,

$$u' = \frac{u}{\varepsilon} \tag{13.39}$$

In other words the cross-sectional area which determines the interstitial velocity for a given volumetric flow rate is proportional to the inter-particle voidage. Kozeny further suggested that the equivalent pore space diameter d_B is given by

$$d_B = \frac{\varepsilon}{S_B} \tag{13.40}$$

where S_B is the particle surface area per unit bed volume which comes into contact with the fluid passing through the bed. In turn this quantity is related to the specific surface by the fraction of the bed occupied by particles $(1 - \varepsilon)$. Thus

$$S_B = S(1 - \varepsilon) \tag{13.41}$$

Substituting each of these assumptions into Equation (13.38), and further assuming that the equivalent pore space length is proportional to the bed depth, results in the Carman–Kozeny equation

$$u = \frac{\varepsilon^3 \Delta P}{K S^2 (1 - \varepsilon)^2 \mu L} \tag{13.42}$$

which can be used to predict fluid velocity or flow rate as a function of pressure drop for a bed of incompressible particles, that is, where the bed voidage is constant. The dimensionless constant K is known as Kozeny's constant and has a value of approximately 5.0 although strictly it is a function of both intra-particle porosity and particle shape.

A comparison can now be made between the Darcy and Kozeny relationships. Thus from Equations (13.37) and (13.42) the permeability of the bed is

$$\beta = \frac{\varepsilon^3}{K S^2 (1 - \varepsilon)^2} \tag{13.43}$$

In other words permeability depends upon the geometry of the bed, that is, bed voidage and specific surface, which in turn is a function of particle size. Permeability has dimensions of m^2 and for fine particles values in the range 10^{-10}–$10^{-12}\,m^2$ can be expected. Note that for spheres $S = 6/d$. Bed permeability is often expressed in terms of the specific resistance α where

$$\alpha = \frac{1}{\beta} \tag{13.44}$$

Thus α has dimensions of m^{-2} and values of the order of 10^{10}–$10^{12}\,m^{-2}$ for fine particles. This definition is used in the analysis of filtration.

Example 13.13

Coffee particles, which may be assumed to be spheres $400\,\mu m$ in diameter, are to be dried in a stream of warm air. If the bulk and particle densities are 618 and $1{,}030\,kg\,m^{-3}$, respectively, calculate the permeability of the bed of coffee particles.

The inter-particle voidage is given by Equation (13.18) and therefore

$$\varepsilon = 1 - \left(\frac{618}{1030}\right), \qquad \varepsilon = 0.40$$

For spheres $S = 6/d$. Now assuming Kozeny's constant to be equal to 5, the permeability of the bed becomes

$$\beta = \frac{(0.40)^3 (400 \times 10^{-6})^2}{180(1 - 0.40)^2}\,m^2 \;\text{ or }\; \beta = 1.58 \times 10^{-10}\,m^2$$

For large particles (greater than about 600 or 700 μm) the Carman–Kozeny relationship is inadequate and predicts far too low a pressure drop. Therefore Ergun suggested a semi-empirical equation for the pressure drop per unit bed depth, containing two terms. Ergun's equation may be expressed as

$$\frac{\Delta P}{L} = \frac{150(1 - \varepsilon)^2 \mu u}{\varepsilon^3 d^2} + \frac{1.75(1 - \varepsilon)\rho_f u^2}{\varepsilon^3 d} \tag{13.45}$$

The first term represents the pressure loss due to viscous drag (this is essentially the Carman–Kozeny equation) whilst the second term represents kinetic energy losses, which are significant at higher velocities (kinetic energy being proportional to velocity squared). Equation (13.45) is valid in the range $1 < Re < 2{,}000$ where the Reynolds number is defined by

$$Re = \frac{u\rho}{S(1 - \varepsilon)\mu} \tag{13.46}$$

13.4. FLUIDISATION

13.4.1. Introduction

Fluidisation is a technique which enables solid particles to take on some of the properties of a fluid. For example, fluidised solids will adopt the shape of the container in which they are held and can be made to flow, under pressure, from an orifice or overflow a weir. Solids may be fluidised either by a liquid or by a gas. These phenomena give rise to a series of characteristics (for example good mixing and good heat transfer) which are exploited in a wide range of food processing operations such as freezing, drying, mixing and granulation.

Consider a bed of particles, say of a size similar to sand. When a fluid is passed upwards through the particles, the bed remains packed at low fluid velocities; the particles do not move. However, if the fluid velocity is increased sufficiently, a point will be reached at which the drag force on a particle will be balanced by the net weight of the particle. The particles are suspended in the upward moving fluid and move away from one another. This is the point of incipient fluidisation at, and beyond which, the bed is said to be fluidised. The superficial fluid velocity in the bed at the point of incipient fluidisation is called the minimum fluidising velocity u_{mf}. At velocities in excess of that required for minimum fluidisation one of two phenomena will occur.

(i) The bed may continue to expand and the particles will space themselves uniformly. This is known as particulate fluidisation and in general occurs when the fluidising medium is a liquid (Figure 13.9).

(ii) Alternatively, the excess fluid may pass through the bed in the form of bubbles. This is called aggregative fluidisation and usually occurs where the fluidising medium is a gas. This type of behaviour gives rise to the analogy of a boiling liquid (Figure 13.10).

A fluidised bed requires a distributor plate which supports the bed when it is not fluidised, prevents particles from passing through and promotes uniform fluidisation by distributing the fluidising medium evenly. The nature of the distributor plate influences the number and size of bubbles formed in aggregative fluidisation. Several types of plate are possible including porous or sintered ceramics and metals, layers of wire mesh and drilled plates. The pressure drop across the plate should be high to promote even gas distribution and is usually some fraction of bed pressure drop, often up to 50%. Fluidised beds can be operated either in batch or continuous mode. In continuous operation the solids are fed via screw conveyors, weigh feeders or pneumatic conveying lines and are withdrawn via standpipes or weirs.

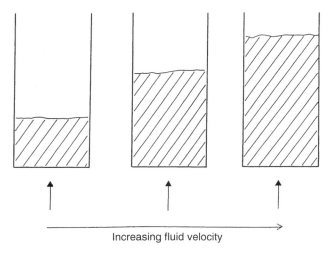

Increasing fluid velocity

Figure 13.9. Particulate fluidisation.

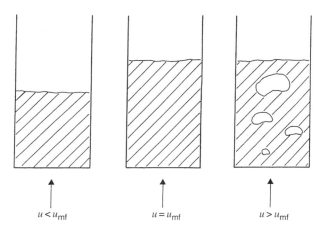

Figure 13.10. Aggregative fluidisation.

13.4.2. Minimum Fluidising Velocity in Aggregative Fluidisation

The behaviour of a fluidised bed and its effectiveness as a mixer, drier or freezer depends crucially upon the superficial gas velocity in the bed relative to the minimum fluidising velocity. It is essential that the minimum fluidising velocity is known; it may be determined experimentally or may be calculated from elementary fluid mechanics concepts. In addition much effort has been expended on producing accurate semi-empirical relationships to predict u_{mf}.

The relationship between bed pressure drop and superficial fluidising velocity is shown in Figure 13.11. As the gas velocity increases, the pressure drop increases in the fixed bed, or packed bed, region and then levels out as the bed becomes fluidised. Ideally, the pressure drop then remains constant as the weight of the particles is supported by the fluid. However if the velocity is then reduced marked hysteresis is observed. This is because the bed voidage remains at the minimum fluidising value whereas with increasing gas velocity considerable vibration of the particles takes place, the voidage is lower, and the pressure drop correspondingly is slightly greater. In practice too there will be a maximum pressure drop through which the curve passes because of particle interlocking. Also, as the velocity is reduced, the transition between the fluidised and fixed curves is gradual rather than sudden.

(*a*) *Experimental measurement* The standardised procedure to measure minimum fluidising velocity is to fluidise the bed of particles vigorously for some minutes and then reduce gas velocity in small increments, recording the bed pressure drop each time. This may be done with a simple water manometer with one leg open to atmosphere and one leg connected to a narrow tube placed in the bed. The data are then interpreted as in Figure 13.12; u_{mf} corresponds to the intersection of the straight lines representing the fixed and fluidised beds.

(*b*) *Carman–Kozeny equation* The Carman–Kozeny expression for minimum fluidising velocity is derived by substituting for the pressure drop in Equation (13.42). The pressure drop across a fluidised bed of bed height h can be obtained by treating the fluidised solids as a fluid with a density equal to the difference between the solid and fluid densities. Thus the pressure drop is given by Equation (2.13) but includes the term $(1 - \varepsilon)$ to take account of the

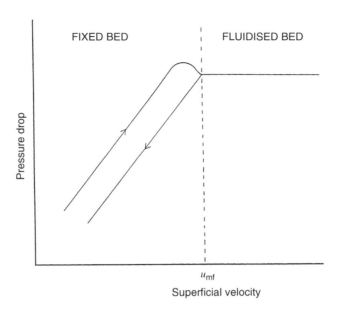

Figure 13.11. Pressure drop as a function of superficial gas velocity.

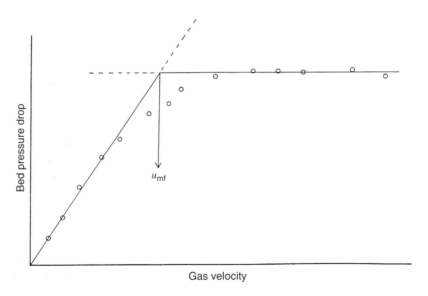

Figure 13.12. Measurement of minimum fluidising velocity: aggregative fluidisation.

fraction of the bed volume occupied by solids. Hence at minimum fluidising velocity

$$\Delta P_{mf} = (\rho_s - \rho_f)(1 - \varepsilon_{mf})gh_{mf} \qquad (13.47)$$

and therefore

$$u_{mf} = \frac{\varepsilon_{mf}^3(\rho_s - \rho_f)g}{KS^2(1 - \varepsilon_{mf})\mu} \qquad (13.48)$$

Now for $K = 5$, and for spherical particles where the specific surface is equal to $6/d$, the minimum fluidising velocity becomes

$$u_{mf} = \frac{\varepsilon_{mf}^3 (\rho_s - \rho_f) g d^2}{180(1 - \varepsilon_{mf})\mu} \tag{13.49}$$

At minimum fluidisation the drag force acting on a particle due to the flow of fluidising gas over the particle is balanced by the net weight of the particle. The former is a function of surface area and the latter is proportional to particle volume. Consequently the surface-volume mean diameter is the most appropriate particle size to use in expressions for minimum fluidising velocity. Note that Equation (13.49) suggests that u_{mf} is proportional to the difference in density between particle and fluid, proportional to the square of particle diameter and inversely proportional to fluid viscosity.

Example 13.14

A food powder is to be dried in a 0.5 m diameter fluidised bed using air at 50°C. It is found that minimum fluidising conditions are obtained when the bed pressure drop is 6,000 Pa for a bed height of 0.50 m. Using the Carman–Kozeny relationship, determine the minimum fluidising velocity if the surface-volume mean particle diameter is 180 μm and the particle density is 2,300 kg m^{-3}.

For air at 50°C the density is 1.1 kg m^{-3} and the viscosity is 1.98×10^{-5} Pa s. The voidage at minimum fluidising velocity is obtained from Equation (13.47) and hence

$$6000 = (2300 - 1.1)(1 - \varepsilon_{mf})\, 9.81 \times 0.50$$

from which the voidage is

$$\varepsilon_{mf} = 0.468$$

Thus the Carman–Kozeny equation gives the minimum fluidising velocity as

$$u_{mf} = \frac{(0.468)^3 (2300 - 1.1)(180 \times 10^{-6})^2\, 9.81}{180(1 - 0.468) 1.98 \times 10^{-5}}\ \text{m s}^{-1}$$

and

$$u_{mf} = 0.0395\ \text{m s}^{-1}$$

(c) *Ergun equation* The Carman–Kozeny equation works well for fine particles. However for large particles, for example peas in a fluidised bed freezer, the minimum fluidising velocity is high and the kinetic energy losses are significant. In these circumstances the Carman–Kozeny expression vastly over estimates u_{mf} and the Ergun equation must be used.

Writing Ergun's equation for minimum fluidising conditions gives

$$\frac{\Delta P}{h_{mf}} = \frac{150(1 - \varepsilon_{mf})^2 \mu u_{mf}}{\varepsilon_{mf}^3 d^2} + \frac{1.75(1 - \varepsilon_{mf})\rho_f u_{mf}^2}{\varepsilon_{mf}^3 d} \tag{13.50}$$

and substituting for pressure drop from Equation (13.47) gives

$$(\rho_s - \rho_f)g = \frac{150(1 - \varepsilon_{mf})\mu u_{mf}}{\varepsilon_{mf}^3 d^2} + \frac{1.75\rho_f u_{mf}^2}{\varepsilon_{mf}^3 d} \tag{13.51}$$

This expression is unwieldy but can be simplified considerably. Multiplying through by $\rho_f d^3 / \mu^2$ results in

$$\frac{\rho_f(\rho_s - \rho_f)d^3 g}{\mu^2} = \frac{150(1 - \varepsilon_{mf})u_{mf}d\rho_f}{\varepsilon_{mf}^3 \mu} + \frac{1.75\rho_f^2 d^2 u_{mf}^2}{\varepsilon_{mf}^3 \mu^2} \tag{13.52}$$

which can be put into the form

$$Ga = \frac{150(1 - \varepsilon_{mf})}{\varepsilon_{mf}^3} Re_{mf} + \frac{1.75 Re_{mf}^2}{\varepsilon_{mf}^3} \tag{13.53}$$

This is a quadratic equation where the Galileo number Ga is defined by

$$Ga = \frac{\rho_f(\rho_s - \rho_f)d^3 g}{\mu^2} \tag{13.54}$$

and the Reynolds number at minimum fluidisation by

$$Re_{mf} = \frac{\rho_f u_{mf} d}{\mu} \tag{13.55}$$

Example 13.15

A novel method for germinating tomato seeds includes a relatively rapid drying stage in a gas–solid fluidised bed. The seeds have a flat, irregular disc-like shape but may be assumed to have a diameter of 2 mm. The solid density of the seeds is $1{,}600\,kg\,m^{-3}$. Determine the minimum fluidising velocity of the seeds in air at 300 K if the pressure drop across a 0.2 m deep bed at minimum fluidisation is 1568 Pa.

At 300 K the density and viscosity of air are $1.177\,kg\,m^{-3}$ and $1.846 \times 10^{-5}\,Pa\,s$, respectively. The voidage at minimum fluidisation is obtained from Equation (13.47), thus

$$1568 = (1 - \varepsilon_{mf})(1600 - 1.177)\,9.81 \times 0.20$$

from which

$$\varepsilon_{mf} = 0.50$$

The Galileo number can be calculated from Equation (13.54) to give

$$Ga = \frac{1.177(1600 - 1.177)\,9.81(2 \times 10^{-3})^3}{(1.846 \times 10^{-5})^2}, \qquad Ga = 4.333 \times 10^5$$

Substituting Ga and ε_{mf} into the Ergun equation yields the quadratic expression

$$4.333 \times 10^5 = 600\,Re_{mf} + 14\,Re_{mf}^2$$

from which the positive root is $Re_{mf} = 155.8$. Clearly the negative root has no physical significance and can be ignored. Thus

$$155.8 = \frac{1.777 u_{mf} 2 \times 10^{-3}}{1.846 \times 10^{-5}}$$

and the minimum fluidising velocity is then

$$u_{mf} = 1.22\,m\,s^{-1}$$

Example 13.16

Compare the Carman–Kozeny and Ergun relationships for determining the minimum fluidising velocity in air of the following particles: (a) surface-volume mean diameter = 600 μm, particle density = 2,400 kg m^{-3}; (b) surface-volume mean diameter = 9 mm, particle density = 1200 kg m^{-3}. In each case assume that the bed voidage at minimum fluidisation is 0.45 and that the density and viscosity of air are 1.1 kg m^{-3} and 2 × 10^{-5} Pa s, respectively.

For the 600 μm diameter particles the Carman–Kozeny equation gives the minimum fluidising velocity as

$$u_{mf} = \frac{(0.45)^3(2400 - 1.1)(600 \times 10^{-6})^2 9.81}{180(1 - 0.45)2 \times 10^{-5}} \text{ m s}^{-1}, \qquad u_{mf} = 0.389 \text{ m s}^{-1}$$

The Galileo number is

$$Ga = \frac{1.1(2400 - 1.1)\, 9.81(600 \times 10^{-6})^3}{(2 \times 10^{-5})^2}, \qquad Ga = 1.398 \times 10^4$$

which on substitution into the Ergun equation gives

$$1.398 \times 10^4 = 905.4\, Re_{mf} + 19.20\, Re_{mf}^2$$

The positive root is $Re_{mf} = 12.25$ and therefore

$$12.25 = \frac{1.1 u_{mf} 600 \times 10^{-6}}{2 \times 10^{-5}}$$

from which

$$u_{mf} = 0.371 \text{ m s}^{-1}$$

In other words there is good agreement between the Carman–Kozeny and Ergun models for the 600 μm particles. However for the 9 mm diameter particles the Carman–Kozeny equation gives

$$u_{mf} = \frac{(0.45)^3(1200 - 1.1)(9 \times 10^{-3})^2 9.81}{180(1 - 0.45)2 \times 10^{-5}} \text{ m s}^{-1} \text{ or } u_{mf} = 43.8 \text{ m s}^{-1}$$

This is an unrealistically high velocity. In contrast, using the Ergun model, $Ga = 2.358 \times 10^7$ and

$$2.358 \times 10^7 = 905.4\, Re_{mf} + 19.20\, Re_{mf}^2$$

This gives a Reynolds number of $Re_{mf} = 1084.9$ and therefore

$$1084.9 = \frac{1.1 u_{mf} 9 \times 10^{-3}}{2 \times 10^{-5}}$$

from which

$$u_{mf} = 2.192 \text{ m s}^{-1}$$

Example 13.16 shows that for large particles the kinetic energy losses far outweigh the losses due to viscous drag. For large food particles, such as peas and other vegetable pieces in a fluidised bed freezer or drier, the Carman–Kozeny equation is wholly inadequate to predict minimum fluidising velocity and the Ergun equation must be used. However this is a little cumbersome and it is possible to ignore the first term in Equation 13.53 which then reduces to

$$Re_{mf}^2 = \frac{\varepsilon_{mf}^3 Ga}{1.75} \tag{13.56}$$

For the 9 mm diameter particles of Example 13.16, Equation (13.56) gives the minimum fluidising velocity as $2.238 \, \text{m s}^{-1}$ which is a good approximation to that predicted by the full Ergun equation.

(d) *Semi-empirical correlations*   Probably the most useful and accurate of the many semi-empirical equations available is that due to Leva:

$$u_{mf} = 0.0079 \frac{d^{1.82}(\rho_s - \rho_f)^{0.94}}{\mu^{0.88}} \tag{13.57}$$

This equation allows the prediction of minimum fluidising velocity from a knowledge of the mean particle diameter, the particle density, the density of fluidising medium and the viscosity of fluidising medium (SI units).

13.4.3. Gas–solid Fluidised Bed Behaviour

(a) *Influence of gas velocity*   As the superficial gas velocity is increased beyond the minimum fluidising velocity a greater proportion of gas passes through the bed in the form of bubbles, the bubbles grow larger, particle movement is more rapid and there is a greater degree of 'turbulence'. Fluidising gas velocity is the single most important variable affecting the behaviour of a bed of given particles and it is expressed usually as either:

(i) multiples of u_{mf}

for example $u/u_{mf} = 3$ implies that the gas velocity is three times that required for minimum fluidisation, or as

(ii) excess gas velocity, $u - u_{mf}$

for example $u - u_{mf} = 1.2 \, \text{m s}^{-1}$ implies that the gas velocity is $1.2 \, \text{m s}^{-1}$ *greater* than that required for minimum fluidisation.

As an approximation, it may be assumed that all the gas over and above that required for minimum fluidisation flows up through the bed in the form of bubbles. This is the assumption of the 'two-phase theory;' in other words the bed consists of a bubble phase and a dilute or lean phase of fluidised solids. If the total volumetric flow of gas is Q then, according to the two-phase theory,

$$Q = Q_{mf} + Q_B \tag{13.58}$$

where Q_{mf} is the volumetric flow of gas at minimum fluidisation and Q_B is the volumetric bubble flow rate. In practice, proportionately more gas flows interstitially (i.e., between the particles) as the velocity is increased than at u_{mf}. In addition, there is a limited interchange of gas between the bubble phase and the dilute phase. As the gas velocity is increased further the very smallest particles are likely to be carried out of the bed in the exhaust stream. This is because at any realistic fluidising gas velocity the terminal falling velocity of the very smallest particles will be exceeded. The loss of bed material in this way is known as elutriation and will increase as u/u_{mf} increases. Further increases in gas velocity result in greater elutriation and a more dilute concentration of the solids remaining in the bed. Eventually all the particles will be transported in the gas stream at the onset of pneumatic conveying.

(*b*) *Geldart's classification*   Geldart suggested classifying fluidised particles into four groups. This classification is shown diagrammatically in Figure 13.13 in the form of a plot of the density difference between particle and fluid against mean particle size. Group A particles are typically between 20 and 100 μm in diameter with a particle density less than 1,400 kg m^{-3}. These particles exhibit considerable bed expansion as the fluidising velocity increases and collapse only slowly as the velocity is decreased. In other words they tend to retain the fluidising gas. On the other hand the larger and denser group B particles (40–500 μm, particle density in the range 1,400–4,000 kg m^{-3}) form freely bubbling fluidised beds at the minimum fluidising velocity. This is the classical fluidised bed behaviour; group B particles can be defined by

$$(\rho_s - \rho_f)^{1.17} \geq 9.06 \times 10^5 \tag{13.59}$$

Groups A and B are the most frequently encountered classes of particles, giving stable fluidisation.

Group C, with particle diameters below 30 μm, consists of cohesive powders which display a tendency to agglomerate and are very difficult to fluidise. This group is characteristic of a number of food materials such as flour or very fine spray dried particles. Large particles with a mean diameter greater than 600 μm and a density above 4,000 kg m^{-3} are classed as Group D. Such particles display very little bed expansion and generally give unstable fluidisation;

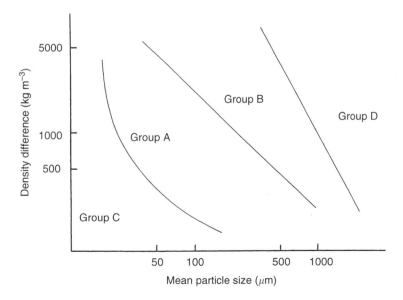

Figure 13.13. Geldart's classification of fluidised particles.
Source: Adapted from Geldart, D., Powder Tech., 7 (1973) 285.

channelling of the gas is prevalent. Food examples include seeds and vegetable pieces. Group
D particles can be defined by

$$(\rho_s - \rho_f)d^2 \geq 10^9 \tag{13.60}$$

13.4.4. Bubbles and Particle Mixing

The size of bubbles at the bottom of the bed is determined by the nature of the distributor
and they grow as they rise through the bed. Small bubbles are overtaken by larger ones and
coalescence takes place at the base of the larger bubble. Consequently, the number of bubbles
at any bed cross-section decreases with bed height. Small bubbles are approximately spherical
but with a slight indentation at the base. However the size of the indentation increases with
bubble diameter and the bubbles take on the characteristic shape shown in Figure 13.14. As
gas velocity increases the nature of the bubbles changes. Especially in deep beds, the bubbles
grow to the size of the bed container and push plugs of material up the bed as they rise. The
particles then stream past the slugs at the bed walls on their downward path. This is known as
slugging and is to be avoided in food processing applications.

Particle mixing in a fluidised bed is brought about solely by the movement of bubbles.
As a bubble rises in the bed (Figure 13.15) it gathers a wake of particles and then draws up
a spout of particles behind it. The wake grows as the bubble rises and a proportion of it may
be shed before the bubble reaches the bed surface. Growth and shedding of the wake may be
repeated several times in the life of a single bubble. Overall a quantity of particles equal to one
bubble volume is moved through a distance of 1.5 bubble diameters by a single rising bubble.
Because of the large numbers of bubbles present in a fluidised bed, the particle mixing pattern
is highly complex and extremely rapid.

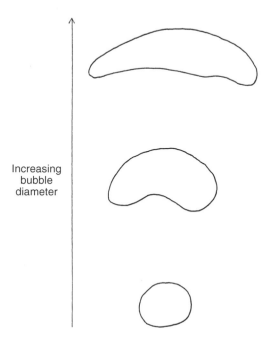

Increasing
bubble
diameter

Figure 13.14. Shape of bubbles in a fluidised bed.

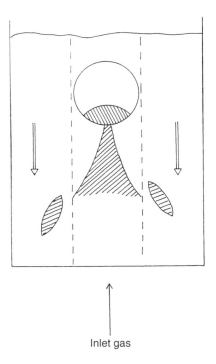

Inlet gas

Figure 13.15. Particle mixing in a fluidised bed.

Based upon these observations, Rowe derived an expression for the average circulation time t around a bed in terms of excess gas velocity and bed height at minimum fluidisation.

$$t = \frac{h_{\mathrm{mf}}}{0.6(u - u_{\mathrm{mf}})} \tag{13.61}$$

Although this expression may not accurately predict circulation time, and in any case particles do not follow a simple pre-determined circuit around the bed, it serves to illustrate the importance of excess gas velocity in determining particle mixing rates.

The rapid particle movement in a fluidised bed leads to its use in particle mixing. However the mixing mechanism is opposed by segregation, where the concentration of one component increases at the bottom of the bed. Particle size has only a slight influence upon segregation (unlike in mechanical mixers). However, segregation increases markedly with the difference in density between two components. When the density ratio reaches a value of about seven almost no mixing occurs in a fluidised bed. Three main mechanisms can be identified by which particles move in a bed and thus bring about either mixing or segregation:

i   particles rising in bubble wakes
ii   large, dense particles falling through bubbles
iii small, dense particles percolating downwards through the interstices of the dense phase of the bed.

13.4.5. Heat and Mass Transfer

High rates of heat and mass transfer are possible in fluidised beds and these characteristics are responsible for their use in drying, freezing and granulation. Heat may be added to the bed either from hot inlet gas or from heating elements in the bed wall or immersed coils. Similarly heat can be removed by using coils in which a coolant circulates. In the case of fluidised bed freezers the inlet air is at a temperature well below the relevant freezing point.

High rates of heat transfer are due to a number of reasons. First, the presence of particles in a fluidised bed increases the heat transfer coefficient by up to two orders of magnitude, compared with the value obtained with gas alone at the same velocity. The bed particles are responsible for the transfer of heat and, because of the high rate of particle movement (and very short residence times close to the heat transfer surface), the bulk bed temperature comes very close to that of the heat transfer surface. Particle to particle heat transfer coefficients are of the order of $400\ \mathrm{W\ m^{-2}\ K^{-1}}$. Second, the volumetric particle heat capacity is about 1,000 times greater than that of a gas and therefore approximates to that of a liquid. Third, the very high specific surface of the particles results in high heat fluxes.

The temperature drop between hot inlet gas and the bed temperature takes place over only a few particles diameters just above the gas distributor plate. Bed temperatures are uniform, because of the rapid particle mixing, and thus close bed temperature control is possible. Equally it is possible to use high inlet temperatures without exposing temperature sensitive particles to thermal damage.

13.4.6. Applications of Fluidisation to Food Processing

(*a*) *Drying*   Fluidised bed driers exploit the rapid particle mixing and high heat transfer coefficients which characterise aggregative fluidisation. Consequently high drying rates are possible together with close temperature control, which avoids overheating and thermal damage, and uniform moisture content. The enthalpy of vaporisation is supplied by the fluidising gas which is usually air. However submerged heating elements can also be used especially if the gas velocity has to be reduced in order to retain fine particles in the bed.

Various drying regimes have been used including high temperature short time drying for fruits and vegetables, in which temperatures up to $150°C$ are combined with residence times of a few minutes. The ability to control particle residence time is another advantage of a fluidised bed. In a plug flow (PF) unit the fluidised bed is rectangular (aspect ratios between 5 and 40 are common) with the feed inlet at one end and a weir arrangement at the other over which dried solids flow out to the next process stage. The depth of the fluidised bed (usually less than about 10–15 cm) is controlled by the weir height and therefore, for a constant feed rate, the average residence time in the drier can be controlled. It is an assumption of plug flow that very little longitudinal mixing takes place and therefore the PF arrangement allows the particles to experience a series of different processing conditions as they move through the bed. The provision of separate compartments in the plenum chamber allows different gas velocities and temperatures to be used. For example, at the bed inlet it may be desirable to use high velocities and temperatures to cope with a high moisture content whilst more moderate conditions may be needed as the moisture content is reduced. Cooling may take place immediately prior to discharge.

It is essential that the wet food particulates in the food are able to be fluidised. Too high a moisture content means that the particles may agglomerate to a size which cannot be fluidised. The loss of fluidisation in this way is disastrous for a drier; clumps of wet material are likely to adhere to the internal surfaces of the bed and to block the distribution plate and when this happens the shut-down of the plant is inevitable. The material and energy balance over the bed must be satisfied: sufficient heat must be supplied to evaporate the moisture in the food and the exhaust gas from the bed must not become saturated with water vapour. If the exhaust gas does become saturated a phenomenon known as quenching occurs which in turn leads to defluidisation. A number of techniques have been used to assist the flow of fluidised solids through driers and to prevent excessive agglomeration of wet material. Gravity flow is normally used but it is possible to mount the entire fluidised bed on a spring chassis and vibrate it. Alternatively stirrers placed in the bed can be used to break up agglomerated particles or a pulsed air flow may be employed.

An alternative to the PF arrangement is a well-mixed bed or continuous stirred tank reactor (CSTR) design where the outlet moisture content is equal to the average moisture content in the bed. This allows the wet solids to be fed into relatively dry material which helps to maintain fluidisation. However it is difficult to obtain very low product moisture contents with CSTR operation. Various combinations of fluidised bed drier may be used, for example a well-mixed unit followed by a plug flow bed designed to give a lower final moisture content. Fluidised bed drying has been used successfully for peas, carrots, nuts, strawberries, instant potato, starch, soybean, grain and other products. Spouted bed drying has been used successfully for large particles where the gas flow rates for fluidisation would be excessive.

(*b*) *Freezing* Fluidised bed freezing is a modification of air-blast freezing which exploits higher heat transfer coefficients and therefore allows the use of more compact equipment. The solids to be frozen are fluidised by refrigerated air at temperatures of $-30°C$ or below. The particles are frozen independently and very rapidly to give a free flowing Individually Quick Frozen (IQF) product. Foods frozen in a fluidised bed have an attractive appearance because water on the surface of the particle is frozen giving a glazed look which also helps to protect the food.

One of the largest volume products frozen in this way is frozen peas. A high processing capacity is needed for the short pea season and the peas are then placed in long term bulk storage to be followed by packaging at a later date. Peas are frozen by direct fluidised freezing in which the peas are in direct contact with the cold air stream. Other applications include beans, sweetcorn, sprouts, cauliflower, carrot, diced meat, shrimps, and a variety of fruits including those which are difficult to freeze in other ways such as strawberries and soft fruits. However the necessary fluidising velocity often imposes a limit on the size of the product. Immersion freezing is used for packaged foods which are frozen in bed of inert particles. In direct freezing gravity feed and vibration are used as well as a technique in which a wire mesh conveyor belt passes through the bed on which food blocks are placed and fluidised solids circulate around them. This latter method is used for example with fish fingers.

(*c*) *Fermentation* Gas–solid fluidised beds can be used as bioreactors for fermentation reactions. This technique has distinct advantages over conventional fermentation systems and has considerable potential for exploitation. For example aqueous glucose solutions can be atomised within a bed of yeast particles with the latent heat for vaporisation of both ethanol and extraneous water being supplied in the inlet gas. The fluidised bed allows a uniform temperature in the fermenter and the easy addition and removal of heat, substrate, nutrients and gases avoiding the usual high dilution of micro-organisms. Pure ethanol is condensed directly from the exhaust gases of the fluidised bed without the need for costly downstream processing whilst the removal of product in this way reduces the inhibition of the yeast by the ethanol. A number of volatile products can be manufactured in this way, including a range of flavour compounds, using a variety of micro-organisms. An alternative use for a fluidised bed biore-actor involves the cultivation of micro-organisms with the advantage that subsequent drying of the product takes place in the bed thus reducing the overall plant volume.

The other major applications of aggregative fluidisation are mixing, the mechanisms of which are covered in section 13.4.4 and granulation which is covered in detail in section 13.6.3. In addition fluidisation has been used to blanch vegetables in hot air, for cooking (e.g. the pre-cooking of rice) and for roasting coffee.

13.4.7. Spouted Beds

A spouted bed requires a cylindrical container with an inverted conical base and a vertical pipe entering at the apex of the cone. If coarse solids are placed in the container and gas is introduced through the pipe, the solids will begin to 'spout' as in Figure 13.16. Particles are carried at high velocity upwards in the central jet in a dilute phase and at the bed surface the particles form a kind of fountain and rain down onto the clearly defined surface. They then travel down the bed in the annulus at much lower velocities than those in the central spout.

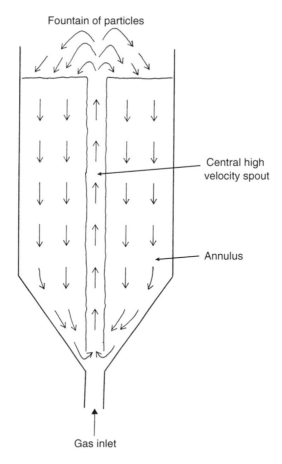

Figure 13.16. Spouted bed.

On reaching the gas inlet point the particles are re-entrained and thus complete a well defined and regular cycle.

Spouted beds can exist only with large particles, usually greater than 1–2 mm in diameter. Particle motion is much more regular than in a fluidised bed although there is a distribution of cycle times with particles re-entering the spout at various points up the bed. The pressure drop down the bed is not uniform; it is small at the base and reaches a maximum at the surface where it approaches the pressure drop needed to support the solids. If the gas velocity is sufficient to fluidise the particles then the spouted bed becomes unstable; thus there is a maximum spoutable depth. The applications of spouted beds include drying, particularly the drying of wheat, granulation and the coating of seeds and large particles.

13.4.8. Particulate Fluidisation

Particulate fluidisation, where the fluidising medium is usually a liquid, is characterised by a smooth expansion of the bed. Liquid/solid fluidised beds are used in continuous crystallisers, as bioreactors in which immobilised enzyme beads are fluidised by the reactant solution and in physical operations such as the washing and preparation of vegetables. The empirical Richardson–Zaki equation describes the relationship between the superficial fluid velocity and bed voidage:

$$\frac{u}{u_0} = \varepsilon^n \tag{13.62}$$

where u_0 is the superficial velocity at a voidage of unity and n is an index which depends on the particle Reynolds number. Figure 13.17 shows the form of a logarithmic plot of velocity

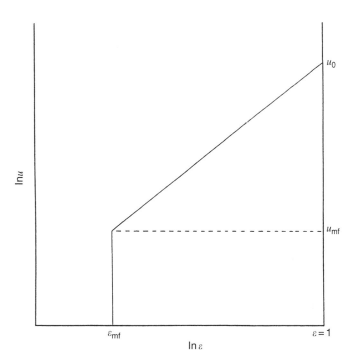

Figure 13.17. Measurement of minimum fluidising velocity: particulate fluidisation.

against voidage. The vertical line represents the packed bed where the bed voidage does not change with velocity. The inclined section of the plot represents the fluidised bed, described by Equation (13.62), and has a gradient equal to n. The minimum fluidising velocity is found at the intersection of the two lines. The value of n is usually in the range 2.4–5.0, with smaller values at higher Reynolds numbers. There is a degree of similarity between the fluidisation of a particle and the settling of a particle in a fluid. For sedimentation

$$u_0 = u_t \qquad (13.63)$$

that is, the velocity at a voidage of unity is equal to the terminal falling velocity. However in particulate fluidisation a wall effect exists and

$$\ln u_0 = \ln u_t - \frac{d}{D} \qquad (13.64)$$

where D is the bed diameter. Thus as the bed diameter increases relative to the particle diameter fluidisation behaviour approaches that of a sedimenting particle.

Example 13.17

Carrot pieces are to transported hydraulically. In laboratory tests to determine the minimum fluidising velocity, 10 kg of 1 cm cubes of carrot were fluidised in a column of 0.25 m diameter. The following data were generated:

Superficial velocity (m s^{-1})	Bed height (m)
0.00245	0.353
0.00409	0.353
0.00674	0.353
0.0111	0.353
0.0235	0.388
0.0388	0.494
0.0639	0.749
0.105	2.042

Calculate the minimum fluidising velocity if the density of carrot is $1{,}050 \, \text{kg m}^{-3}$.

The total volume of particles in the bed is equal to the mass of carrot pieces divided by the carrot density, which is equal to $\frac{10}{1{,}050} \, \text{m}^3$. The total bed volume is equal to $\pi(0.25)^2 h/4 \, \text{m}^3$, where h is the bed height at any instant. Thus, from Equation (13.17), the bed voidage is given by

$$\varepsilon = 1 - \frac{10 \times 4}{\pi(0.25)^2 h \, 1050}$$

This allows bed voidage to be calculated as a function of fluid velocity as follows:

Superficial velocity (m s^{-1})	Bed voidage
0.00245	0.450
0.00409	0.450
0.00674	0.450
0.0111	0.450
0.0235	0.500
0.0388	0.607
0.0639	0.741
0.105	0.905

On a plot of $\ln u$ against $\ln \varepsilon$ the lines for fixed and fluidised beds intersect at $u_{mf} = 0.0183$ m s^{-1}. The gradient of the line gives $n = 2.5$.

13.5. TWO-PHASE FLOW: PNEUMATIC CONVEYING

13.5.1. Introduction

Two-phase flow, the flow of gas/solid, liquid/solid or gas/liquid mixtures, is important in many areas of food processing. Gas/solid flow is exploited in pneumatic conveying which is a method of transporting powders and particulate materials, ranging from fine powders to pellets several thousand microns in diameter, in a gas stream over long distances either from bulk carriers to storage hoppers or within factories between process stages. The former may use a portable system mounted on a bulk delivery tanker but the latter is more likely to be a permanent installation including fans, pipework, valves, filters and silos. The basic principle of pneumatic conveying is that particles are carried in an air stream with a velocity which exceeds their terminal falling velocity. Pneumatic conveying lines can be either vertical or horizontal with runs of several hundred metres through ducting with a minimum diameter of about 0.1 m. The air stream is under pressure and therefore difficulties arise in introducing the solids into the air and in separating solids from the air at the destination; this requires the use of carefully designed rotary valves. For food applications the air must be filtered and the pipework and valves manufactured from stainless steel.

13.5.2. Mechanisms of Particle Movement

The characteristics of gas/solid flow depend upon the gas velocity and the solids loading, usually expressed as the solid/gas mass ratio. In horizontal conveying, as the gas velocity is reduced, and as the solid/gas mass ratio increases, the two-phase system passes through the following stages:

i   A uniform distribution of particles exists over the pipe cross section.
ii Segregation begins, with the larger particles settling at the bottom of the duct.
iii Particles settle at the bottom before accelerating and, forming 'dunes.'
iv The dunes are stationary and the particles move from dune to dune, as they are conveyed in the gas above the dunes.

v   A continuous bed is formed on the bottom of pipe, the bed increases in depth along
the pipe length and moves slowly forward.

vi   The bed remains stationary and the pipe becomes blocked.

In vertical lines, of course, the distribution of particles over the pipe cross-section remains
constant.

13.5.3. Pneumatic Conveying Regimes

Three broad regimes in pneumatic conveying may be identified. First, dilute phase or lean
phase conveying. Here the volume fraction of solids is less than 0.05, that is, the voidage is
greater than 0.95 and therefore the particles are usually in full suspension. The solids/gas mass
ratio is less than 5. Gas velocities generally are in the range 20–30 m s^{-1} although are sometimes
higher than this. For example flour is conveyed with gas velocities of about 18–20 m s^{-1} and
wheat at 25 m s^{-1}. Second, the moving bed in which the volume fraction of solids is less than
or equal to 0.60 (i.e., the voidage is greater than 0.40). This regime can exist only in horizontal
conveying. The third identifiable regime is dense phase conveying. Here slugs of solids with
a voidage of about 0.50 alternate with slugs of gas. The overall solids/gas mass ratio is greater
than 50, velocities are much lower than in dilute phase conveying (about 3 m s^{-1}) and therefore
gas usage is considerably less but the risk of blockage is greater. Dense phase conveying has
the distinct advantage that separation of gas and solids is easier and in addition there is less
erosion of pipe bends and less particle attrition. However the possibility of choking imposes a
limitation on the system. When this occurs unsteady pressure gradients are generated and it is
difficult to predict the required pressure drop. Metering of solids into dense phase conveying
lines is essential and they require about 40% more power than lean phase conveying systems.

13.5.4. Pneumatic Conveying Systems

The pressure system (Figure 13.18a) is used for free flowing solids over a wide range of
particle size, at high solids/gas mass ratios, which are fed into a pressurised line via a rotary
valve; the valve prevents the back flow of air. The air is moved by a positive displacement
blower and the system pressure drop is about 0.3 bar. This kind of arrangement is suitable for
long horizontal runs which feed several destinations. Gas and solids are separated either by
filter or by cyclone attached to the receiving silo.

In vacuum systems (Figure 13.18b) material is collected from several points and conveyed
to a single receiver which is upstream of the fan. The air stream is always at a pressure below
atmospheric. Such systems do not require a rotary valve at the solids inlet. They are used particularly for fine solids and smaller solids/gas mass ratios. However vacuum conveying is suitable
only for relatively short distances (300 m maximum) and to vertical runs with few bends.

Combined systems (Figure 13.18c) attempt to combine the advantages of both the pressure
and vacuum arrangements. A vacuum is used to move material over relatively short distances
to a separator. The air then enters the suction side of the fan and the solids are fed into the
discharge from the fan via a rotary valve. In this way particulate solids from several sources
can be conveyed to several destinations.

13.5.5. Safety Issues

The handling of fine powders always creates the potential for dust explosions and this
is true particularly of pneumatic conveying; milling, grinding and spray drying are also

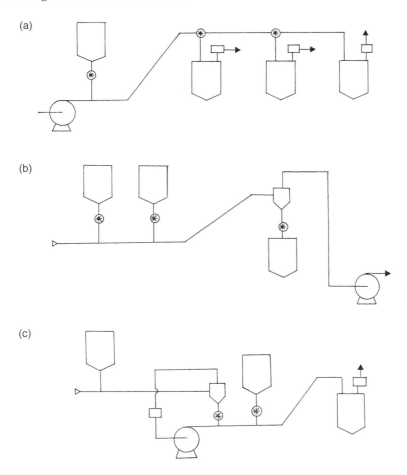

Figure 13.18. Pneumatic conveying systems: (a) pressure, (b) vacuum, and (c) combined vacuum-pressure.

hazardous. A dust explosion is simply a very rapid combustion reaction where the rate of reaction is increased dramatically because of the very high specific surface of finely divided solids. Flour, sugar, instant coffee, custard powder, dried milk, and soup powder are examples of food materials which present an explosion risk. A dust explosion will occur if the concentration of dust is within the explosive limits and a source of ignition exists. Static electricity is very often a source of ignition. The general precautions which can be adopted include the provision of explosion relief panels, the creation of a slight negative pressure in silos and the earthing of process equipment. Pneumatic conveying systems should have under and over pressure sensors in order to close down the system if the explosive limit is approached.

13.6. FOOD PARTICLE MANUFACTURING PROCESSES

13.6.1. Classification of Particle Manufacturing Processes

The manufacture of particles with specific properties is of increasing importance in food processing. Particle size is often, but by no means always, the most important of these properties

(a)

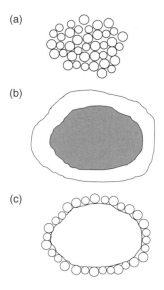

(b)

(c)

Figure 13.19. Granule morphology.

and food particles may range in size from a few microns to several millimetres, and sometimes larger. There is no comprehensive theory which covers size enlargement techniques and therefore the operation of such processes relies heavily on empirical knowledge. There are numerous terms in use to describe powder production methods and some terms have different meanings in different industries. The word *granulation* is probably the best all-embracing term and may be taken to mean 'to produce granules,' either by producing assemblies of smaller particles or by breaking large pieces into smaller ones. This wide definition of granulation then allows size reduction to be considered as a particle production process. Granules may take a number of forms (Figure 13.19) including agglomerates, layered particles and coated particles.

There are many reasons for producing powders of a specified size and structure and these may include one or more of the following:

i   to improve the flow of the product as it leaves a packet or container;
ii to improve product appearance;
iii to improve material handling in intermediate process steps, for example flow in and out of process equipment;
iv to impart a particular particle structure or property to the final product, for example, to aid dissolution or reconstitution when placed in water.

This latter requirement gives rise to the term *instantising*. Rapid dissolution favours small particles because of their high specific surface but these may not be wetted readily when placed in water. An agglomerated structure (Figure 13.19a) allows the larger particle to sink and disperse more readily in water. Water is then taken up into the void spaces of the agglomerate, the solid bridges dissolve and rapid dissolution of the primary particles then follows. Agglomerated products include instant milk powder, starch, coffee, instant soups, dry pudding mixes, cocoa, and sugar.

Table 13.3 is an attempt to classify various granulation processes. Although by no means comprehensive, this table shows that these process fall into four major groups.

TABLE 13.3
Classification of Particle Production Processes

Size reduction	Agitation methods	Drop formation	Compaction and extrusion
			Screw extrusion
Milling	Agglomeration	Spray drying	Extrusion cooking
Crushing	Granulation	Prilling	Roll compaction
Grinding	Pelletisation	Spray cooling	Tabletting
Communition	Instantization	Pastillisation	Briquetting
Shredding		Globulation	
Cutting			
Slicing			
Dicing			
	1. Mechanically or pneumatically agitated solids	1. Melt or slurry	1. Compression of powder/binder mixture
	2. Addition of binder in liquid form	2. Break-up of liquid phase	2. Cutting or breakage of solid pieces to give pre-determined size
	3. Phase change from liquid to solid binder	3. Drying or solidification	
Particle size is a function of energy input	Granule size is a complex function of a large number of variables	Granule size is a function of droplet size and droplet interaction	Granule size is a function of die size and energy input to size reduction step

(*a*) *Agitation methods*   In this group of processes a liquid binder is added to particles which are agitated in some kind of mixing device. In food systems the binder is an aqueous solution of, for example, lactose, dextrose, carboxy methyl cellulose, gelatine or a food gum or alternatively it may be a molten fat. Thus it is possible to use any powder mixer (fluidised bed, rolling drum or spouted bed) as a granulator. The solids being granulated may be a mixture of powders which are held together by the binder or may consist of a single component. The binder will usually undergo a change of phase either by solidifying a melt or by the removal of a solvent leaving the solute behind as the 'adhesive.' In the latter case it is usually necessary to have a drying stage; in fluidised bed granulation both particle growth and drying can be achieved in the same equipment. Very fine powders can be agglomerated without binder but the product granules are weak and the final size is limited. In many food applications the particle surface is wetted with a fine spray of water, or by the injection of steam, which allows particles to fuse together.

(*b*) *Drop formation methods*   Here, a melt (e.g., in prilling, spray cooling or globulation) or a solution (e.g., in spray drying) is broken up into small droplets. Spray drying requires the addition of thermal energy whereas in the other processes the melt solidifies in an air stream or on a cooled continuously moving belt. In spray drying and spray cooling the droplets are of the order of 100 μm diameter; it is important to stress that spray drying, although covered as a drying operation in chapter twelve, should be seen also as a particle formation process. Spray cooling employs a conventional spray drying tower and the molten liquid is atomised and then solidifies in a counter-current stream of cold air. Small particles (100–$200\,\mu$m) can be produced at throughputs equivalent to those achieved in spray drying; of the order of $40\,\mathrm{t\,h}^{-1}$ in a 9 or 10 m diameter tower.

In traditional prilling towers the molten droplets are larger, of the order of 5 mm, and solidify while falling through stationary air in a tower. The tower may be several tens of metres high and the droplets are formed by using small holes drilled in a plate. Belt cooling or pastillisation produces pastilles by solidifying a molten liquid on a steel belt which moves continuously and is cooled from underneath by sprays of cold water. Particles have a minimum size of 2–3 mm. In each of these operations the product particle size is determined largely by the size of the liquid droplets.

(c) *Compaction and extrusion*   In this group of processes powders (sometimes containing a wide distribution of particles sizes) or soft solids are forced together under great pressure and formed into pre-determined shapes. Often a liquid binder is included in a powder feed. In screw extrusion a screw rotates inside a pressurised barrel and forces the feed material through a die. The die may contain one or more orifices which define the granule or pellet size. The extrudates are cut into pellets by rotary cutter blades. Subsequently a spheroniser may be used to produce a rounded granule. A binder is usually required to plasticise the solid feed. A wide definition of extrusion would extend to the production of pasta, confectionery shapes and sausages. In the extrusion cooking of corn-based snack foods the moisture in the feed flashes off as the very high pressure generated in the heated barrel of the extruder is reduced to atmospheric at the exit and a considerable expansion of the product occurs.

In roll-type extrusion (Figure 13.20) the feed material is nipped between counter-rotating rollers and forced through holes in the periphery of one or both rollers. The extrudates are cut to the required length by adjustable knives. In roll compaction (Figure 13.21) the feed is screw-conveyed to two counter-rotating rollers which compact it and produce a sheet (about 1 mm thick) which is then milled to give irregular-shaped granules. Granule size then depends upon the degree of communition. Only very low liquid contents can be tolerated. Scale-up is achieved by increasing the diameter and length of the rollers. Compaction has a wide range of applications including the production of dried soup mixes, animal feeds and briquetting of sugar cubes.

(d) *Size reduction*   Size reduction includes a number of apparently disparate operations such as shredding, cutting, milling and grinding and is used on a very wide range of food materials including vegetable matter. Size reduction is used where the product specification requires a given size range, where an increase in surface area is required to aid drying, dissolution, blanching and the rate of extraction or leaching. In addition size reduction facilitates mixing and reduces bulk volume which in turn facilitates bulk solids handling.

13.6.2. Particle–Particle Bonding

An understanding of the nature of the bonds between particles in an agglomerate is fundamental to an understanding of granulation. The strength of inter-particle bonds determines not only granule strength (itself an important property) but also granule size. The effect of different bonding mechanisms and of the primary particle size on granule strength is shown in Figure 13.22.

(a) *Intermolecular forces*   The attractive force between particles is inversely proportional to the seventh power of the separation distance of the particles. Surface roughness

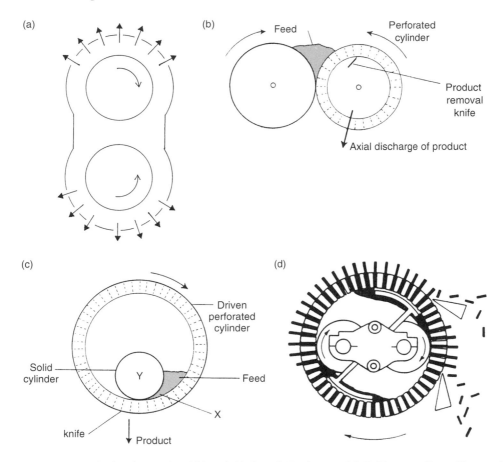

Figure 13.20. Methods of extrusion ((b) and (c) from J. Benbow and J. Bridgwater, Paste Flow and extrusion, Oxford University Press, 1993, © John Benbow and John Bridgwater 1993, reproduced by permission of Oxford University Press; (d) from P.J. Sherrington and R. Oliver, Granulation, Heyden, 1981, © John Wiley & Sons Ltd., reproduced with permission).

increases the effective separation distance and consequently van der Waals forces are not usually significant in granulation systems.

(*b*) *Electrostatic forces* For dry particles, electrostatic forces can be of the same order of magnitude as van der Waals forces.

(*c*) *Liquid bonds* Adsorbed liquid layers on the particle surface have the effect of smoothing out surface roughness and decreasing particle separation distances. Consequently, van der Waals forces may increase in magnitude significantly. Of greater significance are the mobile liquid bonds which are present in a granule initially when the solids to be granulated are contacted with liquid binder. They are usually a prelude to the formation of permanent solid bridges. Increasing the quantity of liquid changes the nature of the bonds and influences the overall granule strength. An assembly of particles containing bonds at the contact point between individual particles is described as being in the pendular state (Figure 13.23a). Increasing the liquid content of the agglomerate gives rise to the funicular state (Figure 13.23b) and finally the capillary state (Figure 13.23c) in which the inter-particle voidage is saturated

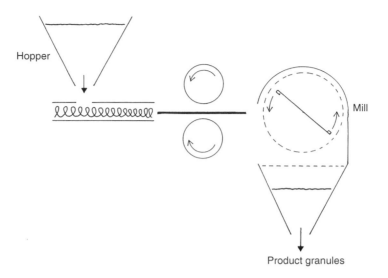

Figure 13.21. Roll compaction.

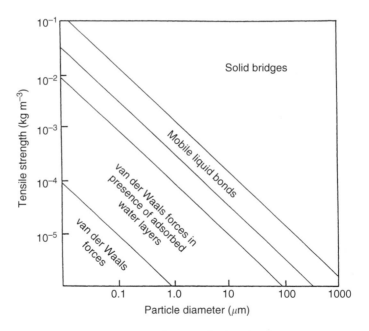

Figure 13.22. Tensile strength of agglomerates.

with liquid. This is amenable to theoretical analysis; the tensile strength of a capillary state granule is given by

$$\sigma = \frac{8(1 - \varepsilon)\gamma}{\varepsilon d} \tag{13.65}$$

where ε is the inter-particle voidage within the granule, γ is the surface tension of the liquid and d is the diameter of the constituent particles.

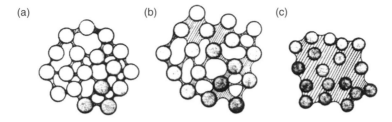

Figure 13.23. Mobile liquid bonds: (a) pendular; (b) funicular; (c) capillary (from P.J. Sherrington and R. Oliver, Granulation, Heyden, 1981, © John Wiley & Sons Ltd., reproduced with permission).

(*d*) *Solid bridges* Whilst the analysis of the strength of moist agglomerates and particles bound by liquid is well developed, solid bridges between particles do not lend themselves readily to theoretical treatment. The strength of crystalline bridges depends not only on the amount of material present, but also upon its structure. A finer crystal structure results in stronger bonds and there is some correlation between bond strength and higher drying temperatures. By assuming that all the material available for forming solid bridges is distributed uniformly over all points of contact between constituent particles in the granule and that the material has a constant tensile strength, the strength of an agglomerate can be defined by

$$\sigma = \varepsilon \theta f \tag{13.66}$$

in which θ is the intrinsic tensile strength of the bridge and f is the fraction of the void volume filled with binder. Little more can be said from a theoretical point of view. For a given concentration of binder, particle size and granule size, granule strength is a function of the structure and physical properties of the binder used and further information can only be obtained by experiment. However experimental measurements are difficult and tedious and it is difficult to measure the strength of the solid bridge independently because it cannot easily be cast into a form amenable to standard tests.

(*e*) *Growth mechanisms in granulation* In agglomeration processes granules grow by the successive addition of primary particles to an agglomerate. This occurs when two particles or two granules are brought into contact with sufficient liquid binder present to hold the two species together. In tumbling beds of powder (e.g. the pan granulator), two agglomerates colliding will be kneaded together by the tumbling action of the mixer and, because of their surface plasticity, form an approximately spherical granule. This sequence of events, forming a nucleation stage, continues until the granules are sufficiently large that the torque tending to separate them is too great to allow a permanent bond. Subsequent growth occurs by a 'crushing and layering' mechanism (Figure 13.24) in which the smallest and weakest granules are crushed by larger ones and the material redistributed around the surface of the large granule in a uniform layer. A distinction must be drawn here between the layering of smaller particles around a larger granule and the deposition of solute which provides a coating around a primary particle.

13.6.3. Fluidised Bed Granulation

Fluidised bed granulation (Figure 13.25) is a term that has been applied to processes which produce granules or dry powder from a solution or slurry in a fluidised bed to which

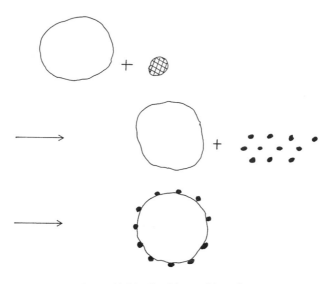

Figure 13.24. Crushing and layering.

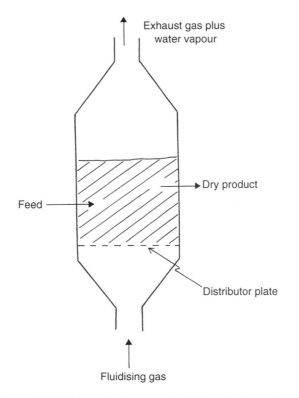

Figure 13.25. Schematic diagram of a fluidised bed granulator.

sensible heat is applied. Growth of bed particles, creation of new particles and drying of the product may all take place. Heat for evaporation of the solvent or for the removal of moisture from bed particles can be supplied either in the fluidising air or through the bed walls, and the wet feed material may be introduced under, or sprayed onto, the bed surface.

An excess of liquid feed, either over the whole bed or in a localised region, produces excessive and uncontrollable particle agglomeration and leads to a loss of fluidisation called 'bed quenching.' The term 'dry quenching' is used when defluidisation is the result of the excessive formation of dry granular material and 'wet quenching' where failure is caused by excessive free liquid. For a given bed outlet temperature, the liquid feed rate must not exceed that which will saturate the outlet air stream. If this condition is not obeyed, the bed material will become increasingly over-wet. Continued operation under these conditions will lead rapidly to wet quenching and the failure of the process. Despite the apparent incompatibility of free liquid and fluidised solids, the use of a fluidised bed for granulation offers several advantages over more traditional methods such as spray drying or pan granulation. In comparison with a spray drier a fluidised bed represents a large reduction in plant volume for the same throughput. Closer control of the physical properties of the product, such as particle size, flow characteristics, bulk density, is possible; a fluidised bed relies not only on the fine atomisation of the feed liquid but also its interaction with existing bed particles. It is possible to form particles in a fluidised bed of a size larger by an order of magnitude.

The successful operation of a fluidised bed granulator depends upon the balance between two opposing factors. First, the binding mechanism which results in particles joining together to form larger ones because of the presence of liquid in the fluidised layer. This is a function of the quantity and physical properties of the liquid feed: its adhesiveness with the solid surface and the strength of the resultant solid bridges. Second, the 'disruptive' forces which result from the abrasive action of, and solids circulation within, the fluidised bed which tends to break down, or prevent the formation of, particle–particle bonds. The magnitude of this effect depends upon the size and nature of the bed particles and the fluidising gas velocity.

Solvent in the feed liquid is evaporated in a well-defined zone close to the spray nozzle, and from the surface of the bed particles with which it inevitably comes into contact resulting in agglomeration. The extent to which clumps of agglomerated particles remain intact determines the outcome of the fluidised bed granulation process. Bed quenching results if insufficient break-down takes place; and break-down into smaller agglomerates, to an equilibrium size, will give a product of agglomerated granules. Further reduction and tearing apart of smaller agglomerates ultimately produces a single bed particle with associated binder, in other words a layered or coated granule. Thus an increase in the excess gas velocity causes a bed to move from quenching through agglomeration to layered growth. Smaller bed particles are more likely to form permanent bonds, and to quench, because of their smaller inertia. With larger initial particles, the mean diameter of product granules decreases and it is possible to achieve layered growth under conditions which would otherwise lead to quenching.

It is possible to operate a fluidised bed granulator in either a batch or a continuous mode. Batch operation produces a continuous increase in bed weight and therefore, if attrition and particle breakdown effects are not dominant, a continuous increase in bed particle size. This necessitates a gradual increase in the volumetric air flow through the bed, to compensate for the increasing minimum fluidising velocity. With continuous operation it is desirable to maintain a stable particle size distribution. In order that granules do not grow to be too large, seed particles or nuclei must be added to the bed, together with the removal of large particles.

13.6.4. Other Particle Agglomeration Methods

(*a*) *Spouted bed granulation*   Spouted beds are characterised by large particles, high inlet gas velocities and regular and ordered particle motion. Solution may be sprayed into the bed at the gas inlet or onto the bed surface. The high voidage, high temperature zone near the gas inlet allows very rapid evaporation of solvent and, together with the effect of large bed particles, results in layered growth with very low rates of agglomeration. Thus spouted beds can be used for coating and have found application in the treating of seed with pesticide coatings.

(*b*) *Pan granulator*   The pan or dish granulator (Figure 13.26) consists of a shallow dish, with a diameter between 0.5 and 6 m, inclined at about 50° to the horizontal and rotated at speeds in the range 10–40 rpm. The speed of rotation is about 75% of that at which solids would be centrifuged. Solids are fed into the pan continuously, climb around the periphery to the '11 O'clock' position and then fall across a pan diameter. Liquid binder is atomised and sprayed onto the solids which are falling down the pan surface. Baffles and scrapers aid the flow of solids and prevent wet material sticking to the wall.

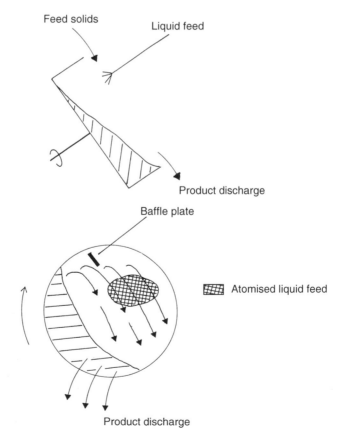

Figure 13.26. Pan granulator.

The pan is a classifying device. Small particles describe large circles at the base of the pan; as they increase in size particles move closer to the surface of the powder bed, describing smaller and smaller circles, until they discharge over the rim. Granule growth is influenced largely by the quantity of binding liquid used and the residence time (which depends upon the angle of inclination and the pan depth). Size decreases with residence time, but not markedly, due to increased exposure to abrasion. Pans are suitable for producing agglomerated granules up to 1–2 mm in diameter. Coating or layered growth is possible with larger particles: sugar and chocolate coatings are applied in a pan followed by drying either with radiant heat or with warm air.

(*c*) *Drum granulator* This consists of an open ended cylinder, with a length two or three times the diameter, inclined at about 5° to the horizontal. Diameters are between 1 and 3 m. The drum rotates slowly (10–30 rpm) and solids are carried around the periphery and fall away when they reach the '11 O'clock' position. Liquid feed is sprayed onto this falling curtain of solids at various points along the drum length. Particles describe a helical path as they travel through the drum before discharging over a rim. Again granule size is influenced largely by residence time and by binder feed rate. The drum is used extensively for high tonnage applications with throughputs up to $50 \, \text{t} \, \text{h}^{-1}$ at a diameter of 3 m.

13.7. SIZE REDUCTION

13.7.1. Mechanisms and material structure

A number of size reduction mechanisms can be identified which are relevant to specific materials and therefore allow, to some degree, a match to be made between size reduction equipment and the food material to be processed. Compression is used for the coarse reduction in size of brittle, hard materials and compressive forces tend to produce relatively few fines. Impaction is relevant to coarse, medium or fine reduction of hard materials whereas attrition, brought about by shear forces, tends to produces very small particles from softer materials. Cutting, slicing and dicing are very different mechanisms which are used to produce definite sizes and shapes from tough, ductile or softer materials and result in very few fine particles. It is important to realise that a number of these mechanisms may operate in a given piece of equipment.

The structure and composition of the material to be processed greatly influence the size reduction mechanisms that can be employed and the equipment used. For example, a crystalline structure (such as sugar) will break along fracture planes which require compression (using a crushing technique) to bring about size reduction. If there are no fracture planes present then new cracks must be developed using impaction. On the other hand a fibrous structure, such as vegetable matter, suggests the need for cutting or shredding. Similarly, cutting is appropriate for ductile materials such as flesh foods.

The presence of moisture can present problems in size reduction operations. Even small quantities of moisture on the surface of fine particles inevitably leads to the agglomeration of fines and therefore a size increase, although such agglomerates will be weak. More seriously, too high a moisture content may lead to the rapid blockage of a mill. Equally, moisture can be useful in suppressing dust and preventing dust explosions and this is exploited in wet milling techniques for example in the milling of corn.

Size reduction is a very inefficient operation with efficiencies as low as 1% or less. Most of the mechanical work is converted to heat energy because of particle-particle friction and friction between particles and surfaces and therefore cooling may be needed to reduce the risk of oxidation of the material. Dust explosions which can have devastating consequences are a very real possibility.

13.7.2. Size reduction equipment

There is a very wide range of size reduction equipment available and the following passages describe briefly the essential details of only a few of the most important types.

(a) *Hammer mill*   A standard hammer mill is shown in Figure 13.27a and consists of a series of hinged or fixed hammers attached to a shaft which rotates at high speed; peripheral speeds up to $100\,\mathrm{m\,s^{-1}}$ are possible. The hammer mill employs impaction primarily, although attrition also takes place. The solids to be milled are impacted in the small gap between the hammer and a toughened breaker plate on the inside casing of the mill. Consequently

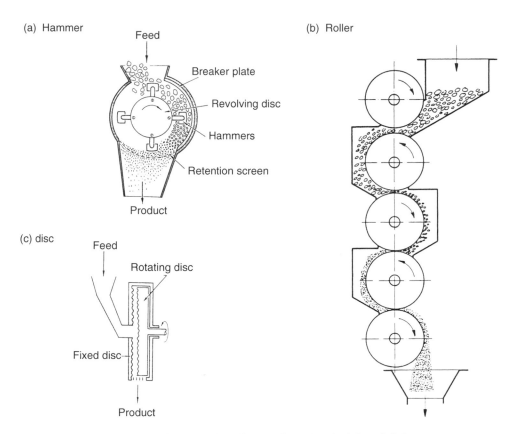

Figure 13.27. Types of mill: (a) hammer; (b) roller; (c) disc ((a) and (c) from J.G. Brennan, J.R. Butters, N.D. Cowell and A.E.V. Lilly, Food Engineering Operations, 3rd ed., Elsevier, 1990, reprinted with permission; (b) from H.A. Leniger and W.A. Beverloo, Food Process Engineering, Reidel Publishing, 1975, with permission).

such devices are used primarily for hard and brittle materials such as pepper, sugar, dried milk, dried vegetables and spices, but hammer mills also find applications with fibrous and vegetable matter.

(*b*) *Crushing rolls* These employ a compression mechanism and are often used in intermediate crushing stages. The rolls are approximately 0.6 m in diameter, rotate up to 300 rpm and are designed to nip large lumps of material. Smooth rolls are employed for producing flaked materials such as corn and for semi-liquids such as chocolate. However the surface may be corrugated for coarse grinding of seeds and grain. A series of rollers may be employed (Figure 13.27b) with increasingly smaller gaps between the rollers, and greater pressure exerted, to give progressively finer and finer grinding.

(*c*) *Disc mills* Disc mills (Figure 13.27c) utilise either a rotating disc which grinds material between itself and a stationary plate, or two counter-rotating discs, and employ largely attrition. The pin mill is a variation on this geometry in which the discs have pins or teeth and material is fed axially and moves radially outwards. The mechanism employed now is impaction at very high peripheral velocities up to 160 m s^{-1}. Disc mills have found application to sugar, nuts, cocoa, nutmeg, cereals, and rice.

(*d*) *Tumbling mills* The ball mill consists of a horizontal rotating drum (with differing sizes of freely moving metal or ceramic balls, about 25–150 mm in diameter, impacting the material to be crushed. Ball mills are used for fine grinding. In the rod mill balls are replaced with rods equal in length to the drum. Attrition occurs in addition to impaction. Rod mills overcome the problems of ball mills clogging with sticky material.

13.7.3. Operating methods

Batch operation is limited to devices such as the ball mill; most other milling equipment is used continuously. However an important distinction must be made between open circuit and closed circuit milling. In open circuit crushing the material passes through the mill once, with no rework of any oversize material, and results in a wide particle size distribution. The largest particles in the output stream must be made to be smaller than the upper limit for the product. This is wasteful of energy and gives too many fines. Closed circuit grinding (Figure 13.28) includes a classification step to remove the undersize particles as product and then recycle the oversize back to the mill. The operation, although requiring a greater investment, is much more efficient and gives a more homogeneous product.

13.7.4. Energy requirement for size reduction

The energy E required to reduce the size of particulate solids depends upon the energy absorbed by the solids and the mechanical efficiency of the process μ_{m} which takes account of frictional losses. Therefore by definition

$$\mu_{m} = \frac{\text{energy absorbed by solid}}{E} \qquad (13.67)$$

The expression for E can be expanded to give

$$E = \frac{e_{s}(A_{P} - A_{F})}{\mu_{m}\mu_{c}} \qquad (13.68)$$

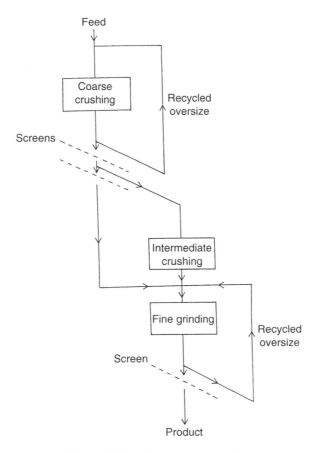

Figure 13.28. Closed circuit grinding.

where e_s is the surface energy per unit area, A_P and A_F are the surface areas per unit mass of the product and feed, respectively, and μ_c is a crushing efficiency. The latter is likely to be very small, of the order of 1%. Using the definitions of specific surface and sphericity, the surface area per unit mass, for non-spherical particles, becomes

$$A = \frac{6}{\phi \rho_s x} \tag{13.69}$$

where ρ_s is the density of the solids and x is the particle size. Now, across n size fractions of the particle size distribution, the total mass specific surface is

$$A = \sum_{i=1}^{i=n} \frac{6 \, \omega_i}{\phi \, \rho_s x_i} \tag{13.70}$$

and assuming that sphericity and density are constant for all size fractions

$$A = \frac{6}{\phi \rho_s x_{3,2}} \tag{13.71}$$

Substituting into Equation (13.68) gives the energy input as

$$E = \frac{e_s}{\mu_m \mu_c} \frac{6}{\phi \rho_s} \left(\frac{1}{x_P} - \frac{1}{x_F} \right) \tag{13.72}$$

where x_F and x_P are the surface-volume mean particle sizes of the feed and product respectively. Equation (13.72) suggests that energy input is a function of the initial and final size of the particles.

Three empirical relationships, due to Rittenger, Kick and Bond, respectively, have been suggested for determining E. Each model assumes that the energy input per unit mass dE required to change the size of material by a small amount dx is proportional to the particle size, that is

$$\frac{dE}{dx} = -c x^{-m} \tag{13.73}$$

where m and c are constants.

Rittenger assumed that $m = 2$, that is, the energy input for size reduction is proportional to the new surface created. Consequently

$$\frac{dE}{dx} = -c x^{-2} \tag{13.74}$$

On integration this yields

$$\int_0^E dE = -c \int_{x_F}^{x_P} x^{-2} \, dx \tag{13.75}$$

which can be evaluated to give

$$E = K_R \left(\frac{1}{x_P} - \frac{1}{x_F} \right) \tag{13.76}$$

where the constant c has been replaced by Rittenger's constant K_R which has units of $J \, m \, kg^{-1}$. Rittenger's equation tends to apply for particles which do not deform before breakage, in other words for brittle materials and for fine grinding. It suggests that the energy required is proportional to the increase in surface area per unit mass.

Kick assumed the value of m in Equation (13.73) to be unity. Hence

$$\int_0^E dE = -c \int_{x_F}^{x_P} x^{-1} \, dx \tag{13.77}$$

On integration this gives

$$E = K_K \ln \left(\frac{x_F}{x_P} \right) \tag{13.78}$$

where K_K is Kick's constant, which has units of $J \, kg^{-1}$. Note the difference in units for K_R and K_K. In Kick's law the energy input is proportional to the size reduction ratio. In other words the same energy input is required for a reduction in size from 1 cm to 1 mm as for a reduction from 100 to 10 μm. In using either of the models due to Rittenger and Kick the relevant constant must be obtained by experiment using both the same equipment and the same material.

The third empirical model is due to Bond who proposed that the work input is proportional to the square root of the surface/volume ratio of the product; this is equivalent to putting $m = 1.5$ in Equation (13.73) and thus

$$\int_0^E dE = -c \int_{x_F}^{x_P} x^{-3/2} dx \qquad (13.79)$$

On integration this gives

$$E = 2c \left[x^{-1/2} \right]_{x_F}^{x_P} \qquad (13.80)$$

and

$$E = 2c \left[\frac{1}{\sqrt{x_P}} - \frac{1}{\sqrt{x_F}} \right] \qquad (13.81)$$

which simplifies to

$$E = \frac{2c}{\sqrt{x_P}} \left[1 - \frac{1}{\sqrt{q}} \right] \qquad (13.82)$$

where $q = x_F/x_P$. Bond put the constant equal to $5 E_i$, where E_i is known as the work index and is defined as the energy required to reduce unit mass of material from an infinite size to a size where 80% of the material is below $100\ \mu m$.

$$E = E_i \left[\frac{100}{x_P} \right]^{1/2} \left[1 - \frac{1}{\sqrt{q}} \right] \qquad (13.83)$$

where x_p has units of μm. The work index is material specific and must be determined by experiment. Hard and brittle materials have a work index in the range 4×10^4–$8 \times 10^4\ J\,kg^{-1}$.

Example 13.18

Grain is milled at a rate of $10\,t\,h^{-1}$ and the power required for this operation is 67.5 kW. Assuming that Bond's law best describes the relationship between energy required and change in particle size, determine the work index for the grain and thus find the total power requirement to mill down to a distribution where 80% passes $100\ \mu m$.

Initial distribution		Final distribution	
Sieve size (μm)	Mass fraction	Sieve size (μm)	Mass fraction
6,730	0.00	605	0.00
4,760	0.05	425	0.08
3,360	0.15	300	0.12
2,380	0.70	212	0.65
1,680	0.10	150	0.11
		100	0.04

From the tabulated data, 80% of the grain in the initial distribution passes a $3,360\,\mu$m sieve and therefore $x_F = 3,360\,\mu$m. Similarly, the final distribution gives $x_P = 300\,\mu$m. Thus

$$q = \frac{3360}{300}, \qquad q = 11.2$$

The power requirement to achieve this degree of size reduction is 67.5 kW. Dividing this figure by the mass flow rate of grain in $kg\,s^{-1}$ gives the energy input per unit mass. Thus

$$E = \frac{67.5 \times 3600}{10 \times 10^3}\,kJ\,kg^{-1} \quad \text{or} \quad E = 24.3\,kJ\,kg^{-1}$$

The work index can be found from Equation (13.83). Hence

$$24.3 = E_i \left[\frac{100}{300}\right]^{1/2} \left[1 - \frac{1}{\sqrt{11.2}}\right]$$

from which

$$E_i = 60.0\,kJ\,kg^{-1}$$

For a final grain size distribution where $x_p = 100\,\mu$m, then $q = 33.6$. Substituting this figure and the work index into Equation (13.83) gives the required energy input as

$$E = 60.0 \left[\frac{100}{100}\right]^{1/2} \left[1 - \frac{1}{\sqrt{33.6}}\right] \frac{10^4}{3600}\,kW, \qquad E = 137.9\,kW$$

Example 13.19

Calculate the power requirement for the problem in Example 13.18 using both Rittinger's law and Kick's law.

As already calculated in Example 13.18, the energy input per unit mass is $E = 24.3\,kJ\,kg^{-1}$. From Equation (13.76) Rittinger's constant is given by

$$K_R = \frac{E}{(1/x_P - 1/x_F)}$$

Thus, putting E in $J\,kg^{-1}$ and x_F and x_P in metres,

$$K_R = \frac{24.3 \times 10^3}{[(1/300 \times 10^{-6}) - (1/3360 \times 10^{-6})]}$$

$$K_R = 8.0\,J\,m\,kg^{-1}$$

The energy required to reduce the grain size from 3,360 to $100\,\mu$m is then

$$E = 8.0 \left(\frac{1}{100 \times 10^{-6}} - \frac{1}{3360 \times 10^{-6}}\right) J\,kg^{-1}$$

$$E = 7.762 \times 10^4\,J\,kg^{-1}$$

The power input is now $77.62 \times 10^4/3600$ or $215.6\,\text{kW}$.

Rittinger's law tends to overestimate the energy input required for size reduction whereas Kick's law usually gives an underestimate. From Equation (13.78)

$$K_K = \frac{24.3 \times 10^3}{\ln(3360/300)}\,\text{J kg}^{-1}, \qquad K_K = 1.006 \times 10^4\,\text{J kg}^{-1}$$

Now substituting for $x_P = 100\,\mu\text{m}$ the energy requirement is

$$E = 1.006 \times 10^4 \ln\left(\frac{3360}{100}\right)\,\text{J kg}^{-1}, \qquad E = 3.535 \times 10^4\,\text{J kg}^{-1}$$

and the power input becomes $35.35 \times 10^4/3600$ or $98.2\,\text{kW}$.

NOMENCLATURE

A	Surface area per unit mass
c	Constant
c_D	Drag coefficient
d	Particle diameter
d_B	Equivalent diameter of void spaces
d_e	Equivalent spherical diameter
D	Bed diameter
e_s	Surface energy per unit area
E	Energy required for size reduction
E_i	Work index
f	Fraction of agglomerate voids filled with binder
F	Drag force
g	Acceleration due to gravity
Ga	Galileo number
h	Fluidised bed height
k	Volume shape factor
K	Kozeny's constant
K_K	Kick's constant
K_R	Rittenger's constant
K'	Constant in Darcy's equation
L	Bed Depth
L_B	Equivalent length of void spaces
m	Particle mass; constant
m'	Mass of displaced fluid
n	Number of size fractions in a particle size distribution; index
N	Number of particles of size x
p	Parameter representing characteristic of a distribution
q	Parameter representing characteristic of a distribution; ratio of feed to product particle size
Q	Total volumetric gas flow rate

Q_B Volumetric bubble flow rate
Q_{mf} Volumetric gas flow rate at minimum fluidisation
Re Reynolds number
S Specific surface
S_B Particle surface area per unit bed volume in contact with fluid
t Time; average particle circulation time
u Relative velocity between sphere and fluid; superficial velocity
u_{mf} Minimum fluidising velocity
u_t Terminal falling velocity
u' Interstitial velocity
u_0 Superficial velocity at a voidage of unity
V Volume
x Particle size
x_l Lower limit of size distribution
x_u Upper limit of size distribution

GREEK SYMBOLS

α Specific resistance
β Bed permeability
γ Surface tension
ΔP Bed pressure drop
ε Inter-particle voidage
θ Intrinsic tensile strength of solid bridge
μ Viscosity
μ_c Crushing efficiency
μ_m Mechanical efficiency
ρ density
ρ_B bulk density
σ tensile strength
ϕ Sphericity
ω Mass fraction

SUBSCRIPTS

f   Fluid
F   Feed
mf   Minimum fluidising conditions
P   Product
S   Solid

PROBLEMS

13.1. For the particle size distribution in Example 13.2, determine the number–volume and weight–moment mean particle diameters.

13.2. Particle size analysis of a sample of spray dried milk gave the following data:

Size band (μm)	Number of particles
40–80	10
80–120	30
120–160	50
160–200	20
200–240	10

Calculate the number–volume and surface–volume mean particle diameters.

13.3. Calculate the equivalent spherical particle diameter of a cylindrical particle 70 μm diameter and 350 μm in length.

13.4. Calculate the sphericity of the particle in problem 13.3.

13.5. Determine the number–length, number–surface and weight–moment mean diameters from the following sieve data:

Sieve size (μm)	Mass of particles (g)
1400	0
1000	0
710	3
500	9
355	17
250	33
180	18
125	10
90	8
63	2
45	0
<45	0

13.6. A food powder has a particle density of 1,228 kg m^{-3} and a bulk density of 700 kg m^{-3}. What is the inter-particle voidage of the powder bed?

13.7. Using Stokes' law, determine the terminal falling velocity of a 75 μm droplet of oil rising in water at 20°C. The oil density is 800 kg m^{-3}.

13.8. Find the viscosity of a 50% aqueous sucrose solution if the glass beads of Example 13.8 take 179 s to fall through a distance of 0.5 m. The solution density is 1,232 kg m^{-3}.

13.9. Determine the limiting diameter of a particle of density $2,650\,\mathrm{kg\,m^{-3}}$ settling according to Stokes' law in water at $20°C$.

13.10. Assuming flow in the Stokes regime, calculate the terminal settling velocity of water droplets of $0.4\,\mathrm{mm}$ diameter in an oil of density $800\,\mathrm{kg\,m^{-3}}$ and viscosity $1.5 \times 10^{-3}\,\mathrm{Pa\,s}$. Take the density of water to be $1000\,\mathrm{kg\,m^{-3}}$.

13.11. What will be the terminal falling velocity of a 2 mm steel ball in oil? Data: density of oil $= 900\,\mathrm{kg\,m^{-3}}$; density of steel $= 7,870\,\mathrm{kg\,m^{-3}}$; viscosity of oil $= 40.5\,\mathrm{mPa\,s}$.

13.12. Determine the diameter of seeds of density $1,500\,\mathrm{kg\,m^{-3}}$ which will settle in water at $20°C$ with a terminal falling velocity of $0.10\,\mathrm{m\,s^{-1}}$.

13.13. Spray dried milk particles are to be agglomerated in a fluidised bed. In small scale tests the pressure drop across the bed was measured as a function of decreasing superficial gas velocity. Determine the minimum fluidising velocity.

Superficial velocity ($\mathrm{cm\,s^{-1}}$)	Pressure drop (Pa)
0.04	200
0.08	400
1.08	540
1.48	740
1.80	860
2.00	900
2.24	940
2.70	1,000
3.02	1,000
3.40	1,020
3.80	980
4.08	1,000
4.60	1,000

13.14. Calculate the minimum fluidising velocity of a food particulate (solids density $1,400\,\mathrm{kg\,m^{-3}}$) fluidised with air at $52°C$. The cumulative percentage oversize curve approximates to a straight line between $100\,\mu m$ (at 0% mass) and $500\,\mu m$ (at 100% mass). At minimum fluidisation conditions a bed pressure drop of $2.25\,\mathrm{kPa}$ is required for a bed height of $0.30\,\mathrm{m}$.

13.15. Potato cubes ($8\,\mathrm{mm} \times 8\,\mathrm{mm} \times 8\,\mathrm{mm}$) are to be frozen in a fluidised bed with air at a temperature of $-23°C$. In laboratory tests at incipient fluidisation the pressure drop across a $0.25\,\mathrm{m}$ deep bed was found to be $1.4\,\mathrm{k\,Pa}$. Determine a suitable equivalent particle diameter and hence estimate the minimum fluidising velocity. The potato density is $1,100\,\mathrm{kg\,m^{-3}}$.

13.16. Calculate the minimum fluidising velocity of the particles in Example 13.14 using the Leva equation.

FURTHER READING

T. Allen, *Particle Size Measurement*, Chapman and Hall, London (1981).

J. S. M. Botterill, *Fluid-Bed Heat Transfer*, Academic Press, New York (1975).

C. Capes, *Particle Size Enlargement*, Elsevier, Amsterdam (1979).

J. F. Davidson and D. Harrison, *Fluidisation*, Academic Press, New York (1971).

H. A. Leniger and W. A. Beverloo, *Food Process Engineering*, Reidel, Dordrecht (1975).

M. Loncin and R. L. Merson, *Food Engineering; Principles and Selected Applications*, Academic Press, New York (1979).

P. J. Sherrington and R. Oliver, *Granulation*, Heyden (1981).

Mixing and Separation

14.1. MIXING

14.1.1. Definitions and Scope

The mixing and/or agitation of liquids, solids and (to lesser extent) gases is one of the commonest of all operations in the food processing industries. Of the possible combinations of these states, those of principal interest are liquid/liquid mixtures, solid/solid mixtures, and liquid/solid mixtures or pastes. However it is important at this early stage to define exactly what is meant by the terms 'agitation' and 'mixing' and it is perhaps easiest to do this by considering liquid/liquid systems. The agitation of a liquid may be defined as the establishment of a particular flow pattern within the liquid, usually a circulatory motion within a container. On the other hand mixing implies the random distribution, throughout a system, of two or more initially separate ingredients.

There are a number of reasons for agitating liquids amongst which may be listed: the suspension of solids within the liquid; the dispersion of a gas within the liquid; the dispersion of a second liquid as droplets (i.e., the formation of an emulsion); the promotion of heat transfer from a heat transfer surface to the bulk liquid; and the mixing of two or more liquids. Now the reasons for mixing (and this applies to all possible combinations of the three states of matter) are: to bring about intimate contact between different species in order for a chemical reaction to occur (this can include the dissolution of solids in a liquid and the extraction of a solute from either liquid or solid phases); and to provide a new property of the mixture which was not present in the original separate components. An example of the latter might be the inclusion of a specific proportion in a food mixture of a given component for nutritional purposes.

It should be clear from the foregoing that mixing is brought about by agitation. However it would be tedious to continue to use both words according to their precise meaning and therefore throughout this chapter the term 'mixing' will be used to mean both the random distribution of components and the means of bringing about that randomness, that is, the mechanisms of agitation.

There are perhaps three criteria by which the performance of a mixer should be assessed. These are:

   i   the degree of mixedness achieved,
   ii the time required to bring about mixing, and,
   iii the power consumption required.

Each of these criteria will be addressed in turn.

14.1.2. Mixedness

The degree of uniformity of a mixed product is measured by the analysis of spot samples removed from the mixer. Local values of the concentration of a component are then compared to average or to expected values. Mixing is complete when one component is randomly distributed in another, using standard deviation or variance as the measure of the difference between the measured local values and the expected value. The variance s^2 is defined by

$$s^2 = \sum \frac{(x_i - \bar{x})}{n - 1} \tag{14.1}$$

where x_i are the local values, $\bar{x}$ is the average or expected value and n is the population or number of samples. For the case where $\bar{x}$ is the average of the experimental values s^2 is an estimate only of the true variance, and is known as the experimental variance.

For an unmixed system of two separate components the variance, s_0^2, is given by

$$s_0^2 = p(1 - p) \tag{14.2}$$

where p is the proportion of one of the components. As mixing proceeds the variance will decrease until, for a uniform distribution, the variance is zero. However, a random distribution tends towards a limiting value of variance s_∞^2 which is defined as being acceptable. For liquids, and for gases, mixing usually takes place on a molecular scale and therefore almost perfect mixing is possible. However, even with gases and liquids, s_∞^2 is not equal to zero, rather it will be very close to zero after a reasonable time. Solids mixing is rather different; mixing does not occur on a molecular scale because the smallest solid particles are several orders of magnitude larger than the largest molecules and the best possible mixing which can be obtained is a predetermined degree of randomness. Another way of expressing this difference is to refer to point uniformity and overall uniformity (Figure 14.1). In each of the diagrams in Figure 14.1 there are equal numbers of black and white squares. In liquid mixing point uniformity is possible but in the mixing of solids overall uniformity only can be achieved.

A further way of looking at this problem is to consider the concept of the 'scale of scrutiny'. In turn this is related to the sampling procedures which are adopted. The sampling of a mixture must be completely random in order to give statistical validation to the procedure. The sampling technique, that is, the method used to remove physically the sample from the mixture, must neither enhance mixing nor promote segregation of the mixed components and thus give a false picture. However the size of the sample which is removed is equally important. Where overall uniformity rather than point uniformity exists, the scale of scrutiny is the sample size at which mixing is considered to be 'good'. Scale of scrutiny has been defined variously as

> the minimum size of the regions of segregation which would cause the mixture to be regarded as imperfectly mixed

and

> establishing a scale of scrutiny appropriate to the end use of the mixed product fixes the size or volume of the samples that should be used to assess the mixture quality.

For example it would be inappropriate to judge the effectiveness of mixing by removing 1 kg samples from a mixer if the final product was to be placed into 100 g packets.

(a)

(b)

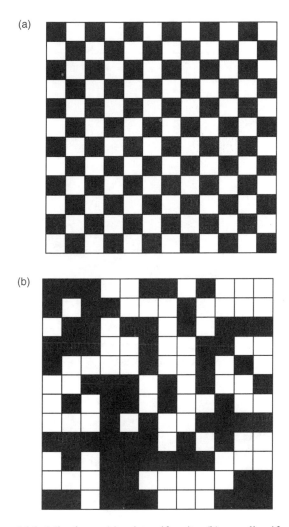

Figure 14.1. Mixedness: (a) point uniformity; (b) overall uniformity.

14.1.3. Mixing Index and Mixing Time

In general, variance does not correlate well with batch mixing time and it is necessary to use a mixing index M, a dimensionless fractional measure of variance or standard deviation, which can be correlated with time. Mixing indices can be defined in a number of ways as shown in Table 14.1.

Definitions (a) and (d) are based on variance, whilst definitions (b) and (e) are based on standard deviation. To an extent, which definition of mixing index is chosen for a given application depends entirely upon its ability to correlate mixedness with time by a simple linear relationship. Definition (c) tends to be used rather less often than the others. It should be noted that, for each of definitions (a), (b) and (c), as s approaches s_∞, M approaches a value of zero

TABLE 14.1
Definitions of Mixing Index

Mixing index	Equation no.
(a) $M = \dfrac{s^2 - s_\infty^2}{s_0^2 - s_\infty^2}$	(14.3)
(b) $M = \dfrac{s - s_\infty}{s_0 - s_\infty}$	(14.4)
(c) $M = \dfrac{\ln s - \ln s_\infty}{\ln s_0 - \ln s_\infty}$	(14.5)
(d) $M = \dfrac{s_0^2 - s^2}{s_0^2 - s_\infty^2}$	(14.6)
(e) $M = \dfrac{s_0 - s}{s_0 - s_\infty}$	(14.7)

corresponding to a mixed sample. As s approaches s_0 then M approaches a value of unity, which represents an unmixed sample. The reverse is true for definitions (d) and (e).

Using definition (a), Equation (14.3), the driving force for mixing, that is, the difference between the variance at a given time t and at the end point, is $s^2 - s_\infty^2$. Thus a rate equation, for the rate of change of variance, can be written as

$$\frac{d(s^2)}{dt} = -k(s^2 - s_\infty^2) \tag{14.8}$$

where k is a rate constant representing the rate at which mixing proceeds. Integrating Equation (14.8) between a variance of s_0^2 at $t = 0$ and s^2 at time t gives

$$\int_{s_0^2}^{s^2} \frac{d(s^2)}{s^2 - s_\infty^2} = -k \int_0^t dt \tag{14.9}$$

and

$$\ln\left[\frac{s^2 - s_\infty^2}{s_0^2 - s_\infty^2}\right] = -kt \tag{14.10}$$

or, substituting from definition (a) of mixing index,

$$M = \exp(-kt) \tag{14.11}$$

The time required to bring about a given degree of mixing can now be found by plotting the logarithm of experimental values of the mixing index against time and extrapolating to the desired final value of M. A linear relationship may be obtained from any of the mixing indices in Table 14.1 and the existence of a straight line validates the choice of mixing index.

A major difference between solids and liquids mixing can now be seen; for liquids, for which s_∞^2 can approximate to zero, the task is to determine an acceptable value of s^2 (and hence of M) based upon the product specification and the normal distribution (Table 14.2).

TABLE 14.2
Normal Distribution Curve: Percentage Area under
the Curve for a Given Standard Deviation s either
side of Mean $\bar{x}$

x	Area %	x	Area %
$\bar{x} \pm 2.00\,s$	95.45	$\bar{x} \pm 2.55\,s$	98.92
$\bar{x} \pm 2.05\,s$	95.96	$\bar{x} \pm 2.60\,s$	99.07
$\bar{x} \pm 2.10\,s$	96.43	$\bar{x} \pm 2.65\,s$	99.20
$\bar{x} \pm 2.15\,s$	96.84	$\bar{x} \pm 2.70\,s$	99.31
$\bar{x} \pm 2.20\,s$	97.22	$\bar{x} \pm 2.75\,s$	99.40
$\bar{x} \pm 2.25\,s$	97.56	$\bar{x} \pm 2.80\,s$	99.49
$\bar{x} \pm 2.30\,s$	97.86	$\bar{x} \pm 2.85\,s$	99.56
$\bar{x} \pm 2.35\,s$	98.12	$\bar{x} \pm 2.90\,s$	99.63
$\bar{x} \pm 2.40\,s$	98.36	$\bar{x} \pm 2.95\,s$	99.68
$\bar{x} \pm 2.45\,s$	98.57	$\bar{x} \pm 3.00\,s$	99.73
$\bar{x} \pm 2.50\,s$	98.76	$\bar{x} \pm 4.00\,s$	99.994

Area %	x	Area %	x
20.00	$\bar{x} \pm 0.253\,s$	80.00	$\bar{x} \pm 1.282\,s$
30.00	$\bar{x} \pm 0.385\,s$	90.00	$\bar{x} \pm 1.645\,s$
40.00	$\bar{x} \pm 0.524\,s$	95.00	$\bar{x} \pm 1.960\,s$
50.00	$\bar{x} \pm 0.674\,s$	99.00	$\bar{x} \pm 2.576\,s$
60.00	$\bar{x} \pm 0.842\,s$	99.90	$\bar{x} \pm 3.291\,s$
70.00	$\bar{x} \pm 1.036\,s$	99.99	$\bar{x} \pm 3.891\,s$

For solids it is not possible for s_∞^2 to reach zero and therefore the product specification is used to determine a target variance s_∞^2. Thus by definition, when this target is reached M equals either unity or zero, respectively.

It should be noted that

$$\frac{s^2 - s_\infty^2}{s_0^2 - s_\infty^2} = 1 - \frac{s_0^2 - s^2}{s_0^2 - s_\infty^2} \tag{14.7}$$

and therefore, based on definition (e) in Table 14.1,

$$1 - M = \exp(-kt) \tag{14.8}$$

Example 14.1

In a full scale mixing trial two liquids were mixed in the proportions 1:4. Samples removed over a period and analysed for the minor component, generated the following data:

Time (min)	Variance
1	0.084
2	0.044
4	0.012
6	3.24×10^{-3}
10	2.41×10^{-4}

Determine the mixing time to ensure that 99.5% of all samples withdrawn from the mixer will contain ±5% of the mean concentration of the minor component.

The solution to this problem will be based upon the mixing index defined in Equation (14.3). Now, the proportion of one component is 0.20 and therefore

$$s_0^2 = 0.20 \times 0.80, \qquad s_0^2 = 0.16$$

Note that, in a binary system, it is immaterial which fraction is used to determine the initial variance. For liquids, mixing takes place at a molecular scale and therefore $s_\infty^2 \cong 0$. Using the experimental values of variance as a function of time together with Equation (14.3) it is possible to construct the following table:

t (min)	M
1	0.525
2	0.275
4	0.0750
6	0.0203
10	1.506×10^{-3}

Now $M = \exp(-kt)$ and therefore a plot of the logarithm of M against time produces a straight line, the gradient of which gives the rate constant k as -0.65 min^{-1}. In order to find an acceptable value of mixing index use must be made of the normal distribution (Table 14.2). 99.5% of all samples are captured by 2.86 standard deviations either side of the mean. Thus $2.86s$ equates to 5% of the mean concentration of the minor component. This mean concentration is 0.20 when expressed as a fraction. Therefore

$$2.86s = 0.05 \times 0.20, \qquad s = 3.496 \times 10^{-3}$$

Consequently the target variance s^2 is 1.222×10^{-5} and the target mixing index is

$$M = \frac{1.222 \times 10^{-5} - 0}{0.16 - 0} \quad \text{or} \quad M = 7.641 \times 10^{-5}$$

Now, from Equation (14.11),

$$t = \frac{\ln(7.641 \times 10^{-5})}{-0.65} \text{ min}, \qquad t = 14.6 \text{ min}$$

Example 14.2

A minor ingredient was added to a powdered food base to give a 99.6% probability that each 500 g packet contained 25 ± 1 g of the minor component. Mixing took place in a double-cone blender and samples were withdrawn during a mixing trial to generate the following data, based upon variance from the mean value of the mass fraction of the minor ingredient:

Time (min)	Variance
1.5	0.0428
3.0	0.0394
4.5	0.0337
5.0	0.0309
6.0	0.0237

What should the mixing time be?

The mass fraction of the minor ingredient is $\frac{25}{500}$ or 0.05 and therefore the initial (unmixed) variance is

$$s_0^2 = 0.05 \times 0.95 \ \text{ or } \ s_0^2 = 0.00475$$

The acceptable variability in each sample is 1 g in 25 g, that is, 4%, or when expressed as a fraction, 0.04. A 99.6% probability that each packet contains 25 ± 1 g of the minor component equates to an area under the normal distribution curve bounded by 2.90 standard deviations either side of the mean. Thus $2.90\,s$ is equal to 4% of 0.05, that is,

$$2.90s = 0.04 \times 0.05, \qquad s = 6.89 \times 10^{-4}$$

This value defines acceptable mixing and thus the final variance is

$$s_\infty^2 = 4.756 \times 10^{-7}$$

Using this value in Equation (14.6) gives the following values of mixing index with time:

t (min)	M
1.5	0.099
3.0	0.171
4.5	0.291
5.0	0.349
6.0	0.501

A plot of the logarithm of M against time gives a straight line. Because mixing is complete when $s^2 = s_\infty^2$ then the target is $M = 1$ and it is possible to read the required mixing time directly from the graph at $M = 1$. Hence the mixing time is 7.9 min.

14.1.4. Mixing of Liquids

The mixing of liquids is a widespread food processing operation. Most commonly this is achieved in an agitated tank but liquids may be mixed whilst flowing in a pipeline by means of a static mixer. This consists of a pipe section containing baffles which induce rapid changes of direction and high degrees of turbulence in the fluid. They find application in chocolate refining, the blending of fruit pieces into yoghurts and the carbonation of mineral waters. In yoghurt manufacture the relatively low shear avoids the breakdown of fruit particles. A large

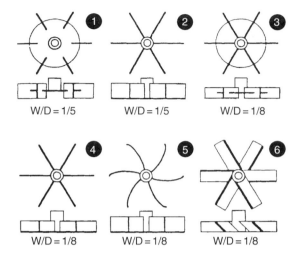

W/D = 1/5   W/D = 1/5   W/D = 1/8

W/D = 1/8   W/D = 1/8   W/D = 1/8

Figure 14.2. Turbines for liquid mixing (from P.J. Fryer, D.L. Pyle and C.D. Rielly (eds.), Chemical Engineering for the Food Industry, Chapman and Hall, 1997, with permission).

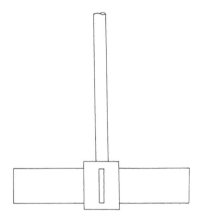

Figure 14.3. Paddle impeller.

number of different types of impeller are in use (see Figures 14.2–14.6); different impellers impart different flow patterns to the liquid and they must be matched to the rheology of the liquid and to the desired shear rate. Mixing vessels usually have rounded bottoms, rather than flat ones, to prevent the formation of dead spaces. The liquid depth is normally equal to the tank diameter. When an impeller rotates in a liquid the liquid is likely to swirl in a mass and a vortex will form (Figure 14.7a). This is undesirable; the vortex may well draw air from the surface down to the impeller with the possibility either of unwanted dissolution of air or a waste of energy as the impeller rotates partly in air at the expense of agitating the liquid. Consequently baffles are fitted to the tank which consist of vertical strips of metal running the full depth of the inside surface of the tank. Normally four baffles are used, their width being about 10% of the tank diameter. Baffles minimise vortex formation, prevent swirling of the liquid, and result in more rapid mixing (Figure 14.7b).

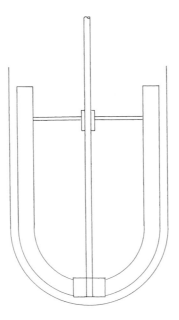

Figure 14.4. Anchor impeller.

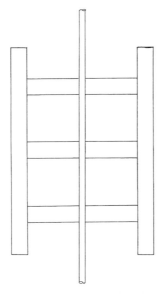

Figure 14.5. Gate impeller.

The impellers used in liquid mixing fall into three main groups.

(*a*) *Propellers* Marine propellers, usually with three blades and a diameter of about one third of the tank diameter, are used for agitating low viscosity fluids and rotate at high speeds of the order of 10–25 Hz. The resultant flow is axial (Figure 14.8a) with strong vertical

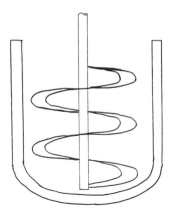

Figure 14.6. Helical impeller.

(a)

(b)

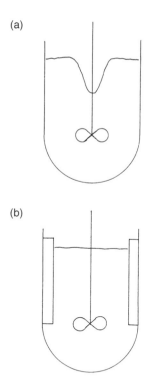

Figure 14.7. Vortex formation: (a) without baffles, and (b) with baffles.

currents, which are particularly useful for keeping solids in suspension. The upper viscosity limit for the use of propellers is about 5 Pa s.

(*b*) *Turbines*   Turbines (Figure 14.2) take a number of forms but usually have six or eight blades. The blades may be straight, curved, flat, or pitched; disc turbines have blades mounted orthogonally to a horizontal disc. The ratio of turbine diameter to tank diameter is

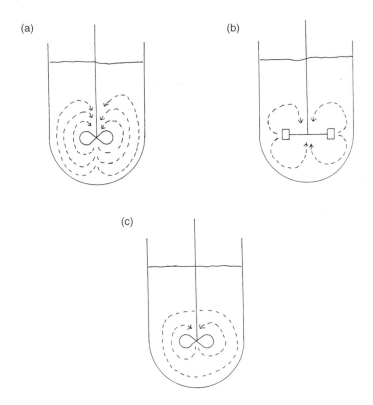

Figure 14.8. Flow patterns in liquid mixing: (a) axial flow, (b) radial flow, and (c) shear-thinning.

usually about 0.6. Turbines can be used to agitate more viscous liquids and the flow leaving the impeller is radial giving an overall improved agitation pattern (Figure 14.8b).

(c) *Impellers for highly viscous or non-Newtonian fluids* Smaller impellers can be very ineffective with non-Newtonian liquids. Consider a shear-thinning liquid being mixed with a propeller as in Figure 14.8c. Shear rates are not uniform throughout a mixing vessel and close to the propeller shear rates are high and the apparent viscosity of the liquid will be low. Further away, close to the vessel walls, shear rates are considerably lower and the apparent viscosity will remain much higher than at the centre. Consequently there will be a well mixed region close to the impeller whilst the bulk of the liquid is poorly mixed. To overcome this problem very viscous and non-Newtonian fluids are agitated with impellers such as paddles, anchors, gates, or helical ribbons (Figures 14.3–14.6). Each of these impeller types has a larger relative diameter than either turbines or propellers and is designed to sweep the bulk of liquid in the tank more effectively. They rotate more slowly at speeds of 1–2 Hz. Paddles consist of either two or four blades and the impeller diameter is between 0.5 and 0.8 of the tank diameter. More complex designs such as the anchor, or gate, occupy up to 95% of the tank diameter and are able to scrape liquid layers away from the vessel surface, thus avoiding the poor mixing described above. Helical ribbons are still more complex impellers which give improved mixing by imparting a significant vertical velocity component in addition to the radial component.

14.1.5. Power Consumption in Liquid Mixing

In order to scale up power requirements for agitated vessels, from experiments with small tanks to large industrial tanks, it is necessary to use correlations based upon dimensional analysis and subsequent experiment using a wide range of fluids with different densities and viscosities. The technique is exactly that which is used to correlate heat transfer coefficients with a series of variables and which was described in chapter seven. For such scale-up rules to be valid there must be similarity between different-sized vessels. Three kinds of similarity are needed: geometrical, kinematic, and dynamic. Geometrical similarity is perhaps the easiest to achieve; corresponding dimensions at the two scales must have the same ratio in all cases. Referring to Figure 14.9, these ratios are: D_T/D, Z/D, W/D, H/D, and B/D, where D is the impeller diameter, D_T is the vessel diameter, H the depth of liquid, Z the height of the impeller from the bottom of the vessel, W the impeller blade depth, and B is the baffle width. In addition the number of baffles and the pitch of propellers must be constant. Kinematic similarity implies that paths of motion within the liquid must be alike at different scales and that the ratios of

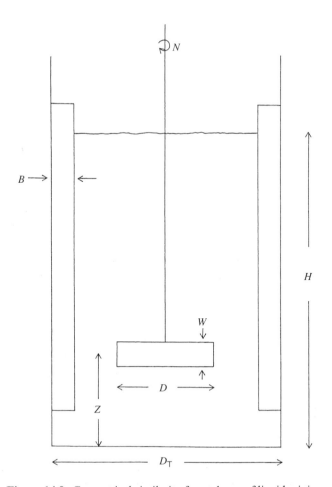

Figure 14.9. Geometrical similarity for scale-up of liquid mixing.

velocities at corresponding points must be equal. Finally, dynamic similarity means that the ratio of shear stresses at corresponding points must be equal. This is particularly important in the mixing of food liquids which are non-Newtonian; variations in the shear regime at different scales may mean that very different apparent viscosities may exist with significant implications for the effectiveness of mixing.

Now, assuming that the power input P required to agitate a liquid is dependant upon the density of the liquid, ρ, its viscosity or apparent viscosity, μ, the impeller diameter, D, the speed of rotation of the impeller, N (measured in Hz) and the acceleration due to gravity, g, dimensional analysis produces a correlation of the form

$$N_{\mathrm{p}} = a\,Re^x\,Fr^y \tag{14.14}$$

where the constant a and the indices x and y are determined by experiment. In Equation (14.14) N_{p} is the power number defined by

$$N_{\mathrm{p}} = \frac{P}{\rho N^3 D^5} \tag{14.15}$$

and may be thought of as analogous to a drag coefficient or friction factor. Fr is the Froude number defined by

$$Fr = \frac{N^2 D}{g} \tag{14.16}$$

The Froude number is the ratio of inertial to gravitational forces and the latter are important only when vortices are formed at the liquid surface. Baffles are designed to prevent vortices and therefore the Froude number may usually be ignored in baffled vessels. The Reynolds number is written in a slightly different format to that which has been met so far. The characteristic length is the impeller diameter and the relevant velocity is that at the tip of the impeller blade, equal to the product ND. Thus

$$Re = \frac{\rho N D^2}{\mu} \tag{14.17}$$

The Reynolds number is, of course, the ratio of inertial to viscous forces and indicates the degree of turbulence. For mixing vessels of this kind laminar flow occurs when $Re < 10$ and turbulent flow for $Re > 10^4$. The transition region equates to an unusually wide range of Reynolds number: $10 < Re < 10^4$.

Correlations of the form of Equation (14.14) are usually presented graphically as in the example of Figure 14.10 for the power consumption of disc turbines. At low Reynolds number the relationship between Re and power number for different impeller types form a single curve which is steeply inclined indicating that in the laminar region the power consumption is strongly dependant upon the apparent viscosity of the liquid. In the transition and turbulent regions the impeller geometry and the resultant flow pattern influence the power drawn, however the power number is much more nearly constant with increasing Reynolds number than in the laminar region.

For Newtonian fluids the Reynolds number is a function of dynamic viscosity. However non-Newtonian fluids require an apparent viscosity based upon an average shear rate $\dot{\gamma}$ which characterises the shear experienced by the bulk of the liquid in the vessel. The simple correlation

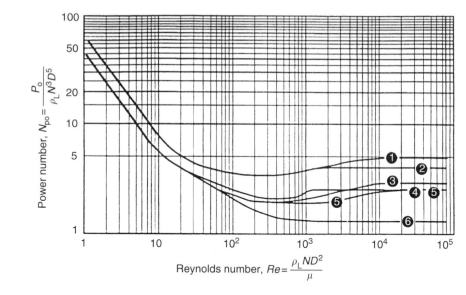

Figure 14.10. Power number as a function of Reynolds number for turbines (from P.J. Fryer, D.L. Pyle and C.D. Rielly (eds.), Chemical Engineering for the Food Industry, Chapman and Hall, 1997, with permission).

due to Metzner and Otto is usually used for both Bingham and shear-thinning power law fluids. Thus

$$\dot{\gamma} = \text{constant} \times N \qquad (14.18)$$

where the constant has a value in the range 10–13, but is usually assumed to be 12. An apparent viscosity can now be found using Equation (6.63). For shear-thickening power law fluids the average shear rate is obtained from the Calderbank and Moo-Young equation

$$\dot{\gamma} = 38N \left(\frac{D}{D_T} \right)^{0.5} \qquad (14.19)$$

Example 14.3

During the preparation of a jam filling a 25% fruit concentrate is to be agitated in a baffled vessel of 1.2 m diameter with a conventional disc turbine, 0.4 m diameter, which rotates at 600 rpm. The concentrate is a shear-thinning power law fluid with a consistency coefficient of 70.0 Pa s^n and a flow behaviour index of 0.35. The concentrate density is 1,280 kg m^{-3}. What power will the turbine draw?

In order to calculate the apparent viscosity of the concentrate an average shear rate must be specified. The rotation speed N is $\frac{600}{60} = 10$ Hz and therefore, using the correlation of Otto and Metzner, the average shear rate is

$$\dot{\gamma} = 12 \times 10 \, \text{s}^{-1} \quad \text{or} \quad \dot{\gamma} = 120 \, \text{s}^{-1}$$

The apparent viscosity is now

$$\mu_a = 70 \times 120^{(0.35-1)} \, \text{Pa s}, \qquad \mu_a = 3.12 \, \text{Pa s}$$

This allows the Reynolds number to be calculated as

$$Re = \frac{1280 \times 10(0.4)^2}{3.12}, \qquad Re = 657.2$$

From curve 1 in Figure 14.10 the power number is $N_p = 3.5$ and thus, using the definition of power number in Equation (14.15),

$$P = 3.5 \times 1280(10)^3(0.4)^5$$

that is, the turbine has a power requirement of

$$P = 45.9\,\text{kW}$$

14.1.6. Correlations for the Density and Viscosity of Mixtures

Empirical relationships are available for both the density and viscosity of binary mixtures. Density can be found from

$$\rho = v_1\rho_1 + v_2\rho_2 \tag{14.20}$$

where v is the volume fraction and the subscripts 1 and 2 represent the continuous and dispersed phases, respectively. This relationship suggests that density is an additive property but the correlations for viscosity are inevitably more empirical in nature. For an unbaffled vessel

$$\mu = \mu_1^{v_1}\mu_2^{v_2} \tag{14.21}$$

and for a baffled vessel

$$\mu = \frac{\mu_1}{v_1}\left(1 + \left[\frac{1.5\mu_2 v_2}{(\mu_1 + \mu_2)}\right]\right) \tag{14.22}$$

Note that in Equations (14.21) and (14.22) μ represents both dynamic and apparent viscosity where appropriate.

Example 14.4

A new thickening agent is to be manufactured from a blend of xanthan gum and guar gum in equal volumetric proportions. A 1.0 m diameter baffled vessel fitted with a 0.60 m diameter standard Rushton turbine is available. The impeller rotates at 420 rpm. Determine the power requirement for adequate agitation. Rheological data for both xanthan gum (shear-thinning) and guar gum (shear-thickening) is given in the table.

	Xanthan (continuous phase)	Guar (dispersed phase)
$K\,(\text{Pa}\,\text{s}^n)$	30.0	10.0
$n\,(\text{—})$	0.50	1.40
Density $(\text{kg}\,\text{m}^{-3})$	1100	1200

This problem involves the mixing of two power law liquids and both a mean density and a mean apparent viscosity must be used for the Reynolds number. For a shear-thinning fluid the average shear rate from the Otto and Metzner equation, at a rotation speed of 7 Hz, is $\dot{\gamma} = 84\,\text{s}^{-1}$. Therefore xanthan gum has an apparent viscosity of

$$\mu_a = 30 \times 7^{(0.5-1)}\,\text{Pa s} \quad \text{or} \quad \mu_a = 11.3\,\text{Pa s}$$

For a shear-thickening liquid Equation (14.19) must be used to find $\dot{\gamma}$. Thus

$$\dot{\gamma} = 38 \times 7 \left(\frac{0.6}{1.0}\right)^{0.5}, \qquad \dot{\gamma} = 206\,\text{s}^{-1}.$$

The apparent viscosity of the guar gum is therefore

$$\mu_a = 10 \times 206^{(1.4-1)}\,\text{Pa s} \quad \text{or} \quad \mu_a = 84.3\,\text{Pa s}$$

The mean density of the mixture is $1{,}150\,\text{kg m}^{-3}$ and the mean apparent viscosity is given by

$$\mu = \frac{11.3}{0.5}\left(1 + \left[\frac{1.5 \times 84.3 \times 0.5}{(11.3 + 84.3)}\right]\right)\,\text{Pa s}$$

and therefore $\mu = 37.5\,\text{Pa s}$.

The Reynolds number is now

$$Re = \frac{1150 \times 7(0.6)^2}{37.5}, \qquad Re = 77.3$$

This gives a power number of $N_p = 3.9$ (from curve 1 in Figure 14.10). Consequently

$$P = 3.9 \times 1150(7)^3(0.6)^5 \quad \text{or} \quad P = 119.6\,\text{kW}$$

14.1.7. Mixing of Solids

A theory for solids mixing, as is the case generally for the processing of particulate solids, is not available as it is for liquids. However a number of mixing mechanisms can be identified and to a limited extent these can be related to mixing equipment. The solids mixing mechanisms which can be identified are diffusion, convection, and shear.

Particles diffuse under the influence of a concentration gradient in the same way that molecules diffuse. They move by inter-particle percolation, that is, in the void spaces between other particles under the influence of either gravity or, in higher speed mixers, of centrifugal effects. Ficks's law can be used to describe this phenomenon. Convection describes the movement of groups of particles from one place to another within the mixer volume because of the direct action of an impeller or a moving device within the mixer body. As in convection within fluids this is likely to be a more significant effect than diffusion but diffusional effects will still be present. The shear mechanism operates when slipping planes are formed within the particulate mass, perhaps because of the action of a blade, which in turn allow particles to exploit new void spaces through which particles can then diffuse.

In addition to these mixing mechanisms, segregation acts against mixing to separate components which have different physical properties. Segregation is usually due to gravitational forces, but is heightened when centrifugal effects are present and occurs when particles have the possibility of falling through the spaces between other particles. The degree of segregation is a function of particle size (with smaller particles being more likely to segregate), density, and shape. Thus, larger size differences and larger density differences in a particulate mixture are likely to bring about increased segregation and make mixing more difficult. It is more difficult to quantify the effect of shape, although gross differences in shape are more likely to lead to poor mixing.

14.1.8. Equipment for Solids Mixing

A huge variety of devices for the mixing of solids is available and it is impossible to classify these exhaustively. However four types of mixer suggest themselves.

(*a*) *Pneumatic agitation*   This category includes fluidised bed mixing and to a lesser extent the spouted bed. The mechanisms of mixing were covered in chapter thirteen and do not need to be repeated here. However it is important to emphasise that the density difference between particles is more important than size difference in determining the rate and extent of particle segregation.

(*b*) *Tumbling mixers*   Essentially these are enclosed containers which rotate about a horizontal axis and include the horizontal drum, double-cone, V-cone, Y-cone, and cube. Figure 14.11 shows a Y-cone mixer. These mixers are operated in batch mode being partially filled with solids (up to about 60% by volume). Tumbling mixers are run at a fraction of the critical speed required for centrifugation (at which all the solids would be thrown outwards towards the mixer walls) with a practical maximum speed of about 100 rpm. They are used for free-flowing solids, but small amounts of liquid may be added, and are best suited for particles of similar size and density because strong segregation can occur. Such mixers may have baffles fitted to the inner walls which help to lift solids or alternatively may be fitted with ploughs to assist convection.

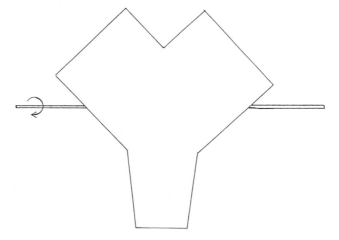

Figure 14.11. Y-cone mixer.

(*c*) *Ribbon mixers*   Essentially a ribbon mixer is a horizontal trough within which a helical ribbon rotates about the horizontal axis. Sometimes two counter rotating ribbons are used and these have the effect of moving solids in opposite directions. It is usually operated in batch mode with mixer volumes up to about $15\,m^3$, but continuous operation is possible with feed rates up to $10\,t\,h^{-1}$. Helical ribbon mixers can be used for slightly cohesive solids, for very thin pastes or for the addition of liquids to solids. Mixing is strongly convective and segregation is far less pronounced than in either tumbling mixers or fluidised beds.

(*d*) *Vertical screw mixers*   A vertical screw mixer consists of an open cone with an orbiting screw-type impeller. The screw turns upon its own axis moving solids from top to bottom of the cone. At the same time the screw, which is pivoted at the base of the cone and on a rotating arm, travels in an orbit close to the internal cone wall. It is generally operated batch-wise with volumes up to $10\,m^3$. Applications include the addition of small quantities of a minor component to large particle masses. Convective mixing is dominant and therefore it is used for segregating mixtures. Although the rate of mixing is low, power consumption is less than for a ribbon mixer.

14.2. FILTRATION

14.2.1. Introduction

There are a number of ways in which solid particles can be separated physically from a liquid on an industrial scale without the use of thermal energy or high temperatures. The simplest method of separating out solids from a suspension is to allow the particles to settle under gravity; this is a very slow process but may well be suited to large volumes of liquid such as waste water. Centrifugation is a much faster separation process but is relatively costly because of the high non-thermal energy demand. Filtration is in some respects a compromise between these two options.

Filtration may be defined as the separation of solid from a suspension of solids in a liquid by using a porous medium which retains the solids and allows the liquid to pass through. It is a mechanical separation and requires far less energy than either evaporation or drying. However, other operations such as drying may follow the filtration stage to give the final desired moisture content of the filtered solids. Filtration may be categorised into those operations where the separated solid is the required product and the liquid which passes through the filter, the filtrate, is unwanted and second, into those operations where the filtrate is the product. The former are examples of cake filtration in which a relatively thin filter medium is used on which particles collect. The particles build up rapidly on the surface of the filter medium and it is this cake of solids which is then responsible for filtration (Figure 14.12). The concentration of solids in the suspension is likely to be high and almost all of the solids will be retained in the filter cake. Cake filtration requires the use of a filter press such as a leaf filter, a plate, and frame press or a rotary vacuum filter. Such equipment is used, for example, in sugar production to remove the last few percent of moisture from the raw juice, to separate vegetable oil from the seeds from which the oil has been extracted and in brewing is used for the filtering of mash.

The second category consists of processes where a clarified filtrate is the product. Examples of this kind of filtration range from deep bed filtration to the use of cartridge filters. Deep bed filtration (Figure 14.13) employs a bed of sand and gravel, decreasing in particle

Suspension

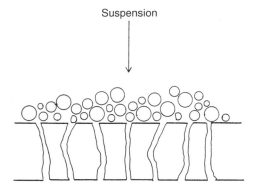

Figure 14.12. Cake filtration.

Suspension

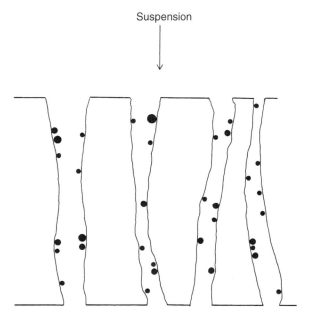

Figure 14.13. Deep bed filtration.

size down the bed, and is used in large throughput water purification. The particles are considerably smaller than the void spaces of the filter bed, the solids concentration is low and the particles penetrate the bed a considerable distance before being trapped by surface forces. Usually there will be no layer of solids on the surface of the bed. Cartridge filters are used extensively in the clarification of both soft drinks and alcoholic beverages, particularly to remove residual yeasts. A cartridge filter consists of a series of thin metal discs which are stacked over a vertical shaft and inserted into a cylindrical casing. The shaft may be hollow or fluted to allow liquid to pass axially along the cartridge; the discs are pressed such that there is a very narrow gap between them. Liquid is forced into the casing and flows inwards between the discs allowing the filtered solids to build up principally at the edge of the discs.

14.2.2. Analysis of Cake Filtration

The analysis of cake filtration depends upon the relationship between the volume of filtrate V which is produced in a given time t and the pressure drop ΔP which is needed to maintain the flow of filtrate through the filter medium. Kozeny's equation [Equation (13.42)] must now be adapted as follows: first, the superficial velocity of the fluid (filtrate) is replaced by the volumetric flow rate per unit cross-sectional area of the bed. Thus

$$u = \frac{1}{A}\frac{dV}{dt} \tag{14.23}$$

Second, the permeability of the bed is expressed in terms of the specific resistance of the filter cake α and therefore

$$\frac{1}{\alpha} = \frac{\varepsilon^3}{K S^2 (1 - \varepsilon)^2} \tag{14.24}$$

Kozeny's equation now becomes

$$\frac{1}{A}\frac{dV}{dt} = \frac{\Delta P}{\alpha\mu L} \tag{14.25}$$

where L is the depth of the cake deposited on the filter medium. However the resistance to the flow of filtrate which is presented by the filter medium must be accounted for by introducing an extra cake depth L_0 which presents a resistance equivalent to the filter medium. Hence

$$\frac{dV}{dt} = \frac{A\,\Delta P}{\alpha\mu(L + L_0)} \tag{14.26}$$

which is known as the general filtration equation. This expression is based upon the assumption that the cake can be modelled as a series of capillaries in which the flow is laminar. If it is assumed that the filter cake is incompressible, that is, that the cake voidage is constant with time even though the mass of the cake increases throughout the filtration operation, then the cake volume LA can be equated to νV and

$$L = \frac{\nu V}{A} \tag{14.27}$$

where ν is the volume of cake deposited per unit volume of filtrate passing through the cake. However if the filter cake is compressible then an increase in pressure drop will result in a denser cake with a higher resistance to flow. If the cake is inelastic then it is important that the pressure drop should not exceed the normal operating pressure because operation will then take place with a higher resistance than is necessary.

Substituting from Equation (14.27) the general filtration equation becomes

$$\frac{dt}{dV} = \frac{\alpha\mu\nu V}{A^2\,\Delta P} + \frac{\alpha\mu L_0}{A\,\Delta P} \tag{14.28}$$

Now, as solids build up in the cake with time and the bed depth L increases, the pressure drop required to maintain a constant flow rate of filtrate will gradually increase. Consequently there are two possible methods of operating a filter press:

   i   constant rate filtration in which ΔP must be increased continually, and
  ii   constant pressure filtration in which the flow rate of filtrate will decrease as the operation proceeds.

The latter is by far the most common mode of operation.

14.2.3. Constant Pressure Filtration

Equation (14.28) can be integrated over the time t taken to collect a volume V of filtrate and therefore

$$\int_0^t dt = \frac{\alpha\mu v}{A^2\,\Delta P}\int_0^V V\,dV + \frac{\alpha\mu L_0}{A\,\Delta P}\int_0^V dV \tag{14.29}$$

from which

$$t = \frac{\alpha\mu v V^2}{2A^2\,\Delta P} + \frac{\alpha\mu L_0 V}{A\,\Delta P} \tag{14.30}$$

Clearly the relationship between t and V is not linear. Now dividing through by V gives

$$\frac{t}{V} = \frac{\alpha\mu v V}{2A^2\,\Delta P} + \frac{\alpha\mu L_0}{A\,\Delta P} \tag{14.31}$$

and a plot of t/V against V will give a straight line as in Figure 14.14. The gradient of this line is $\alpha\mu v/2A^2\,\Delta P$ from which the specific cake resistance α can be obtained if the other quantities are known. Of these the pressure drop will be set, the area will be fixed by the geometry of the filter press and the viscosity of the filtrate can be measured. The volume of cake deposited per unit volume of filtrate v can usually be obtained from a material balance. Once the specific cake resistance has been determined a knowledge of the intercept in Figure 14.14, equal to $\alpha\mu L_0/A\,\Delta P$, allows the equivalent depth of the filter medium L_0 to be calculated. A knowledge of the specific cake resistance and, where it is significant, the equivalent depth of the filter medium are required to find the filtration area in a given application and hence the size of the filter press.

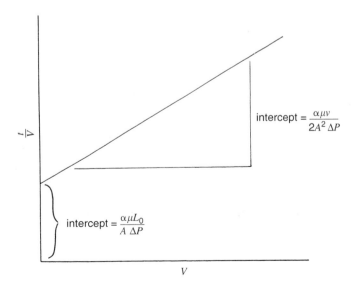

intercept $= \dfrac{\alpha\mu v}{2A^2\,\Delta P}$

intercept $= \dfrac{\alpha\mu L_0}{A\,\Delta P}$

Figure 14.14. Constant pressure filtration.

Example 14.5

A filter press produces 500 l of water from an aqueous food suspension after 10 min and 1,000 l after 35 min. How long will it take to filter 2,000 l and what will the filtration rate be at that stage?

If it is necessary to determine only the relationship between the filtrate volume and the time for collection of that volume, then Equation (14.31) can be simplified to $t = k_1 V^2 + k_2 V$ where k_1 and k_2 are constants. Only two data points are available and therefore it is not necessary to plot the data. Substituting for each data point in turn, and working in SI units, gives

$$600 = k_1(0.5)^2 + 0.5 k_2, \qquad 2100 = k_1(1.0)^2 + k_2$$

Solving these equations simultaneously yields $k_1 = 1,800\,\mathrm{s\,m^{-6}}$ and $k_2 = 300\,\mathrm{s\,m^{-3}}$. Thus the time required to collect 2,000 l is

$$t = 1800(2.0)^2 + (300 \times 2.0)\,\mathrm{s}, \qquad t = 7800\,\mathrm{s} \ \text{ or } \ 130\,\text{min}.$$

14.2.4. Filtration Equipment

A plate and frame filter press consists of a series of ribbed or grooved plates covered on both surfaces by a filter cloth. The plates are usually held vertically in a rack and are separated by alternately placed hollow frames which thus form a series of chambers between alternate plates. The whole assembly is clamped tightly together and the feed slurry is forced into each chamber at pressures up to 1,000 kPa. The filter cake builds up on the cloth at each side of the chamber and the filtrate, after passing through the cloth, flows down the grooves on the plates. After filtration is complete (which is detected by the fall in the filtrate flow rate or by an increase in pressure) the cake can be washed through to remove any soluble impurities. The press is then opened and the solids removed. The plate and frame press has the advantages of being of relatively simple construction with no moving parts. The filter cloths are readily replaced and the capacity can be varied by changing the number and size of the plates and frames. However the great disadvantages are the intermittent operation and labour intensive nature of this equipment.

A rotary vacuum filter is a continuous filter consisting of a perforated horizontal drum with the filter cloth covering the external surface (Figure 14.15). A solid inner cylinder forms an annular space which is divided into a number of sectors each of which is connected separately to the vacuum line by means of a complex rotary valve. The drum is immersed in the slurry to be filtered and is rotated at about 2 rpm. As the drum rotates a given sector picks up slurry which is sequentially filtered and washed. The washed cake is then removed by a scraper blade just before the sector re-enters the feed slurry. The rotary valve allows the filtrate and the wash liquor to be collected separately. The rotary vacuum filter has the great advantage of continuous operation and the speed of rotation can be varied to influence the cake thickness. However the inherent disadvantage is that vacuum operation limits the pressure drop to something less than atmospheric pressure.

Example 14.6

A 35% by mass food suspension was filtered using a small vacuum filter of area 200 cm^2 operated at an absolute pressure of 31.3 kPa. After 938 s the volume of filtrate collected was

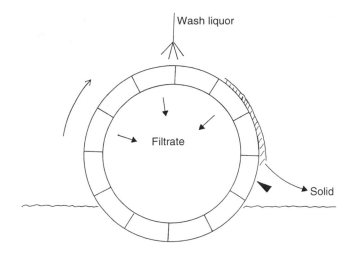

Figure 14.15. Rotary vacuum filter.

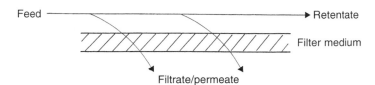

Figure 14.16. Cross-flow filtration.

10 l and a total of 3,686 s was required to collect 20 l. A filter cake was formed with a voidage of 0.45 and a moisture content, on a dry basis, of 20%. Determine the specific cake resistance and the equivalent bed depth of the filter medium. The density and viscosity of the filtrate were 1,000 kg m^{-3} and 0.001 Pa s, respectively and the density of solids 2,600 kg m^{-3}.

Substituting the filtrate volume, and the time required to collect that volume, into the constant pressure filtration equation gives two expressions

$$938 = \frac{\alpha \mu v (0.01)^2}{2 A^2 \, \Delta P} + \frac{\alpha \mu L_0 (0.01)}{A \, \Delta P}$$

and

$$3686 = \frac{\alpha \mu v (0.02)^2}{2 A^2 \, \Delta P} + \frac{\alpha \mu L_0 (0.02)}{A \, \Delta P}$$

which can be solved simultaneously to yield

$$\frac{\alpha \mu v}{2 A^2 \, \Delta P} = 9.05 \times 10^6 \, \text{s m}^{-6}, \qquad \frac{\alpha \mu L_0}{A \, \Delta P} = 3300 \, \text{s m}^{-3}$$

Now taking a basis of 100 kg of suspension, the mass of solids is 35 kg, all of which is retained in the cake. The bulk density of the cake is equal to $\rho_S (1 - \varepsilon)$ where ρ_S is the solids density and ε is the cake voidage. Thus the cake volume is $35/2600(1 - 0.45)$. However the filter cake also contains residual moisture and the mass of water retained is $0.20 \times 35 = 7\,\text{kg}$. Consequently the mass of filtrate is reduced to $58\,\text{kg}$ and the cake volume per unit volume of filtrate v becomes

$$v = \frac{35}{2600(1 - \varepsilon)} \frac{1000}{58} \quad \text{or} \quad v = 0.422$$

Substituting for the filter area, filtrate viscosity and operating pressure drop, the specific cake resistance α is now

$$\alpha = \frac{9.05 \times 10^6 \times 2 \times (200 \times 10^{-4})^2 \times 70 \times 10^3}{0.422 \times 10^{-3}}\,\text{m}^{-2}, \qquad \alpha = 1.2 \times 10^{12}\,\text{m}^{-2}$$

Similarly the equivalent bed depth of the filter medium is

$$L_0 = \frac{3300 \times 200 \times 10^{-4} \times 70 \times 10^3}{1.2 \times 10^{12} \times 10^{-3}}\,\text{m}, \quad \text{or} \quad L_0 = 0.00385\,\text{m}$$

Example 14.7

The suspension in Example 14.6 is to be filtered at the full scale to give a filtrate production rate of $12\,\text{m}^3\,\text{h}^{-1}$. A rotary vacuum filter is to be used which operates at a fixed speed of $0.4\,\text{rpm}$ with 30% of the drum surface immersed in the suspension and at an absolute pressure of $31.3\,\text{kPa}$. The same filter medium is to be used. Suggest suitable dimensions for the filter.

In problems involving rotary filters is sensible to adopt as a basis for the calculation an integral number of revolutions of the drum. The volume of filtrate V collected during one revolution is the product of the time taken to make one revolution and the required filtrate rate. Thus

$$V = 150 \times 3.33 \times 10^{-3}\,\text{m}^3, \qquad V = 0.5\,\text{m}^3$$

Of the time taken for the filter to make one revolution, $150\,\text{s}$, only 30% can be counted as filtering time which is accordingly $45\,\text{s}$.

Substituting these values, together with the specific cake resistance and equivalent bed depth of the filter medium from Example 14.6, into the constant pressure filtration equation results in a quadratic expression for the unknown filter area A

$$45 = \frac{1.2 \times 10^{12} \times 10^{-3} \times 0.422 \times (0.5)^2}{2A^2 \times 70 \times 10^3} + \frac{1.2 \times 10^{12} \times 10^{-3} \times 0.00385 \times 0.5}{A \times 70 \times 10^3}$$

which simplifies to

$$45A^2 - 33A - 904.3 = 0$$

The positive root of this equation gives the area as $4.86\,\text{m}^2$. Sensible dimensions might then be a diameter of $1.0\,\text{m}$ and a length of $1.55\,\text{m}$.

14.2.5. Filter Aids

The particles in the filter cake should be as large as possible in order to maximise the flow rate of filtrate. Consequently flocculating agents are sometimes added to the slurry to promote coagulation of the solids. Filter aids may be added to increase the permeability of the cake. These are high voidage materials such as Kieselguhr or diatomaceous earths. When a thin layer of filter aid is placed on the filter medium this is known as 'pre-coating'. If the filter cake is the valuable product then separation of the filter aid from the product presents a further problem.

14.3. MEMBRANE SEPARATIONS

14.3.1. Introduction

Membrane separation techniques use very thin semi-permeable membranes to separate small particulate or molecular species from a solution or suspension. At the lower end of the filtration size range microfiltration is the term used to describe the removal of small particles, large colloids or microbial cells in the size range 0.1–10 μm. For example, microfiltration has been used extensively in cheese production to remove bacteria from milk and bacteria, moulds, and spores from brine. Below this size range, in the context of food processing, membrane separation implies either ultrafiltration or reverse osmosis.

Ultrafiltration uses a semi-permeable membrane to remove emulsions, colloids or macro-molecules such as proteins, in the range 0.1 μm–5 nm, from solution. In reverse osmosis, which is sometimes referred to as hyperfiltration, small molecules such as dissolved salts less than 5 nm in size are separated from the solvent. Each of these operations is now widespread in the food industry and are used in a range of applications including the concentration of juices, production of protein concentrates, waste water treatment and desalination. The major advantages of membrane separations are the ability to concentrate solutions without the use of thermal energy and the ability to retain volatile components which tend to be lost in evaporation or spray drying.

The solvent (usually water) which passes through the membrane is known as the permeate and is equivalent to the filtrate in conventional filtration. The permeate may contain small dissolved salts in reverse osmosis, whilst larger species are retained, and in ultrafiltration the permeate may contain much larger molecules. The term retentate is used to refer to the liquid stream which is retained on the feed side of the membrane. In most operations the retentate will be recycled across the membrane to increase the degree of concentration.

Ultrafiltration and reverse osmosis are the techniques of most interest in food processing but there are a number of other membrane separation processes in use. Dialysis is a term which is usually applied to the treatment of renal failure, in other words the artificial kidney. However electrodialysis is the application of an electromotive force across a membrane to aid the diffusion of dissolved salts or ionic species. The membrane does not permit the passage of water and therefore electrodialysis has been used to remove salt from liquid foods. A relatively new development is the use of direct osmosis concentration in which the solution to be concentrated and an osmotic agent (a concentrated brine or glucose solution) come into intimate contact with opposite sides of a semi-permeable membrane. Water will flow from the juice to the osmotic solution if the osmotic pressure of the solution is greater than that of the juice.

Conventional filter presses normally use a technique known as dead end filtration in which the suspension to be filtered is pumped up against the filter medium. In contrast to this, membrane separations such as ultrafiltration and reverse osmosis make use of cross flow filtration, illustrated in Figure 14.16, in which the feed stream flows across the membrane surface allowing the permeate to pass through the membrane. This flow arrangement prevents the formation of a filter cake and therefore increases the permeate flux. The feed stream remains as a solution or suspension and this allows the retentate to be recycled. A further advantage of cross flow filtration is that it allows the selective removal of particular particle or molecule sizes in the permeate.

14.3.2. Osmosis and Reverse Osmosis

Figure 14.17a depicts a semi-permeable membrane on either side of which are a solvent and a solution, respectively. For the sake of illustration these may be assumed to water and an aqueous solution, say sodium chloride. Osmosis is a natural phenomenon in which water moves across the membrane into the solution from a high concentration of water to a lower water concentration. The size of the pores in the membrane are such to impede the passage of the solute molecules. If now a pressure is applied to the sodium chloride solution water can be made to flow in the opposite direction, that is from solution to pure water. At a certain pressure Π an equilibrium is established (Figure 14.17b) where the flow of water in each direction is equal, in other words there is no net flow of water and no change in concentration. However if the pressure is increased beyond Π then there will be a net flow of water from the sodium chloride solution. This is the principle of reverse osmosis (Figure 14.17c) in which water can be removed from a solution by the application of a pressure gradient. Reverse osmosis allows only very small molecules to pass through the membrane and therefore the applied pressure

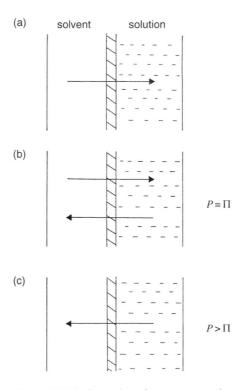

Figure 14.17. Osmosis and reverse osmosis.

must be large enough to overcome the very high osmotic pressure of small solute molecules. Consequently the operating pressures in reverse osmosis can be as high as 100 bar.

A major applications of reverse osmosis is the desalination of sea water to provide drinking water. The pores of the cellulose acetate membranes which are used are sufficiently small to exclude sodium chloride, sodium bromide, calcium chloride, and the other salts found in sea water. In the food industry reverse osmosis is used to concentrate fruit and vegetable juices and is an alternative to the traditional evaporation process in which important flavours are lost.

14.3.3. General Membrane Equation

Crucial to the operation of any membrane process is a knowledge of the rate at which permeate is produced. This is normally expressed as a permeate flux, that is, the mass or volumetric flow rate of permeate per unit area of membrane. Because of the complexity of reverse osmosis and ultrafiltration there is no universally accepted theory which can be used to predict permeate flux. However the general membrane equation is often used to indicate the dependence of permeate flux upon the applied transmembrane pressure, the osmotic pressure difference across the membrane and the resistance provided by the membrane.

$$N = \frac{\Delta P - \Delta \Pi}{\mu \sum R} \qquad (14.32)$$

Figure 14.18 represents the concentration, pressure, and osmotic pressure on each side of the membrane. The permeate flux is directly proportional to the difference between the transmembrane pressure ΔP and the osmotic pressure difference $\Delta \Pi$ where $\Delta P = P_1 - P_2$ and $\Delta \Pi = \Pi_1 - \Pi_2$. Flux is then inversely proportional to the permeate viscosity and the sum of the mass transfer resistances $\sum R$ which may include the resistance of the membrane itself and fouling layers due to deposits from the feed material. In reverse osmosis the osmotic pressure of the feed stream Π_1 is likely to be very large, being proportional to the molar concentration

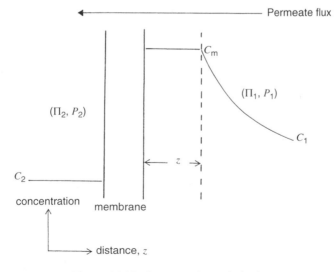

Figure 14.18. Concentration polarisation.

TABLE 14.3
Osmotic Pressure of Orange Juice and Tomato Juice as a
Function of Concentration

Orange juice		Tomato juice	
Concentration (°Brix)	Osmotic pressure (bar)	Concentration (°Brix)	Osmotic pressure (bar)
10.5	14	4.5–5.0	9–10
21.5	30	8–12	16–25
31.5	59	20–24	41–49
42	94	28–29	57–59
		36–37	74–76

of the rejected solute (and therefore inversely proportional to the solute molecular weight). The osmotic pressure of the permeate will be very close to zero. Therefore the transmembrane pressure must exceed $\Delta\Pi$ in order for water to flow from the feed solution and hence the need for the very large transmembrane pressures in reverse osmosis.

14.3.4. Osmotic Pressure

The osmotic pressure Π is a function only of the solution being concentrated, not of the membrane. Table 14.3 gives some examples of the very high osmotic pressures of fruit and vegetable juices at concentrations of industrial interest.

It is notoriously difficult to predict or to measure osmotic pressure, although for dilute solutions the van't Hoff relation can be used. This is based on the form of the ideal gas law

$$\Pi V = nRT \tag{14.33}$$

which in terms of the molar concentration of the solution C becomes

$$\Pi = CRT \tag{14.34}$$

where R is the universal gas constant and is T the absolute temperature of the solution.

Example 14.8

Calculate the osmotic pressure of a 1% by mass sodium chloride solution at 20°C. The solution density is $1{,}007\,\mathrm{kg\,m^{-3}}$.

The molecular weight of sodium chloride is 58.5 and therefore the molar density of the solution, that is, the number of kmoles of sodium chloride per cubic metre, is

$$C = \frac{0.01 \times 1007}{58.5}\,\mathrm{kmol\,m^{-3}} \quad \text{or} \quad C = 0.172\,\mathrm{kmol\,m^{-3}}$$

Now using the van't Hoff relationship the osmotic pressure is

$$\Pi = 0.172 \times 8314 \times 293\,\mathrm{Pa}, \qquad \Pi = 4.19 \times 10^{5}\,\mathrm{Pa}$$

The accepted value for the osmotic pressure of a 1% sodium chloride solution is $8.6 \times 10^{5}\,\mathrm{Pa}$ and therefore the van't Hoff equation underestimates the osmotic pressure by a factor of two.

14.3.5. Ultrafiltration

Ultrafiltration is similar in principle to reverse osmosis but is designed to separate molecules of very much larger molecular weight, of the order of tens of thousand, and allows the passage of molecules at least up to 1,000 molecular weight into the permeate. Therefore, although in some ways the distinction between reverse osmosis and ultrafiltration is not clear cut, the applied pressure difference in ultrafiltration is required to overcome a very much smaller osmotic pressure difference than in reverse osmosis and consequently the operating pressures are of the order of only 1–5 bar. Membranes tend to be characterised by the size of the molecular species which is retained; thus a particular molecular weight cut off is specified. Molecules below this size pass into the permeate and larger molecules remain in the retentate.

Ultrafiltration finds application in the concentration of whole milk in cheese manufacture and in the removal of water from waste whey. It is the principle behind the membrane reactor in which an enzymic reaction proceeds with the enzyme retained behind a high molecular weight cut off membrane but the reaction products are allowed to leave the reactor with the permeate. Alternatively the enzyme may be immobilised in the porous structure of the membrane.

14.3.6. Membrane Properties and Structure

Membrane materials must be film-forming, hydrophilic and have wet strength. Early membranes were made from celluloses and polyamides but have now largely been replaced with polysulphones. More recent developments include the use of inorganic oxides, such as alumina, which give a narrower pore size distribution, and endothelial cells from animal cardiovascular systems mounted on a ceramic support. The self-renewal of these cells helps to prevent the build up of fouling layers. The membrane film is cast onto a support in the form of a sheet or tube and the membrane may be either symmetrical, in which the pores have parallel sides such that the cross section is constant along the pore length or asymmetrical in which the pore cross section increases from the feed side to the permeate side. This design prevents the membrane from being blocked by solute molecules.

Four membrane characteristics may be considered to be important: selectivity, resistance, cost, and permeate flux. First, any membrane must give the required degree of separation over long periods. The ability of a membrane to reject a solute can be expressed as a rejection factor R defined by

$$R = \frac{C_1 - C_2}{C_1} \tag{14.35}$$

where C_1 is the concentration of the species of interest in the feed stream and C_2 is the concentration of the same component in the permeate. Thus when the concentration in the permeate is zero $R = 1$ indicating 100% rejection by the membrane. If the membrane fails to prevent the passage of the relevant solute molecules the permeate concentration will rise so that it equals the feed concentration, that is, $C_1 = C_2$, and therefore $R = 0$. For reverse osmosis systems the rejection factor increases with increasing transmembrane pressure and it is therefore desirable to operate at the maximum economic pressure. For ultrafiltration the increase in R with molecular weight takes the form of the curve in Figure 14.19 and membranes may be designed with either a sharp or a diffuse molecular weight cut off. Second, membranes must be able to withstand high pressures (especially for reverse osmosis) and the resultant mechanical stress, moderate temperatures up to about 50°C; low pH and must also be able to resist microbiological effects.

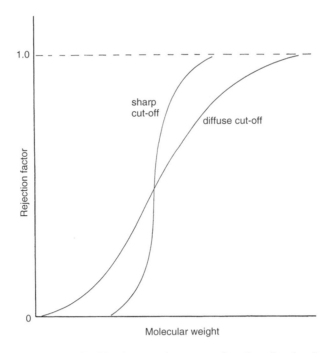

Figure 14.19. Selectivity of ultrafiltration membranes as a function of molecular weight cut-off.

Third, the capital and operating costs must be considered. Whilst membranes are relatively expensive, of the order of several hundreds of pounds per square metre, the total capital cost per unit volume of product is a more important cost indicator and may well be as low as other methods despite the high cost of membranes. Table 12.1 in chapter twelve shows the cost of membrane processes relative to other methods of water removal. It is important to realise that reverse osmosis and ultrafiltration do not incur thermal energy costs; operating costs are largely those associated with the pumping of liquid streams. Finally the permeate flux must be sufficient to give an adequate production rate. The prediction of permeate flux is considered in detail in section 14.3.8.

14.3.7. Membrane Configurations

There are four configurations or geometries which are widely used for industrial equipment: flat sheet systems, tubular membranes, spiral membranes, and hollow fibres.

(*a*) *Flat sheet membranes*   A flat sheet unit resembles a plate and frame filter press and consists of a series of alternate ribbed plates and hollow spacers with membranes placed on either side of each plate (Figure 14.20). The ribs on the plate surface radiate from the centre and form a series of flow channels for the feed stream/retentate. Permeate passes through the membrane and is then in contact with the surface of the spacer which contains a series of small apertures which allows the permeate to drain into the spacer. The plates/spacers are usually circular with a channel length of about 15 cm and a channel depth of 1 or 2 mm. The channel dimensions are chosen to maintain a constant velocity and therefore constant Reynolds number.

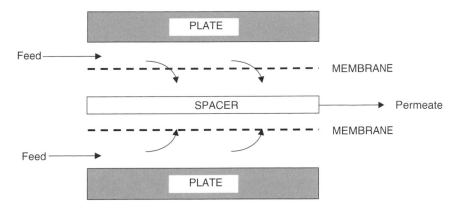

Figure 14.20. Plate and frame membrane configuration.

(*b*) *Tubular membranes* In these units the membrane is cast onto the inside surface of a porous tubular support. The feed solution passes through the tubes and the permeate diffuses through the membrane and the support material in a radial direction. Tubes are up to 2.5 cm in diameter for ultrafiltration but only 1.0 cm in diameter for reverse osmosis in order to withstand the higher pressures required. The tubes, usually about 1 m in length, are mounted in cylindrical cartridges.

(*c*) *Hollow fibre membranes* Hollow fibres are essentially very small diameter tubes, of the order of 25 μm, which are placed in bundles of about 100–150 m^2. Fibres have the advantage of a very high surface area but tend to foul more easily than other configurations. They are commonly used for water treatment.

(*d*) *Spiral membranes* Tubular membranes suffer from the disadvantage that, although they can withstand high pressures, the membrane area is limited. Spiral membranes overcome this and combine a larger area with high operating pressures. Five layers are placed together: a porous sheet which acts as a permeate collector; a layer of membrane; a layer of mesh which generates turbulence in the feed stream; a further membrane; and finally a further permeate collector. This composite sheet is wound into a spiral around a perforated steel tube, the ends of the layers are sealed and the spiral is placed in a cylindrical cartridge; the feed enters between membranes, axially. A membrane module is typically 10 or 15 cm in diameter and 1 m long containing up to 10 m^2 of membrane surface.

14.3.8. Permeate Flux

It might be expected that permeate flux would increase with increasing pressure difference and this is indeed the case until a critical pressure is reached at which the flux levels out (Figure 14.21). Second, at any given transmembrane pressure the flux decreases with time of operation. This decrease begins immediately and can fall to 50% of the original value (Figure 14.22). Three mechanisms may be responsible for these phenomena: fouling, compaction, and concentration polarisation.

Fouling is characterised by an irreversible decline in permeate flux due to the accumulation of macromolecular particles on the membrane surface or the crystallisation of smaller solutes

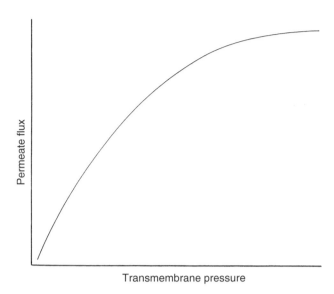

Figure 14.21. Permeate flux as a function of transmembrane pressure.

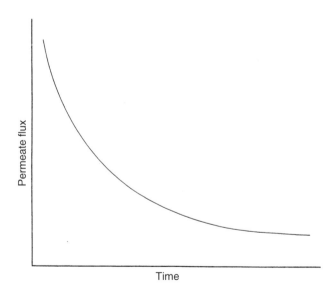

Figure 14.22. Permeate flux decline with time.

within the pores of the membrane. For example during the ultrafiltration of whey the flux increases with decreasing pH; at low pH the solubility of calcium phosphate increases and therefore becomes unavailable as a foulant. In the ultrafiltration of whole milk the concentration of protein within the membrane is also responsible for flux decline. The second possible mechanism, compaction, is due to the mechanical deformation of the membrane under applied pressure leading to mechanical damage of the membrane. At the moderate pressures employed in ultrafiltration the magnitude of compaction and fouling effects are comparable. However the higher pressures used in reverse osmosis increase the importance of compaction.

Probably the most important contribution to flux decline is the formation of a concentration polarisation layer. As solvent passes through the membrane the solute molecules which are unable to pass through the membrane become concentrated next to the membrane surface. Consequently the efficiency of separation decreases as this layer of concentrated solution accumulates. The layer is established within the first few seconds of operation and is an inevitable consequence of the separation of solvent and solute. The consequent decline in permeate flux may be attributed to this increase in concentration. First, the concentration polarisation layer presents a physical barrier and an increased resistance to the flow of solvent. Second, it is responsible for an increase in osmotic pressure on the feed side of the membrane which in turn decreases the driving force for the transfer of solvent. Third, it may also lead to membrane damage as large solute molecules are brought into closer contact with the membrane surface.

14.3.9. Prediction of Permeate Flux

The general membrane equation may now be expanded into the form

$$N = \frac{\Delta P - \Delta \Pi}{\mu (R_{\mathrm{m}} + R_{\mathrm{f}} + R_{\mathrm{p}})} \tag{14.36}$$

where the terms R_{m}, R_{f}, and R_{p} represent the resistance to the flow of permeate presented by the membrane, the fouling layer, and the concentration polarisation layer, respectively.

A fundamental difference between the models based on Equation (14.32) is the method of including the effects of concentration polarisation. In some models the osmotic pressure difference is omitted because the rejected solute molecules are thought to be too large to contribute a significant osmotic pressure whilst others dispense altogether with the term R_{p}. A simple permeability model can be formulated which assumes the resistance of the membrane to be dominant and expresses this as an effective mass transfer coefficient for the membrane k_{e}. Hence

$$N = k_{\mathrm{e}}(\Delta P - \Delta \Pi) \tag{14.37}$$

This effective mass transfer coefficient should not be confused with a liquid film coefficient. However, in the manner of the definition of the film mass transfer coefficient in Whitman's two-film theory, k_{e} may be equated to an effective diffusivity of water through the porous structure of the membrane D_{e} divided by the membrane thickness z. Thus

$$k_{\mathrm{e}} = \frac{D_{\mathrm{e}}}{z} \tag{14.38}$$

Example 14.9

Whey containing 6% solids by mass is to be concentrated to 12% solids in an ultrafiltration module at a feed rate of 4,300 kg h^{-1} and a temperature of 20°C. The permeate contains 0.5% solids, which may be assumed to be lactose. The effective diffusivity of water through the membrane, which is 100μm thick, is 5×10^{-7} m^2 s^{-1}. If the module operates with a transmembrane pressure of 900 kPa, and the osmotic pressure of the whey is 600 kPa, determine the necessary membrane surface area.

An overall material balance for the ultrafiltration module yields

$$4300 = R + P$$

where R and P are the mass flow rates of retentate and permeate, respectively. Substituting this into the component balance for whey solids gives

$$0.06 \times 4300 = 0.12(4300 - P) + 0.005P$$

from which

$$P = 2243.5 \, \text{kg h}^{-1}$$

This must be converted to a molar flow rate and hence, dividing by the molecular weight of water and changing the time base,

$$P = 0.0346 \, \text{kmol s}^{-1}$$

The permeate contains 0.5% lactose. This exerts an osmotic pressure which can be estimated from the van't Hoff relationship. Thus the molar concentration of the permeate C_2 is

$$C_2 = \frac{0.005 \times 1000}{342} \, \text{kmol m}^{-3}$$

and therefore the osmotic pressure Π_2, assuming the permeate density to be $1{,}000 \, \text{kg m}^{-3}$, is

$$\Pi_2 = \left(\frac{0.005 \times 1000}{342} \right) 8314 \times 293 \, \text{Pa}, \qquad \Pi_2 = 3.56 \times 10^4 \, \text{Pa}$$

The osmotic pressure of the whey, the feed stream, is given as $600 \, \text{kPa}$ and therefore the osmotic pressure difference $\Delta\Pi$ becomes $(600 \times 10^3 - 3.56 \times 10^4) \, \text{Pa}$. The permeate flux can be obtained by using the permeability model [Equation (14.37)]. The effective mass transfer coefficient for the transport of water through the membrane is given by the effective diffusivity of water divided by the membrane thickness. However in order for k_e to have units of $\text{kmol N}^{-1} \text{s}^{-1}$ so that the permeate flux is in $\text{kmol m}^{-2} \text{s}^{-1}$ the coefficient must be defined by

$$k_e = \frac{D_e}{zRT}$$

The permeate flux is now

$$N = \frac{5 \times 10^{-7}}{10^{-4} \times 8314 \times 293} [9 \times 10^5 - (600 \times 10^3 - 3.56 \times 10^4)] \, \text{kmol m}^{-2} \text{s}^{-1}$$

and

$$N = 6.89 \times 10^{-4} \, \text{kmol m}^{-2} \text{s}^{-1}$$

The required membrane area A is equal to the permeate flow rate divided by the flux, so that

$$A = \frac{0.0346}{6.89 \times 10^{-4}} \, \text{m}^2, \qquad A = 50.2 \, \text{m}^2$$

Example 14.10

A reverse osmosis unit is to be used to concentrate tomato juice at $25°C$ from 4.5% solids by mass to 8% solids, with an output of $1,285\,kg\,h^{-1}$. The solids content of the feed juice can be assumed to have a mean molecular weight of 234; the juice density is $1,045\,kg\,m^{-3}$. The osmotic pressure of the concentrate is estimated to be 12 bar and the unit operates with a transmembrane pressure of 30 bar. Assuming an arithmetic average osmotic pressure for the feed/retentate, and the permeate to be pure water, show that a membrane area of $72\,m^2$ is sufficient to obtain the required concentration. The membrane is $100\,\mu m$ thick and the effective diffusivity of water is $2.5 \times 10^{-8}\,m^2\,s^{-1}$.

The component balance for the solids in the juice is

$$0.045\,F = 0.08 \times 1285$$

from which the feed flow rate is $F = 2284.4\,kg\,h^{-1}$. Consequently, from the overall material balance, the mass flow rate of permeate is $1,000\,kg\,h^{-1}$. The osmotic pressure of the feed, from van't Hoff, is

$$\Pi_F = \left(\frac{0.045 \times 1045}{234}\right) 8314 \times 298\,Pa \quad \text{or} \quad \Pi_F = 4.98 \times 10^5\,Pa$$

Assuming an arithmetic average osmotic pressure for feed and retentate, Π_1 is then $8.49 \times 10^5\,Pa$. The permeate is pure water and has an osmotic pressure of zero and therefore the permeate flux is given by

$$N = \frac{2.5 \times 10^{-8}}{10^{-4} \times 8314 \times 298}[30 \times 10^5 - 8.49 \times 10^5]\,kmol\,m^2\,s^{-1}$$

and $N = 2.17 \times 10^{-4}\,kmol\,m^{-2}\,s^{-1}$. For a membrane area of $72\,m^2$, the permeate flow rate is $2.17 \times 10^{-4} \times 72 \times 3600 \times 18 = 1012.7\,kg\,h^{-1}$. Thus the membrane area is adequate.

Some studies have taken into account only concentration polarisation. For ultrafiltration especially, concentration polarisation strongly influences the flux of solvent and dominates the membrane characteristics with the consequence that mass transfer in the feed stream is the important design consideration, not the membrane characteristics. Thus, the so-called film model assumes that the bulk transport of solute towards the membrane surface is balanced at steady-state by back diffusion of solute away from the membrane. This concept may be understood by referring to Figure 14.18.

The thickness of the concentration polarisation layer may become constant because of two opposing factors. First, the convective mass transfer of solute towards the membrane which is due to the bulk motion of the permeate. This is equal to NC_S/C_W where C_S and C_W are the concentrations of solute and solvent, respectively. Second, the build up of large solute molecules near the membrane surface produces a concentration gradient with a high concentration of solute near the membrane surface and a lower concentration in the bulk feed stream. Therefore back diffusion of solute away from the membrane takes place and this flux is equal to $-D(dC_S/dz)$ where D is the diffusivity of the rejected solute in the feed solution. At equilibrium therefore

$$N\frac{C_S}{C_W} = -D\frac{dC_S}{dz} \tag{14.39}$$

and this differential equation can be integrated across the membrane thickness z and between the solute concentration in the feed stream C_1 and that at the membrane surface C_m. Hence

$$N \int_{0}^{z} dz = -D \int_{C_m}^{C_1} \frac{C_W \, dC_S}{C_S} \qquad (14.40)$$

which results in

$$N = \frac{D}{z} \ln \left(\frac{C_m}{C_1} \right) \qquad (14.41)$$

As in the classic Whitman two-film theory which was covered in chapter eight, a liquid film mass transfer coefficient k can be substituted for D/z and therefore

$$N = k \ln \left(\frac{C_m}{C_1} \right) \qquad (14.42)$$

Equation (14.42) emphasises the importance of mass transfer in the feed stream. The permeate flux is clearly proportional to the film mass transfer coefficient which characterises the feed stream. Consequently the permeate flux can be maximised by maximising the degree of turbulence in the feed stream which has the effect of reducing the thickness of the concentration polarisation layer. Turbulent flow can be achieved by careful design of the channel dimensions through which the feed stream flows and by maintaining an adequate feed velocity. For tubular membrane systems static mixers or baffles can be employed to create turbulence, for example by placing spheres or rods in the flow channel. For flat sheet or spiral systems a mesh sheet placed between separate membranes sheets has a similar effect. Another technique which is commonly employed is to use oscillating flow. The feed flow is pulsed and each time the flow is stopped the concentration polarisation layer is dispersed. At the instant that flow is restarted the permeate flux returns to its original level only to decline once more. However the average flux obtained is greater than if flux decline is allowed to take its natural course.

Because Equation (14.42) contains no pressure term, the gel polarisation model was advanced to explain the pressure dependence of permeate flux decline. Here it is assumed that concentration polarisation increases with pressure to the point where gelation of the solute occurs leading to increased resistance (as the gel hardens and thickens) and to flux decline. Therefore Equation (14.42) becomes

$$N = k \ln \left(\frac{C_g}{C_1} \right) \qquad (14.43)$$

where C_g is the concentration of solute in the gel layer. A number of limiting flux theories have been suggested which are based on concentration polarisation and which dispense with the concept of a gel layer. For example it has been suggested that there is a dynamic interaction between the membrane and rejected solute in which pore blockage and back diffusion to clear the pores are in equilibrium. The limiting flux with increasing transmembrane pressure arises when the pores are blocked for most of the time.

Many researchers argue that flux decline can best be predicted by the use of semi-empirical models which describe the flux decline curve rather than attempt a mechanistic explanation. Thus Cheryan proposed that because flux decline is due to fouling, and is therefore a function of the cumulative volume of permeate which has passed through the membrane, a simple logarithmic expression may be used

$$N = N_0 V^{-b} \tag{14.44}$$

where N_0 is the initial flux, N is the flux after a volume V of permeate has passed through the membrane and b is a parameter obtained from the gradient of a plot of $\ln(N/N_0)$ against V.

14.3.10. Some Applications of Membrane Technology

(a) *Tomato juice concentration*   As a method of concentrating fruit and vegetable juices, reverse osmosis is more economical than conventional multiple-effect evaporation if the average permeate flux is greater than about $15 \, l \, m^{-2} \, h^{-1}$. There is considerable interest in applying reverse osmosis especially to the concentration of tomato juice. As well as energy savings, reverse osmosis results in improved colour and flavour with none of the browning normally associated with evaporation. There is also excellent retention of organic acids, sugars, free amino acids, and mineral ions.

Tomato juice typically contains about 94.7% water and 5.3% dry solids. The solids content comprises about 1.9% sugars (glucose and fructose), 1.6% dietary fibre, 1.0% protein, 0.5% organic acids, 0.2% potassium, and 0.1% fat. The dietary fibre (pectin and cellulose) consists of insoluble aggregates of tomato cells in suspension which bind many times their own weight in water; the suspended solids are separable by high speed centrifugation. The liquid portion or serum represents about 80% by mass of the original juice and contains about 4.8% total dry solids, which are almost all soluble.

Tomato juice concentration by RO is difficult because of the high pulp content and a higher osmotic pressure than most other juices. Tubular membrane systems are able to handle the suspended solids but the maximum tomato juice concentration attainable in this configuration is about 20°Brix at 70 bar; however this is well below the value of 36–37°Brix required by producers of tomato juice concentrates. Other configurations (plate and frame, spiral-wound, and hollow fibre) have small flow channels which are plugged by the suspended solids in whole tomato juice; thus they are only suitable to concentrate tomato serum. A very recent development is the concept of a high retention-low retention reverse osmosis process which combines membranes with different solute rejection and allows juices to be concentrated to much higher levels than with single membrane systems. This is based on the principle that the osmotic pressure difference can be reduced if the solution to be concentrated is fed through a low retention (LR) membrane which allows a greater quantity of solute to pass through in the permeate. The lost solute is recovered by passing the permeate to a more selective high retention (HR) membrane. A schematic diagram of such a process is shown in Figure 14.23.

The insoluble fibre in whole tomato juice is separated by centrifugation in a continuous centrifuge. Ultrafiltration membranes with a nominal molecular weight cut off between 20,000 and 200,000 then separate the soluble pectin, protein and contaminating bacteria, yeasts and moulds from the serum. The tomato fibre and the ultrafiltration retentate are combined and pasteurised in a scraped-surface heat exchanger. The ultrafiltration permeate, comprising glucose, fructose, organic acids, salts, and some soluble pectin is pre-concentrated in a hollow fibre high retention reverse osmosis module. The permeate from this part of the process is water. The HR retentate comprising glucose, fructose, organic acids, salts, and some soluble pectin is applied to a hollow LR reverse osmosis module. Low retention means that part of the

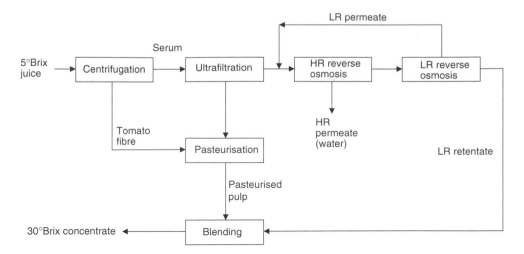

Figure 14.23. Low retention-high retention membrane system.

solute (except the soluble pectin which will be completely retained) passes the membrane and reduces the osmotic pressure that has to be overcome to effect concentration. These solutes are not lost since the LR permeate is recycled to the HR module. With the right combination of LR and HR membranes, it is possible to concentrate solutions to osmotic pressures higher than the applied hydraulic pressure. Finally the pasteurised pulp and concentrated serum are blended to give a concentrated product.

(*b*) *Treatment of waste whey*   Every year some two thousand million litres of whey are produced by the U.K. dairy industry of which the bulk is waste. This waste is a potential source of high quality protein and carbohydrate but the main problem in the processing of whey is the high cost associated with the removal of water. However ultrafiltration offers an efficient and relatively inexpensive recovery method. The value of whey lies in its protein and lactose content; whey proteins have a nutritional value similar to that of whole egg in terms of essential amino acid composition. A large proportion of the whey produced by the dairy industry is used for animal feed but has the potential to be used for human consumption. Yields from whey have always been low because the water content is approximately 94% with about 0.6% protein and 4.5% lactose. Ultrafiltration provides a method of separating and concentrating whey proteins without sacrificing functional properties and is now commercially viable. Following concentration, membrane reactor systems or immobilised enzyme systems can be used to convert the product into a form suitable for use in the food industry. Alternatively evaporation can be used to remove further water or spray drying used to give a solid whey protein concentrate for use, for example, in ice cream manufacture. The use of ultrafiltration in this way greatly reduces the thermal load in evaporation and spray drying.

(*c*) *Cheese manufacture*   Ultrafiltration is commonly used to concentrate milk prior to addition of the starter culture in the manufacture of soft cheeses. This has the advantage of increasing the production rate. Later in the process, in the separation of curds and whey, the loss of fat and protein in the whey can be prevented by using ultrafiltration membranes with a molecular weight cut off of about 30,000.

NOMENCLATURE

a	Coefficient
A	Area
b	Parameter in Equation (14.44)
B	Baffle width
C	Molar concentration
C_1	Feed concentration
C_2	Permeate concentration
C_g	Concentration of solute in gel layer
C_S	Solute concentration
C_W	Solvent concentration
D	Impeller diameter; diffusivity
D_e	Effective diffusivity
D_T	Vessel diameter
Fr	Froude number
g	Acceleration due to gravity
H	Depth of liquid
k	Rate constant
k_e	Effective mass transfer coefficient
K	Consistency coefficient; Kozeny's constant
L	Depth of filter cake
L_0	Equivalent bed depth of filter medium
M	Mixing index
n	Population or number of samples; flow behaviour index; number of moles
N	Rotation speed; permeate flux
N_0	Initial permeate flux
N_p	Power number
P	Power consumption
p	Proportion (fraction)
R	Mass transfer resistance; universal gas constant; rejection factor
Re	Reynolds number
s	Standard deviation
s_0	Initial (unmixed) standard deviation
s_∞	Limiting value of standard deviation
s^2	Variance
s_∞^2	Limiting value of variance
s_0^2	Initial (unmixed) variance
S	Specific surface
t	Time
T	Absolute temperature
u	Superficial velocity
V	Volume; volume of filtrate
W	Impeller blade depth
x	Index
x_i	Local concentration

$\bar{x}$ Average concentration
y Index
z Membrane thickness
Z Height of impeller from vessel bottom

GREEK SYMBOLS

α Specific cake resistance
$\dot{\gamma}$ Shear rate
ΔP Pressure drop across filter press; transmembrane pressure
$\Delta \Pi$ Osmotic pressure difference
ε Voidage
μ Viscosity
μ_a Apparent viscosity
ν Volume fraction; volume of cake deposited per unit volume of filtrate
Π Osmotic pressure
ρ Density

SUBSCRIPTS

1   Feed
2   Permeate
f   Fouling
m   Membrane
p   Concentration polarisation

PROBLEMS

14.1. A minor component is to be added to a powdered food product at 4% by mass. The product specification requires that there be a 99% probability that each 250 g packet contains between 9.5 and 10.5 g of the additive. The variance in the mass fraction of the minor component as a function of time was determined by removing samples from the mixer. Determine the mixing time from the following data:

Time (min)	Variance
1	0.0295
3	0.0271
4	0.0253
7	0.0192
9	0.0131

14.2. An aqueous gum solution using *Xanthomonas campestris* is to be mixed in a standard baffled vessel of 0.9 m diameter using a Rushton turbine at 420 rpm. The xanthan gum may be assumed to be a power law fluid with a flow behaviour index of 0.50, a consistency coefficient of 30.0 Pa s^n and a density of 1,100 kg m^{-3}. Calculate the power drawn by the impeller.

14.3. A preservative is to be mixed into the gum in Problem 14.2 at a concentration of 0.5%. Determine the mixing time to ensure, with 95% confidence, that any sample contains the specified quantity of preservative to ±3%. Samples taken from the vessel and analysed for preservative generated the following data:

Time (min)	Variance
1	2.00×10^{-3}
2	8.06×10^{-4}
3.5	2.06×10^{-4}
6	2.11×10^{-5}

14.4. What will be the effect on power consumption of removing the baffles from the vessel in Example 14.4?

14.5. A large filter press operated at constant pressure gave 100 l of filtrate after 28 s and a further 100 l after a further 44 s. The filter is full when 1,500 l of filtrate have been collected. Calculate the time taken to fill the filter and the final filtration rate.

14.6. A rotary vacuum filter is to be specified for handling a food suspension containing 30% solids. Preliminary tests with a small leaf filter of area $250 \, \text{m}^2$, working at an absolute pressure of 21.3 kPa, yielded 5 l of filtrate after 193 s and 15 l after 1,643 s. Determine the specific cake resistance and the equivalent bed depth of the filter medium if the moisture content of the cake was 10% (dry basis) and the cake voidage 0.4. The rotary vacuum filter is 1.2 m long, has a diameter of 0.8 m and operates at an absolute pressure of 21.3 kPa with 40% of the drum surface immersed in the suspension. If it uses the same filter cloth as the leaf filter and rotates at 0.24 rpm, determine the filtrate production rate and the cake thickness. The solids density and filtrate density are 2,500 and $1,000 \, \text{kg m}^{-3}$, respectively, and the filtrate viscosity is 0.0012 Pa s.

14.7. Determine the membrane area required to remove $125 \, \text{kg h}^{-1}$ of water from a 5% by weight protein solution (average molecular weight of protein $= 22,000$, solution density $= 1,050 \, \text{kg m}^{-3}$) using an ultrafiltration membrane operated at 5 °C and with a mean transmembrane pressure of 500 kPa. The membrane is 90 μm thick and the effective diffusivity of water through the membrane is $8 \times 10^{-8} \, \text{m}^2 \, \text{s}^{-1}$.

14.8. Water is to be removed at a rate of $60 \, \text{kg min}^{-1}$ from a 7.5% (by mass) solution of lactose at 15 °C. A reverse osmosis unit is available which is capable of a maximum transmembrane pressure difference of 60 bar. The membranes to be used are 80 μm thick and the effective diffusivity of water through the membrane is estimated to be $4.9 \times 10^{-8} \, \text{m}^2 \, \text{s}^{-1}$. Determine the membrane area required using a simple permeability model and assuming that the permeate is pure water. The density of a 7.5% lactose solution is $1,028 \, \text{kg m}^{-3}$.

FURTHER READING

M. Cheryan, *Ultrafiltration Handbook*, Technomic (1980).

F. A. Glover, *Ultrafiltration and Reverse Osmosis for the Dairy Industry*, NIRD Technical Bulletin No. 5 (1985).

N. Harnby, M. F. Edwards, and A. W. Nienow, *Mixing in the Process Industries*, Butterworth-Heinemann (1992).

E. Renner and M. H. Abd El-Salam, *Application of Ultrafiltration in the Dairy Industry*, Elsevier, Amsterdam (1991).

K. Scott and R. Hughes, *Industrial Membrane Separation Technology*, Blackie (1996).

V. W. Uhl and J. B. Gray, *Mixing: Theory and Practice*, Academic Press, New York (1966).

Appendix A

(A) LIST OF UNIT PREFIXES

Factor	Prefix	Abbreviation
10^{18}	Exa	E
10^{15}	Peta	P
10^{12}	Tera	T
10^{9}	Giga	G
10^{6}	Mega	M
10^{3}	Kilo	k
10^{2}	Hecto	h
10^{1}	Deca	da
10^{-1}	Deci	d
10^{-2}	Centi	c
10^{-3}	Milli	m
10^{-6}	Micro	μ
10^{-9}	Nano	n
10^{-12}	Pico	p
10^{-15}	Femto	f
10^{-18}	Atto	a

(B) GREEK ALPHABET

	Lower case	Upper case
Alpha	α	A
Beta	β	B
Gamma	γ	Γ
Delta	δ	Δ
Epsilon	ε	E
Zeta	ζ	Z
Eta	η	H
Theta	θ	Θ
Iota	ι	I
Kappa	κ	K
Lamda	λ	Λ
Mu	μ	M
Nu	ν	N
Xi	ξ	Ξ
Omicron	o	O
Pi	π	Π
Rho	ρ	P
Sigma	σ	Σ
Tau	τ	T
Upsilon	υ	Y
Phi	ϕ	Φ
Chi	χ	X
Psi	ψ	Ψ
Omega	ω	Ω

Appendix B

(A) FUNDAMENTAL AND DERIVED SI UNITS

Quantity	SI unit
Length	m
Mass	kg
Time	s
Temperature	K
Electric current	A
Electromotive force	V
Electrical resistance	Ω
Molar mass	kmol
Velocity	$\mathrm{m\,s^{-1}}$
Acceleration	$\mathrm{m\,s^{-2}}$
Force	N
Pressure	Pa
Volume	$\mathrm{m^3}$
Density	$\mathrm{kg\,m^{-3}}$
Viscosity	Pa s
Mass flow	$\mathrm{kg\,s^{-1}}$
Volumetric flow	$\mathrm{m^3\,s^{-1}}$
Work/energy	J
Latent heat	$\mathrm{J\,kg^{-1}}$
Power	W
Heat capacity	$\mathrm{J\,kg^{-1}\,K^{-1}}$
Thermal conductivity	$\mathrm{W\,m^{-1}\,K^{-1}}$
Heat transfer coefficient	$\mathrm{W\,m^{-2}\,K^{-1}}$

(B) CONVERSION FACTORS

Quantity		SI conversion
Length	ft	$0.3048\,\mathrm{m}$
Mass	lb	$0.4536\,\mathrm{kg}$
	tonne	$1000\,\mathrm{kg}$
Temperature difference	°F	$0.5556\,\mathrm{K}$
Temperature scale	°F	$1.8(°C) + 32$
	°C	$K - 273.16$
	°R	$°F + 460$
Volume	litre	$10^3\,\mathrm{m}^3$
	cm^3	$10^{-6}\,\mathrm{m}^3$
	gallon (Imperial)	$4.546 \times 10^{-3}\,\mathrm{m}^3$
	gallon (US)	$3.785 \times 10^{-3}\,\mathrm{m}^3$
Molar mass	gmole	$10^{-3}\,\mathrm{kmol}$
Pressure	bar	$10^5\,\mathrm{Pa}$
	$\mathrm{lb/in}^2$	$6.895\,\mathrm{kPa}$
	$\mathrm{kg/cm}^2$	$98.067\,\mathrm{kPa}$
	atmosphere	$101.325\,\mathrm{kPa}$
	mm mercury	$133.32\,\mathrm{Pa}$
	in water	$249.1\,\mathrm{Pa}$
Work/energy	calorie	$4.1868\,\mathrm{J}$
	BTU	$1055\,\mathrm{J}$
Power	hp	$745.7\,\mathrm{W}$
	BTU/hr	$0.293\,\mathrm{W}$
Density	$\mathrm{g/cm}^3$	$10^3\,\mathrm{kg\,m}^{-3}$
	$\mathrm{lb/ft}^3$	$16.02\,\mathrm{kg\,m}^{-3}$
Viscosity	poise	$0.1\,\mathrm{Pa\,s}$
	lb/(ft h)	$4.13 \times 10^{-4}\,\mathrm{Pa\,s}$
Heat capacity	BTU/(lb °F)	$4.187\,\mathrm{kJ\,kg}^{-1}\,\mathrm{K}^{-1}$
Latent heat	BTU/lb	$2.326 \times 10^3\,\mathrm{J\,kg}^{-1}$
Thermal conductivity	BTU/(hr ft^2 °F/ft)	$1.731\,\mathrm{W\,m}^{-1}\,\mathrm{K}^{-1}$
Heat transfer coefficient	BTU/(hr ft^2 °F)	$5.678\,\mathrm{W\,m}^{-2}\,\mathrm{K}^{-1}$

Appendix C

DERIVATION OF A DIMENSIONLESS CORRELATION FOR FILM HEAT TRANSFER COEFFICIENTS

In convective heat transfer the heat flux is a function of a number of process variables and physical properties: the velocity of the fluid to, or from, which heat is transferred, a length which is characteristic of the geometry (e.g., a pipe diameter), the density, viscosity, thermal conductivity and heat capacity of the fluid, the temperature difference or driving force and the product of the coefficient of thermal expansion and acceleration due to gravity which is important in buoyancy effects responsible for natural convection. Each of these quantities can be broken down into their fundamental dimensions. It is convenient to include heat as a fundamental dimension. These are listed in Table C1.

TABLE C1
Fundamental Dimensions and Variables

	Symbol	Dimensions
(a) Fundamental dimensions		
Heat	Q	
Mass	M	
Length	L	
Time	T	
Temperature	Θ	
(b) Variables		
Heat flux	q	$QL^{-2}T^{-1}$
Velocity	u	LT^{-1}
Characteristic length	L	L
Density	ρ	ML^{-3}
Viscosity	μ	$ML^{-1}T^{-1}$
Heat capacity	c_p	$QM^{-1}\Theta^{-1}$
Temperature difference	ΔT	Θ
Thermal conductivity	k	$QT^{-1}L^{-1}\Theta^{-1}$
Coefficient of thermal expansion $\times$ acceleration due to gravity	βg	$LT^{-2}\Theta^{-1}$

Thus the general relationship may be written as

$$q = f\left(u^a L^b \rho^c \mu^d c_p^w \Delta T^x k^y (\beta g)^z\right) \tag{C1}$$

where a, b, c, etc. are the arbitrary indices. Now Equation (C1) must be dimensionally consistent so that, for example, if length appears as a squared term on the left-hand side then the sum of the indices on the right-hand side must also equal 2. Now balancing the indices for each dimension in turn gives

for heat:	$1 = w + y$	(C2)
for mass:	$0 = c + d - w$	(C3)
for length:	$-2 = a + b - 3c - d - y + z$	(C4)
for time:	$-1 = -a - d - y - 2z$	(C5)
for temperature:	$0 = -w + x - y - z$	(C6)

There are now five equations and eight variables (i.e., indices) and Equations (C2)–(C6) must be solved in terms of three of these. It is appropriate to select a, w, and z. Hence Equation (C2) becomes

$$\underline{y = 1 - w}$$

Equation (C6) becomes

$$x = z + w + y$$
$$x = z + w + (1 - w)$$
$$\underline{x = z + 1}$$

Equation (C3) gives

$$0 = c + d - w$$
$$\underline{d = w - c}$$

Equation (C5) becomes

$$d = 1 - a - y - 2z$$
$$w - c = 1 - a - y - 2z$$
$$c = a - y + y + 2z$$
$$\underline{c = a + 2z}$$

From Equation (C5), Equation (C3) gives

$$\underline{d = w - a - 2z}$$

and Equation (C4) becomes

$$b = -2 - a + 3c + d + y - z$$
$$b = -2 - a + 3(a + 2z) + (w - a - 2z) + (1 - w) - z$$
$$\underline{b = -1 + a + 3z}$$

Substituting these results into Equation (C1) gives

$$q = f\left[\frac{\left(u^a L^a L^{3z} \rho^c \rho^{2z} \mu^w c_p^w \Delta T^z \Delta T k(\beta g)^z\right)}{L \mu^a \mu^{2z} k^w}\right] \tag{C7}$$

The variables are now rearranged so that they form recognisable dimensionless groups, thus

$$q = f\left[\left(\frac{\Delta T\, k}{L}\right)\left(\frac{\rho u L}{\mu}\right)^a \left(\frac{c_p \mu}{k}\right)^w \left(\frac{\beta g\, \Delta T\, L^3 \rho^2}{\mu^2}\right)^z\right] \tag{C8}$$

and

$$\frac{q L}{k \Delta T} = f\left[\left(\frac{\rho u L}{\mu}\right)^a \left(\frac{c_p \mu}{k}\right)^w \left(\frac{\beta g\, \Delta T\, L^3 \rho^2}{\mu^2}\right)^z\right] \tag{C9}$$

However the film heat transfer coefficient h is defined by

$$q = h\,\Delta T \tag{C10}$$

and therefore

$$\frac{q L}{k\,\Delta T} = \frac{h L}{k} \tag{C11}$$

Now, substituting for the Nusselt, Reynolds, Prandtl and Grashof numbers, Equation (C9) becomes

$$Nu = f(Re^a\, Pr^w\, Gr^z) \tag{C12}$$

Appendix D

PROPERTIES OF SATURATED WATER AND WATER VAPOUR

T (°C)	p' (kPa)	h_f (kJ kg^{-1})	h_{fg} (kJ kg^{-1})	h_g (kJ kg^{-1})
0.01	0.6112	0	2500.8	2500.8
25	3.166	104.8	2441.8	2546.6
50	12.33	209.3	2382.1	2591.4
60	19.92	251.1	2357.9	2609.0
70	31.16	293.0	2333.3	2626.3
75	38.55	313.9	2320.8	2634.7
80	47.36	334.9	2308.3	2643.2
85	57.80	355.9	2295.6	2651.5
90	70.11	376.9	2282.8	2659.7
95	84.53	398.0	2269.8	2667.8
100	101.325	419.1	2256.7	2675.8

p' (kPa)	T (°C)	h_f (kJ kg^{-1})	h_{fg} (kJ kg^{-1})	h_g (kJ kg^{-1})
0.6112	0.01	0	2501	2501
10	45.8	192.0	2392	2584
20	60.1	251	2358	2609
50	81.3	340	2305	2645
60	86.0	360	2293	2653
70	90.0	377	2283	2660
80	93.5	392	2273	2665
90	96.7	405	2266	2671
100	99.6	417	2258	2675
150	111.4	467	2226	2693
200	120.2	505	2202	2707
250	127.4	535	2182	2717
300	133.5	561	2164	2725
350	138.9	584	2148	2732
400	143.6	605	2134	2739
450	147.9	623	2121	2744
500	151.8	640	2109	2749

Enthalpy of superheated steam (kJ kg^{-1})

p' (kPa)	T (°C)			
	150	200	300	400
50	2780	2878	3076	3279
100	2777	2876	3075	3278
150	2773	2873	3073	3277
200	2770	2871	3072	3277
300	2762	2866	3070	3275

h_f enthalpy of saturated water.
h_g enthalpy of saturated water vapour.
h_{fg} enthalpy of vaporisation of water.
p' vapour pressure (saturation pressure).
T temperature.

Appendix E

DERIVATION OF LOGARITHMIC MEAN TEMPERATURE DIFFERENCE

Consider a finite element of heat transfer surface dA. The rate of heat transfer between 'hot' and 'cold' fluids is then

$$dQ = U \, dA(T_h - T_c) \qquad \text{(E1)}$$

The heat gained by the 'cold' stream equals that lost by the 'hot' stream and so

$$dQ = m_c c_{p_c} \, dT_c \qquad \text{(E2)}$$

and

$$dQ = -m_h c_{p_h} \, dT_h \qquad \text{(E3)}$$

Rearranging Equations (E2) and (E3) gives

$$dT_c = \frac{dQ}{m_c c_{p_c}} \qquad \text{(E4)}$$

and

$$dT_h = \frac{-dQ}{m_h c_{p_h}} \qquad \text{(E5)}$$

Therefore the differential temperature difference becomes

$$d(T_h - T_c) = -dQ \left[\frac{1}{m_h c_{p_h}} + \frac{1}{m_c c_{p_c}} \right] \qquad \text{(E6)}$$

which on eliminating dQ gives

$$\frac{d(T_h - T_c)}{(T_h - T_c)} = -U \, dA \left[\frac{1}{m_h c_{p_h}} + \frac{1}{m_c c_{p_c}} \right] \qquad \text{(E7)}$$

Now separating the variables and integrating ΔT between ΔT_1 and ΔT_2 and A between 0 and A, gives

$$\int_1^2 \frac{d(T_h - T_c)}{(T_h - T_c)} = -U \left[\frac{1}{m_h c_{p_c}} + \frac{1}{m_c c_{p_h}} \right] \int_0^A dA \qquad \text{(E8)}$$

and

$$\ln\left[(T_h - T_c)\right]_1^2 = -U\left[\frac{1}{m_h c_{p_h}} + \frac{1}{m_c c_{p_c}}\right][A]_0^A \tag{E9}$$

Now on substituting from the enthalpy balance for each fluid

$$m_h c_{p_h} = \frac{Q}{(T_{h_i} - T_{h_o})} \tag{E10}$$

and

$$m_c c_{p_c} = \frac{Q}{(T_{c_o} - T_{c_i})} \tag{E11}$$

and on rearrangement this results in

$$\ln\left(\frac{\Delta T_2}{\Delta T_1}\right) = -U A\left[\frac{(T_{h_i} - T_{h_o})}{Q} + \frac{(T_{c_o} - T_{c_i})}{Q}\right] \tag{E12}$$

When rearranged to give the rate of heat transfer this now falls into the form

$$Q = \frac{-U A\left[(T_{h_i} - T_{h_o}) + (T_{c_o} - T_{c_i})\right]}{\ln(\Delta T_2/\Delta T_1)} \tag{E13}$$

or

$$Q = \frac{U A(\Delta T_2 - \Delta T_1)}{\ln(\Delta T_2/\Delta T_1)} \tag{E14}$$

which is of course

$$Q = U A \, \Delta T_{lm} \tag{E15}$$

where

$$\Delta T_{lm} = \frac{\Delta T_1 - \Delta T_2}{\ln(\Delta T_1/\Delta T_2)} \tag{E16}$$

Appendix F

DERIVATION OF FOURIER'S FIRST LAW OF CONDUCTION

Consider the solid isotropic block depicted in Figure F1. Heat may be conducted through the block in each of three orthogonal directions x, y, and z. The rate at which heat flows into the block plus the generation of heat within the block must equal the sum of the rate at which heat leaves and the rate of accumulation of heat, the latter giving rise to the change in temperature of the solid with time. Thus

$$\text{rate of input} + \text{rate of generation} = \text{rate of output} + \text{rate of accumulation} \tag{F1}$$

Considering for the moment only heat conducted in the x direction, the rate of heat transfer at x is given by

$$Q_x = -k(\Delta y\,\Delta z)\left(\frac{\partial T}{\partial x}\right)_x \tag{F2}$$

where the partial derivative in Equation (F2) acknowledges that heat is transferred in the y and z directions also. Now if heat is generated at a rate of $\dot{Q}$ W m^{-3} then

$$\text{rate of generation} = \dot{Q}(\Delta x\,\Delta y\,\Delta z) \tag{F3}$$

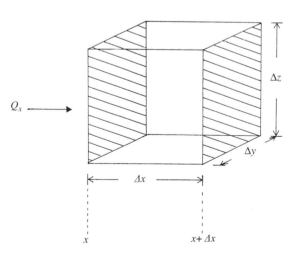

Figure F1. Conduction in three dimensions through an isotropic block.

The rate of heat transfer at $x + \Delta x$ depends upon the local temperature gradient $(\partial T/\partial x)_{x+\Delta x}$

$$Q_{x+\Delta x} = -k(\Delta y \, \Delta z)\left(\frac{\partial T}{\partial x}\right)_{x+\Delta x} \tag{F4}$$

and the rate of heat accumulation is

$$\text{rate of accumulation} = \rho(\Delta x \, \Delta y \, \Delta z)c_\text{p}\left(\frac{\partial T}{\partial t}\right) \tag{F5}$$

where ρ is the density of the solid. Combining Equations (F1)–(F5) and dividing by the volume of the block ($\Delta x \, \Delta y \, \Delta z$) gives

$$\frac{-k(\partial T/\partial x)_x}{\Delta x} + \dot{Q} = \frac{-k(\partial T/\partial x)_{x+\Delta x}}{\Delta x} + \rho c_\text{p}\left(\frac{\partial T}{\partial t}\right) \tag{F6}$$

or

$$\rho c_\text{p}\left(\frac{\partial T}{\partial t}\right) = \frac{k(\partial T/\partial x)_{x+\Delta x} - k(\partial T/\partial x)_x}{\Delta x} + \dot{Q} \tag{F7}$$

In the limit, as Δx approaches zero, this becomes

$$\rho c_\text{p}\left(\frac{\partial T}{\partial t}\right) = k\left(\frac{\partial^2 T}{\partial x^2}\right) + \dot{Q} \tag{F8}$$

which, when dividing through by ρc_p, becomes

$$\left(\frac{\partial T}{\partial t}\right) = \alpha\left(\frac{\partial^2 T}{\partial x^2}\right) + \frac{\dot{Q}}{\rho c_\text{p}} \tag{F9}$$

where α is the thermal diffusivity defined by Equation (F10)

$$\alpha = \frac{k}{\rho c_\text{p}} \tag{F10}$$

If now the flow of heat in the y and z directions is included then

$$Q_y = -k(\Delta x \, \Delta z)\left(\frac{\partial T}{\partial x}\right)_y \tag{F11}$$

and

$$Q_z = -k(\Delta x \, \Delta y)\left(\frac{\partial T}{\partial x}\right)_z \tag{F12}$$

and Equation (F9) becomes

$$\frac{\partial T}{\partial t} = \alpha\left(\frac{\partial^2 T}{\partial x^2} + \frac{\partial^2 T}{\partial y^2} + \frac{\partial^2 T}{\partial z^2}\right) + \frac{\dot{Q}}{\rho c_\text{p}} \tag{F13}$$

This result is more usually expressed in the form of Equation (F14) for the case where there is no internal heat generation and $\dot{Q} = 0$, hence

$$\frac{\partial T}{\partial t} = \alpha\left(\frac{\partial^2 T}{\partial x^2} + \frac{\partial^2 T}{\partial y^2} + \frac{\partial^2 T}{\partial z^2}\right) \tag{F14}$$

REDUCTION TO STEADY STATE

At steady state (i.e., when there is no variation in temperature with time and $\partial T/\partial t = 0$) and for conduction in the x direction only (i.e., both $\partial T/\partial y = 0$ and $\partial T/\partial z = 0$), Equation (F14) reduces to

$$0 = \alpha\left(\frac{\partial^2 T}{\partial x^2}\right) \tag{F15}$$

and, because the thermal diffusivity cannot be equal to zero, it must be the case that

$$\left(\frac{\partial^2 T}{\partial x^2}\right) = 0 \tag{F16}$$

Therefore

$$\frac{\partial T}{\partial t} = \text{constant} \tag{F17}$$

and hence

$$Q = -k(\Delta y\,\Delta z)\left(\frac{dT}{dx}\right) \tag{F18}$$

which is Fourier's second law of heat conduction and was the starting point for the study of conduction in section 7.2.1.

Answers

CHAPTER 2

1. 1.962×10^5 Pa
2. $860.7 \, \mathrm{kg \, m^{-3}}$
3. 18.4 kJ, 613.1 W
4. $\mathrm{MLT^{-1}}$, $\mathrm{ML^2T^{-2}}$, $\mathrm{ML^2T^{-3}}$, $\mathrm{ML^{-1}T^{-2}}$, $\mathrm{MLT^{-2}}$, $\mathrm{ML^2T^{-2}}$, $\mathrm{ML^2T^{-1}}$, $\mathrm{ML^{-2}T^{-2}}$, $\mathrm{ML^{-1}T^{-2}}$, $\mathrm{T^{-1}}$

CHAPTER 3

1. 261.2 K
2. 173 kPa
3. 596 K
4. $0.732 \, \mathrm{m^2}$
5. $-17.9°\mathrm{C}$
6. $1.76 \, \mathrm{kg \, m^{-3}}$
7. 2.48 kg, $1.27 \, \mathrm{kg \, m^{-3}}$
8. 0.349, 0.493
10. $94.3°\mathrm{C}$

CHAPTER 4

1. 0.111, 11.1%, 0.00653, 0.653%
2. 0.365
3. 4.89 kg
4. $0.0556 \, \mathrm{kg \, s^{-1}}$; 37.0% NaCl
5. 14.3%; 1.2 kg
6. $29.46 \, \mathrm{kJ \, kmol^{-1} \, K^{-1}}$
7. $1.95 \, \mathrm{kJ \, kg^{-1} \, K^{-1}}$
8. $31.46 \, \mathrm{kJ \, kmol^{-1} \, K^{-1}}$
9. For each, $c_\mathrm{p} = 1.421 \, \mathrm{kJ \, kg^{-1} \, K^{-1}}$
10. (A) $2{,}545 \, \mathrm{kJ \, kg^{-1}}$
    (B) $2{,}333.3 \, \mathrm{kJ \, kg^{-1}}$
    (C) 3.166 kPa

  (D) $3,277 \, \text{kJ kg}^{-1}$
  (E) 49.4°C
  (F) 60.1°C
  (G) $1.129 \, \text{kg m}^{-3}$
  (H) $2,765.5 \, \text{kJ kg}^{-1}$
  (I) $10.197 \, \text{kPa}$
  (J) $2.026 \, \text{kJ kg}^{-1} \, \text{K}^{-1}$
11. $3.764, 3.847, 3.833$ and $3.848 \, \text{kJ kg}^{-1} \, \text{K}^{-1}$, respectively
12. $228.6 \, \text{kJ}$
13. $29.25 \, \text{kW}; 0.146 \, \text{kg s}^{-1}$
14. $2.51 \, \text{h}$
15. $0.0132 \, \text{kg s}^{-1}$

CHAPTER 6

1. $2.04 \, \text{m s}^{-1}$
2. $0.70 \, \text{kg s}^{-1}$
3. $35,368$, turbulent; 533, laminar
4. $268; 73,333$
5. $0.319 \, \text{m}$ oil; $2.66 \, \text{kPa}$
6. $287 \, \text{Pa m}^{-1}$
7. $0.4 \, \text{m}$
8. $0.405 \, \text{m}$
9. $0.0275 \, \text{m}$
10. $1,892 \, \text{Pa s}^n; 0.203$
11. Herschel–Bulkley; yield stress $140 \, \text{Pa}$, consistency coefficient $66.80 \, \text{Pa s}^n$, flow behavior index, 0.30
12. $84.2 \, \text{kPa}$
13. $0.351 \, \text{Pa s}$
14. $0.0733 \, \text{m} \approx 75 \, \text{mm}$
15. Kelvin–Voigt and Maxwell elements in series, i.e., the Burgers model

CHAPTER 7

1. $7,488 \, \text{W m}^{-2}$
2. $1,167 \, \text{W m}^{-2}; 1,020 \, \text{K}; 570.5 \, \text{K}$
3. $3.39 \, \text{W m}^{-2}$
4. $6.5 \, \text{K}$
5. $0.088 \, \text{m}$
6. $46.5 \, \text{kg}$
7. $135.7 \, \text{W m}^{-1}$
8. 0.843
9. $385.4 \, \text{W}$
10. $84.2 \, \text{W}$
11. $119.7 \, \text{kW}; 22.7 \, \text{K}$

12. $3,685 \, \text{W m}^{-2} \, \text{K}^{-1}$
13. $45.7 \, \text{K}; 55.2 \, \text{K}$
14. $0.60 \, \text{m}$
15. $3.21 \, \text{kg s}^{-1}; 804 \, \text{W m}^{-2} \, \text{K}^{-1}$
16. $2.10 \, \text{kW m}^{-2} \, \text{K}^{-1}; 4.72 \, \text{m}$
17. $1,710 \, \text{W m}^{-2} \, \text{K}^{-1}$
18. $0.52 \, \text{kg s}^{-1}$
19. 30
20. $23,609 \, \text{W m}^{-2} \, \text{K}^{-1}; 157.7°\text{C}$
21. $7,727 \, \text{W m}^{-2} \, \text{K}^{-1}; 10^{-3} \, \text{kg s}^{-1}$
22. $34.7°\text{C}$
23. 0.39
24. $916.5 \, \text{K}$
25. $1,931 \, \text{K}$
26. $82.34 \, \text{kW m}^{-2}$
27. $13.94 \, \text{kW}$

CHAPTER 8

1. $0.0265 \, \text{m s}^{-1}$
2. $4.20 \, \text{kW m}^{-2}; 9.56 \times 10^{-5} \, \text{kmol m}^{-2} \, \text{s}^{-1}; 4.06 \times 10^{-8} \, \text{kmol N}^{-1} \, \text{s}^{-1}$
3. Dissolved oxygen concentration $= 5.55 \times 10^{-5} \, \text{kmol m}^{-3}$
4. $2.19 \times 10^{-7} \, \text{kg s}^{-1}$

CHAPTER 9

1. $0.0062 \, \text{kg kg}^{-1}$ dry air
2. $3.12 \, \text{kPa}$
3. 79.5%
4. $29.5°\text{C}$
5. $0.015 \, \text{kg kg}^{-1}, 46\%, 0.888 \, \text{m}^3 \, \text{kg}^{-1}$
6. $11°\text{C}, 0.0082 \, \text{kg kg}^{-1}, 17.3\%$
7. $10.5°\text{C}, 2.16 \, \text{kg}$
8. $53°\text{C}$
9. $0.0123 \, \text{kg}, 0.0054 \, \text{kg}, 0.0192 \, \text{kg}$

CHAPTER 10

1. $400 \, \text{W m}^{-2} \, \text{K}^{-1}$
2. $1,252 \, \text{s}$
3. $45 \, \text{min}; 9.3°\text{C}$
4. $8,767 \, \text{s}$
5. $106°\text{C}$
6. $12 \, \text{min}$

7.   1 in 10^9
8.   12.8 min
9.   0.0977; 2.455
10. 11.4 min
11. 25 min
12. 30 K; 54 K; 35 min; 1.8
13. 56.9 min; 118.5 min

CHAPTER 11

1.   361.75 kJ kg^{-1}; 50.2 kW
2.   13,885 s
3.   2,062 s
4.   1,607 s
5.   5,466 s

CHAPTER 12

1.   0.977 kg s^{-1}; 14.3 m^2
2.   0.853
3.   0.315 kg s^{-1}; 0.340 kg s^{-1}
4.   5.4 m^2
5.   9.4 K; 4.8 m^2; 0.865
6.   1.45; 65.6 m^2
7.   2.5 kg; 16.7%
8.   19.5 kg
9.   375 min
10. 9.33 × 10^{-6} kg s^{-1}; 14%; 4.5%
11. 104°C

CHAPTER 13

1.   18.60 μm; 22.47 μm
2.   148.4 μm; 160.3 μm
3.   137 μm
4.   0.70
5.   137.5 μm; 156.7 μm; 316.7 μm
6.   0.43
7.   6.13 × 10^{-4} m s^{-1}
8.   0.0155 Pa s
9.   60.7 μm
10. 0.0116 m s^{-1}
11. 0.225 m s^{-1}
12. 1.1 × 10^{-3} m

13. $0.020\,\text{m s}^{-1}$
14. $0.0408\,\text{m s}^{-1}$
15. $2.16\,\text{m s}^{-1}$
16. $0.024\,\text{m s}^{-1}$

CHAPTER 14

1. $12\,\text{min}$
2. $3.21\,\text{kW}$
3. $15\,\text{min}$
4. 8% reduction
5. $2{,}100\,\text{s}; 3.85 \times 10^{-4}\,\text{m}^3\,\text{s}^{-1}$
6. $1.98 \times 10^{12}\,\text{m}^{-2}; 2.64 \times 10^{-3}\,\text{m}; 1.71 \times 10^{-3}\,\text{m}^3\,\text{s}^{-1}; 0.042\,\text{m}$
7. $10\,\text{m}^2$
8. $39.8\,\text{m}^2$

Index